SACRAMENTO PUBLIC LIBRARY
828 "I" STREET
SACRAMENTO, CA 95814
4/2009

The Best of instructables

D0048805

First Edition

By the editors of MAKE magazine and Instructables

O'REILLY®
BEIJING • CAMBRIDGE • FARNHAM • KÖLN • SEBASTOPOL • TAIPEI • TOKYO

The Best of Instructables

by the editors of MAKE magazine and Instructables

Copyright © 2009 O'Reilly Media, Inc. All rights reserved.
Printed in U.S.A.

Published by Make:Books, an imprint of Maker Media,
a division of O'Reilly Media, Inc.
1005 Gravenstein Highway North Sebastopol, CA 95472

O'Reilly books may be purchased for educational, business, or sales promotional use. For more information, contact our corporate/institutional sales department: 800-998-9938 or corporate@oreilly.com.

Print History: October 2008: First Edition

Publisher: Dale Dougherty
Associate Publisher: Dan Woods
Executive Editor: Brian Jepson
Creative Director: Daniel Carter
Lead Editor: Gareth Branwyn
Assistant Editors: Patti Schiendelman,
Eric J. Wilhelm (for Instructables)
Designers: Sutton Long and Gretchen J. Bay
Production Manager: Terry Bronson
Indexer: Patti Schiendelman

The O'Reilly logo is a registered trademark of O'Reilly Media, Inc. *The Best of Instructables*, and related trade dress are trademarks of O'Reilly Media, Inc. The trademarks of third parties used in this work are the property of their respective owners.

Important Message to Our Readers: Your safety is your own responsibility, including proper use of equipment and safety gear, and determining whether you have adequate skill and experience. Chemicals, electricity, and other resources used for these projects are dangerous unless used properly and with adequate precautions, including safety gear. Some illustrative photos do not depict safety precautions or equipment, in order to show the project steps more clearly. These projects are not intended for use by children.

Use of the instructions and suggestions in *The Best of Instructables* is at your own risk. O'Reilly Media, Inc. and the authors disclaim all responsibility for any resulting damage, injury, or expense. It is your responsibility to make sure that your activities comply with applicable laws, including copyright.

ISBN-13: 978-0-596-51952-0

Contents

Preface

When we decided to collaborate with Eric Wilhelm and the folks at Instructables on a best-of book, we knew it was going to be fun and challenging, we had no idea just how fun and how challenging. We also had no idea how deeply interwoven we'd all become with the Instructables community itself.

The book you're holding represents the enthused effort of hundreds of people. As the staffs at MAKE and Instructables were busy choosing our favorite projects, Instructables ran a contest so members could choose their favorites. Thousands of votes were cast. The top 75 can be seen starting on page 304. Projects chosen by both the editors and the community have a special "Contest Winner!" icon throughout the book.

Translating web content to a book is challenging, especially when you're dealing with so many contributors. These are not professional writers or photographers. These are people who are passionate about the things they create and about sharing the results of their effort with others. We tried not to tinker too much with their voice, letting different styles and personalities shine through (glare and all) as much as possible. The authors couldn't have been more generous, taking new pics when possible, correcting errors, proofing layouts, and offering us constant support and encouragement. To all of them we give a resounding Thank you!

A book can only capture some of what the Web has to offer. For each of these projects, a web address is listed at the bottom. Before building a project, go to the Instructable page, check out the additional images and videos, read the comments, and look for updates and corrections.

While this book may represent some of the best projects on Instructables, through this process, we discovered that the best of Instructables is really the community it engenders. The best of Instructables is the willingness of those involved to roll up their sleeves and help out (when you spend your time hanging out on a "how to" site, "can do" is the pervading ethos). And ultimately, the best of Instructables is the itchy excitement one feels while browsing these projects. It's no exaggeration to tell you that it was all we could do, as we edited these pages, to resist the urge to abandon our desktops for our workbenches and start making stuff.

From all of us, editors Gareth Branwyn, Brian Jepson, and Patti Schiendelman; project manager Terry Bronson; designers Sutton Long and Gretchen Bay; art director Daniel Carter; Eric J. Wilhelm and the team at Instructables; and the wonderful worldwide community of Instructable authors, we hope you have as much fun making the projects in *The Best of Instructables* as we did in making the book.

—The Editors

Welcome

Making things by hand is cool again.

You can be a creator, not just a consumer. Whether you're into screen-printing T-shirts, soldering LEDs, modding pre-fab furniture, building robots, or just making dinner, you're a part of the Maker movement that calls Instructables.com its online home.

Instructables is a place where artists, bicyclists, crafters, engineers, modders, cooks, tinkerers, and techies gather to share advice and ideas freely, and post thousands of their fantastic projects. We're excited to finally bring you the best of Instructables in print. We think it's a great introduction to the incredible variety of what you can make and do, and will hopefully inspire you to get started on great projects of your own.

Instructables is *my* biggest project. It all started when I took up kitesurfing while finishing up my Ph.D. in mechanical engineering at MIT. Kitesurfing equipment (modified surfboards, and giant kites to pull you around on them) was still new, and far too expensive for my graduate student budget, so I designed, built, and tested my own gear. I documented the designs and results on my personal website, and got email from lots of people who wanted to ask for advice, share their ideas and pictures, and meet me at the beach. Unfortunately, this took just as long as making the equipment in the first place. It was clear we needed a better way to share our projects online.

Instructables was one of the first ideas to come out of Squid Labs, the innovation research and design firm I started with some of my grad school kitesurfing buddies. We started using the step-by-step format to document our own projects, and liked it so much that we decided to give our friends access. Word spread quickly, and soon people we'd never met were posting fantastic projects. It quickly became clear Instructables was going to be huge, so I spun it off from Squid Labs (see my "How To Start a Business" Instructable—www.instructables.com/id/How_to_Start_a_Business_1 for more information) and built an awesome team to help it keep growing, and to provide a place for all sorts of great projects and discussions.

When someone builds, bakes, or creates something they're proud of, they often put it on the coffee table so visitors will ask "how did you do that?" I think of Instructables as an online coffee table—a place for you to put your projects on display, show off your skills and creativity, and start interesting conversations with the people who stop by to check them out.

I find it deeply satisfying that this book will rest on many homemade coffee tables, and be stacked on crazy custom bookshelves. These awesome projects give you a taste of Instructables, and the sorts of creative things our online community has to offer. Get inspired by these projects, then check out the Instructables website for the full build details. Leave the author a message online, and if you try their project, take a picture and post it in the Comments.

We've had fun putting this book together, and the Instructables community has been amazing—each member contributes, and together we are Instructables. So browse through these fantastic projects, try something new, get some ideas, and post your own Instructables. This is *The Best of Instructables*, Volume 1—I can't wait to see *your* entries in Volume II.

Now turn the page, pick a project, and go make something awesome!

Eric

Eric J. Wilhelm

Home & Garden | Food | Photography | Science | Computers | Electronics

The Best of Instructables
Home & Garden

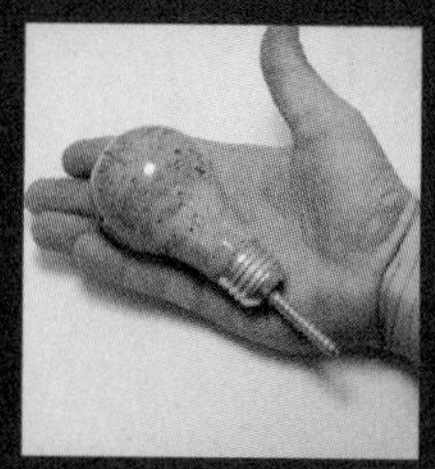

Before we opened Instructables for everyone to use, we loaded it with a bunch of our own projects as examples. The line between my work and my play had always been slightly blurred, but now, it was totally obliterated. Projects that I worked on when I had free time suddenly became mission-critical because we desperately needed content in the system for its launch. Having just moved and left all my furniture behind, I needed bookshelves. So, my sustainably grown hardwood bookshelf project got accelerated, as did many of the surrounding techniques I used and tools I built.

Home projects are some of the most accessible DIY projects. Almost everyone's home was custom-built, and outfitting it properly often means modifying, hacking, and doing it yourself. Even the least handy of us can install a set of shelves, and then aspire to make the next set a little bit nicer, fit closer, be more tailor-made.

In this section, we're showcasing some of the Instructables that really make us wish the authors lived next door—and not just so we could borrow their tools! Cool lights, funky furniture, and energy-saving and waste-reducing ideas abound. After seeing these projects, you'll likely notice an empty wall and wonder just what sort of hidden LED-powered geometric doodad might fit. Be sure to take pictures!

Finally, the section is tied together with a gallery of IKEA hacks. I have something of a love-hate relationship with IKEA: I don't like the way products seem to be positioned as cheap-enough-to-throw-out-when-you-move... But on the other hand, I appreciate that a low-cost piece of furniture is a great starting point for some creative hacking. Hopefully, after you've customized, and made it your own, you'll want to keep it forever.

—Eric J. Wilhelm (ewilhelm)

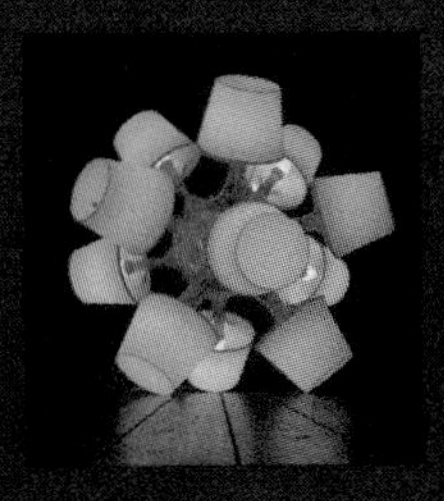

Gallery

IKEA Hacks

Take apart their furniture, re-wire their electronics; you'll never see IKEA the same way again

Modified GALANT Desk Kolby Kirk

/Modified-IKEA-GALANT-desk

This IKEA hacker created endless options for under-desk cable management and suspension of various computer peripherals (router, modem, power strips) by attaching pegboard underneath

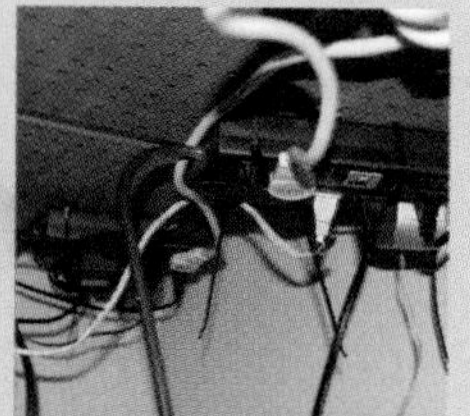

Medium and larger "zip ties" are used to hold the pegboard to the desk frame and the cables and peripherals to the board

Big Lamps from IKEA LAMPAN Lamps

Daniel Saakes

/Big-lamps-from-Ikea-lampan-lamps.

Build Platonic suns with LAMPAN lamps

Use 6, 12, 32 lamps or more!

Power Charging Box with Individual Switches Pedro Rodrigues

/IKEA-Power-Charging-Box-with-individual-switches

Make a home power station for your chargeables.

Bathroom Towel Rail for Your Kitchen

F. Roulier

/ikea-hack--how-to-hack-bathroom-towel-rail-for-yo

Add a shelf and hooks to transform this GRUNDTAL towel rail into a kitchen rack

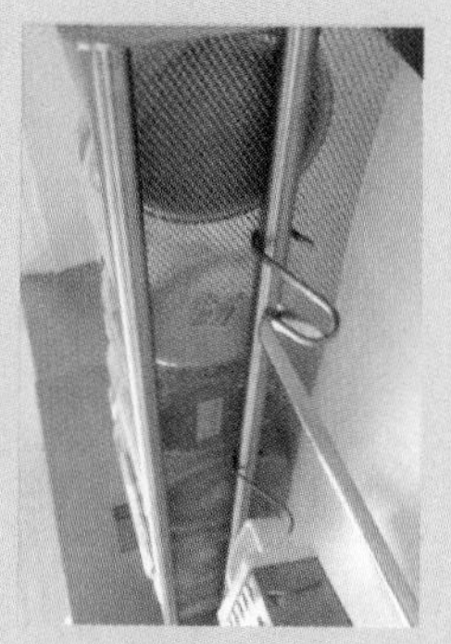

Steel netting makes a great shelf

Make a Mood Lamp

Artur Petrovskyy

/Fun-with-IKEA-lamp

Using a tiny PIC12F683 microcontroller and RGB (Red/Green/Blue) LEDs, you can turn a MYLONIT lamp into a colorful, programmable mood light

TORE Coffee Table Edition Jason Schlauch

/IKEA-Hack:-TORE-Coffee-Table-Edition

Use a TORE file cabinet base as the bones of this slate-topped coffee table

International Clock

Ed Lewis

/International-Clock

A "lazy Susan" bearing and some magnets turn this cheap IKEA RUSCH clock into an International time-keeper—just turn the city of interest to the 12 position to find out the hour there

Ice Straws Paul Jehlen

/IKEA-Ice-Straws

Don't let your cool drink warm up on its way to your mouth—ice straws are the answer!

Place plastic straws into an ice stick tray before filling with water

Tool Box Hack PK Shiu

/IKEA-Tool-Box-Hack

Take an IKEA FIRA mini-chest, attach a pair of KOSING handles (6 for $2), and you have yourself a six-drawer luggable toolchest for home, office, or studio

To view these Instructables, and read comments and suggestions from other IKEA hackers, type in www.instructables.com/id followed by the rest of the address given here. Example: www.instructables.com/id/IKEA-Tool-Box-Hack

Shoe Rack

David Hayward

/Ikea-Hack-STRIPA-Shoe-Rack

STRIPA shelves find new life as a clever wall-hung shoe rack

Scrap Lamp

Adam Kelly/ Killerjackalope

/Ikea-scrap-lamp...

Use ORMEN packaging to make this lamp

Magnetic Refrigerator Lights

Turn your fridge into a canvas for LED art

By John Kowalski

Figure A: Magnetic refrigerator lights in action

With these LED fridge lights, passersby can place the magnetic LEDs any way they wish to create illuminated pictures and messages. It's great for high traffic kitchens and fun for kids and adults alike.

1. What you need

Most of the materials can be found at local hardware and electronics parts stores or online.

Parts:

- Super Shield conductive nickel paint (see Figure B). This can be found at electronics parts stores. It's usually used to add RF shielding to plastic cases. We will be using it because it's electrically conductive.
- 1/4" copper tape used for circuit board repair (optional). If conductive paint can't be found, this may be a possible substitute. It may be a good idea to get some anyway as a way to repair any future scratches or chips in the conductive paint.
- Spray Paint. I used Krylon Fusion For Plastic because it sticks to almost anything, doesn't require a primer, and has a nice finish.
- 10mm LEDs in quantities and colors of your choice. I used 20 LEDs of each—red, green, blue, yellow, and white. These can be bought online from vendors such as www.futurlec.com, www.digikey.com, www.jameco.com, and many others.
- 330Ω surface-mount resistors. Get one for each 2.4 volt LED (Typically red, orange, yellow and sometimes green LEDs are 2.4 volts). The 3.6 volt LEDs (typically blue, white, UV, and true green) do not require resistors.
- One 4.5 volt, 500 mA AC power supply. By using AC, the polarity of the LEDs won't matter. They will light up whichever way they are placed onto the grid. This also reduces power consumption because the LEDs will run at a 50 percent duty cycle.
- 1/8" diameter x 1/16" NdFeB Nickel plated disc magnets. Get two for each LED. These can be found online from vendors such as http://kjmagnetics.com and http://magnet4less.com.
- 1/4" diameter x 1/16" NdFeB Nickel plated disc magnets. I used six—two for attaching the power source to the fridge, and four more for making magnetic jumper wires to bridge the gap between the door and the side of the fridge.
- 5-minute epoxy. Get the kind that you mix from clear and yellow tubes.
- Masking tape
- 1/4" Quilter's tape (see Figure C). This is just masking tape but 1/4" wide, the thinnest tape I could find. You can find this in craft stores. Ideally, you want tape that is just slightly wider than the diameter of the magnets used on the LEDs.
- Solder

Equipment:

- Needle-nose pliers
- Small wire cutters or fingernail clippers
- Soldering iron or gun
- Wire-wrapping tool or other tool with a flat round 1/8" diameter tip. It's really the 1/8" diameter we're going to use as a tool, so you could use a grinded down dollar store screwdriver if that's what's available.
- X-Acto knife
- Wooden toothpick

For additional information, discussion, and more, please visit the Instructables project page:

- The cap from a cheap pen
- Putty/Clay/Plasticene/Play-Doh; this is for holding LEDs in place while you work on them.

2. Paint the fridge

Wash the surface of the fridge first. Once dry, use masking tape to mark off the borders of the grid (display area) and the power traces going to it (see Figure D). Get some newspaper and with more masking tape, cover up all the areas you do not want to get paint on (Figure E).

Apply one even coat of the base paint. I used this base coat to cover any existing scratches on the surface of the fridge and also to help ensure that the conductive paint will not peel off.

Once the base coat is dry to the touch, start applying the 1/4" Quilter's tape to form two separate traces that will power the grid, and then apply all the traces in the grid itself. I used a marker to mark out the vertical spacing between the horizontal lines, then followed the dots as I placed the tape by hand.

Note: Just try to get the spacing right at first and not worry about the edges yet. I marked out dots to keep spacing at 10mm intervals.

The grid should basically look like two interlocked combs, as shown in Figure F. One power line will go to the left side and make "teeth" going back to the right, and the other power line will start at the right side, making teeth go to the left. Neither of the two power lines should ever touch at any point. The grid is basically an open circuit and closed only when the magnetic LEDs will be placed upon it—one magnet of the LED touching one power line and the other magnet touching the other.

As soon as the masking tape is applied, start painting with the conductive paint (see Figure G). This stuff requires caution—you want to make sure that the area is well ventilated. Open all windows and doors, and turn on the vent over the stove.

Keep applying more layers until the can is nearly empty. You may want to save a little bit just in case you ever need to touch up the conductive traces. If you are painting a larger surface area, you may consider using more than one can.

Figure B: Krylon Fusion paint and conductive paint

Figure C: Quilter's tape and masking tape

Figure D: 1. The target area for the LED grid 2. The power line to the grid

Figure E: Add one line of thin masking tape to split the power into two lines

Figure F: The white is where the conductive traces will be and the masking tape is where it won't be conductive. Make sure the top trace is separate from the bottom trace.

Figure G: Our first coat of conductive paint. Keep adding more coats until the can is empty, or nearly empty if you want to keep a bit "just in case."

3. Remove the masking tape

Once the paint is dry to the touch, start removing the newspapers and masking tape, starting from the outside and working your way inside to the finer traces (see Figure H, I, and J).

Be very careful pulling up the masking tape because the conductive paint is so thick that it may pull off along with the tape. I used an X-Acto knife to score the edges in the corners of the fine traces to prevent the paint from being lifted off. Should any conductive traces start lifting, you may be able to press them back down if the paint is still not completely dry. After it is dry, you may try using Krazy Glue to glue down any lifted corners.

Figure H: 1. Try to have thick and even coverage with the conductive paint 2. Peel off the masking tape going from outside-in

Figure I: The thin masking tape gets tricky to remove, so be careful

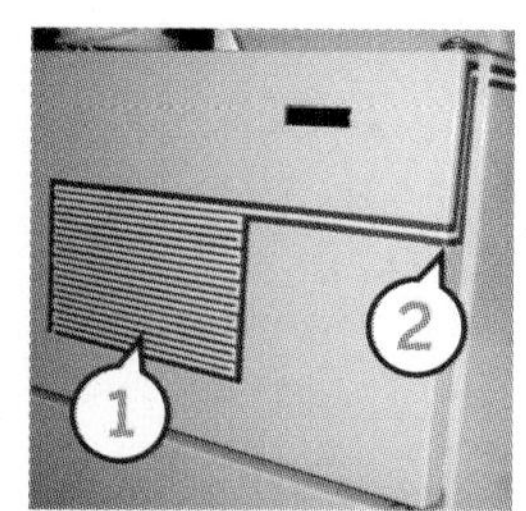

Figure J: 1. Our finished display/LED power grid 2. If you want to hide the power source, you may paint over this part again, just make sure to leave contact points for the power wires and the jump between the door and the side.

Figure K: The power supply connected to the grid with magnets

4. Hook up the power

I used a 4.5 volt / 500 mA AC power adapter with magnets attached (see Figure K) to both wires to power the grid. The wires are glued to the magnets and the tops (not bottoms) of the magnets are covered in hot glue to prevent the two magnets from shorting the circuit should they touch each other. The magnets can be oriented so that the bottoms are magnetically opposed—which should prevent them from attaching to each other if they get pulled from the fridge.

There is a gap in the circuit between the door and the side panel of the fridge. To complete the circuit, I used two small jumper wires with magnets attached to both ends (see Figure L). Make sure to leave enough slack in the wires to cover the gap even when the fridge door is opened all the way.

Now the grid has power. Since the LEDs are not constructed yet, you may test the circuit by placing some magnets onto the traces of the grid and lying some LEDs over the magnets (see Figure M). The pins/wires on LEDs are attracted to the magnets, so no special effort is required for this test.

I measure 15Ω of resistance on each trace going from the back of the fridge to the grid, and about 15Ω more from the side of the grid to the center. When I place a jumper wire in the center of the grid, I measure about 60Ω between both power terminals at the back of the fridge—a full circuit between the power supply, crossing any given point of the grid and back to the power supply, is about 60Ω.

I found that the paint's conductivity increased as the paint dried, so don't be too worried if you get readings higher than that at first. If they're much higher or much lower, you may opt to use a different voltage power supply to compensate or use copper tape to reduce resistance in the traces.

5. Assemble the 3.6 volt LEDs (blue, white, and green)

Bend the pins on the LEDs to lie flat, then use the wire-wrapping tool (or similarly tipped tool) as a shape to bend the wires around so that they encircle the tool. Try to make it so the outside of the circle shapes reaches the outside edges of the bottom of the LED and both circles are on opposite sides of the LED. Figure N shows LEDs in various stages of assembly.

Once the wires are bent, use nail clippers or small wire cutters to cut off excess wire. You only need one looping of the wire to form a circle with each LED pin. Next, use needle-nose pliers to adjust the loop so that the inside of the circle is just a little bit smaller than the diameter of the magnets. This will add a bit of spring to the wires so that they clamp over the magnets when you push them in. (Don't add the magnets just yet.)

Before attaching the magnets to the LEDs I used modeling clay to hold the LEDs upside-down (see Figure O). Get your magnets ready. Mix up a small batch of epoxy. Apply a small bit of epoxy into the center of each circle of wire. Apply just enough for the magnets to hold without having the epoxy overflow over the bottom of the LED.

Figure L: These magnets/ wires jump the gap between the door and the side panel

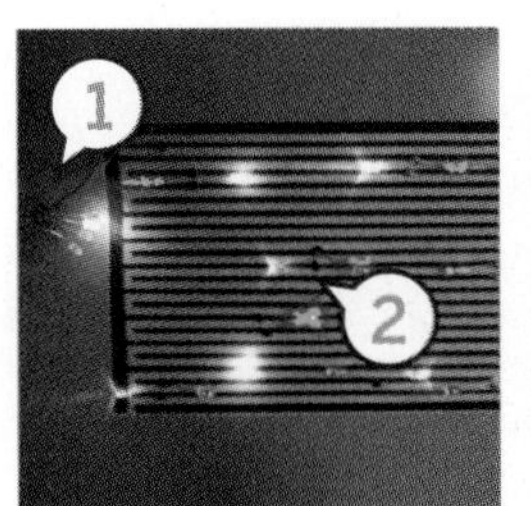

Figure M: 1. A test to work out that 2.4 volt LEDs could coexist with 3.6 volt LEDs on the same grid 2. I didn't have the final LEDs at this point, so I just placed magnets and ordinary 5mm LEDs on the grid to test

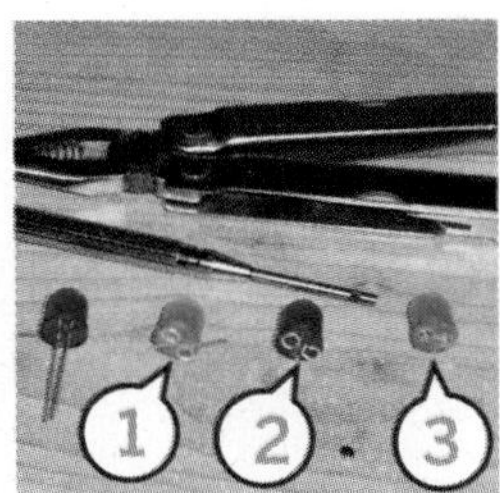

Figure N: 1. Bend the pins into loops 2. Cut off excess wire and adjust the size of the loops 3. Finished LED. (Also pictured: Leatherman, wire-wrapping tool, magnets)

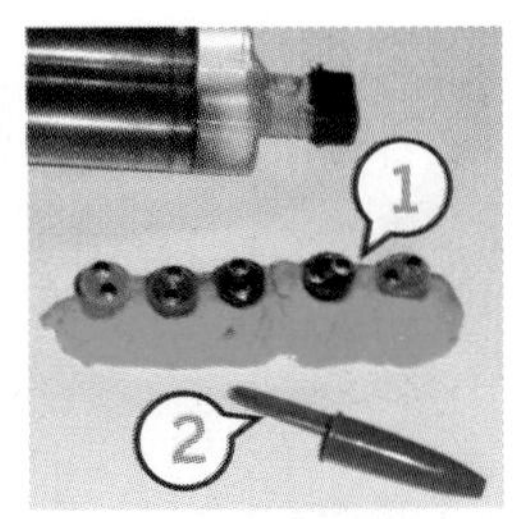

Figure O: 1. Some LEDs after the magnets have been epoxied into the bent wires 2. A pen cap is perfect for pushing magnets into their slots

Now place magnets onto one side of each LED. They will most likely just jump onto the top edge of the circle of wire as the wires will attract the magnets. Once one row is placed, use the cap from a cheap pen to push the magnets into the center of the circle and press down on them until they "click" into place. If they don't "click," or are too hard to push into the circle, you may need to adjust the diameter of the looped wire on your next batch of LEDs.

Add magnets to the second row of pins in the same way. If any epoxy gets onto the pen cap, just wipe it clean with a tissue. This will prevent covering the tops of the magnets with epoxy.

For the next minute or two, use the pen cap to level off the tops of the magnets. Then let them sit for 20 minutes or so as shown in Figure P. Do not move the magnets now because it will potentially break its contact with the looped wire.

After 20 minutes, use a toothpick to carefully scrape any epoxy that may have covered the top surface of the magnets. Don't try placing the magnets onto your fridge until several hours have passed as the metal surface of the fridge will pull at the magnets and break the contact with the LED wires.

You can test your LEDs before the epoxy has completely hardened with a 3V battery attached to small wires—simply touch the wires to the magnets to see the LEDs light up. Try both ways because LEDs will only light up if power is hooked up in the correct polarity.

6. Assemble magnets to the 2.4 volt LEDs (red and yellow)

The 2.4 volt LEDs start similar to the 3.6 volt LEDs, but the wires are bent in a different way to allow for a resistor to be added to the bottom of the LED.

Bend the longer wire of the LED to lie flat and point straight out. Next bend the shorter wire to lie flat around 40 degrees below the first wire. Bend the first wire to come back towards the LED. Make the bend overhang the edge of the LED a bit to make it easier to cut later. Next, use the wire-wrapping tool to curl the bent back wire into a circle. Do the same with the next wire. Try to keep the loops evenly apart, with the edges of the circles going right up to the edges of the LED. Figure Q shows the steps.

Use nail clippers or a wire cutter to cut off excess wire. You only need the wire to make one loop around. Use needle-nose pliers to tighten the inside diameter of the circles to be just slightly smaller than the diameter of the magnets so that the wires will clamp around the magnets when they are pushed into place later.

Place the LEDs in the modeling clay and use epoxy to attach the magnets just like with the 3.6 volt LEDs (see Figure R).

When it comes to testing the LEDs to check that they light up, add an extra resistor to the battery because 2.4 volt LEDs will burn out without a current limiting resistor (see Figure S).

Let the epoxy harden for about two hours, but do not place these LEDs onto your fridge because without the resistor (next step), they will burn out.

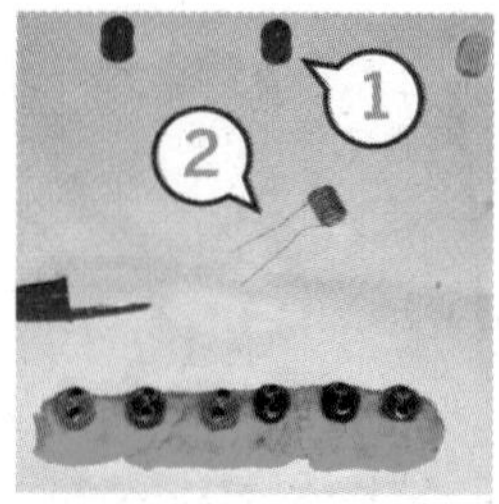

Figure P: 1. Finished LEDs waiting for epoxy to harden completely 2. Small batteries with wires for testing assembled LEDs

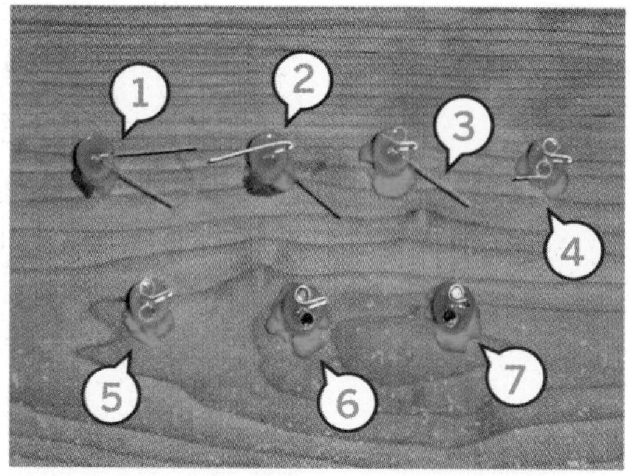

Figure Q: 1. Bend wires flat—long wire straight out, short wire at an angle 2. Bend the long wire back 3. Twist the long wire into a loop 4. Twist the short wire into a loop 5. Clip off excess wire 6. Attach magnets inside wire loops with epoxy 7. After epoxy hardens, clip off bent wire (from 2.) and solder in the resistor

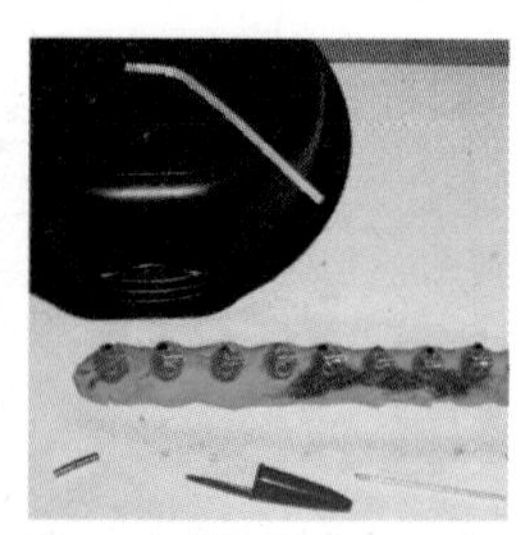

Figure R: The first row of magnets about to go in (epoxy has already been applied, magnets are stacked in the lower left)

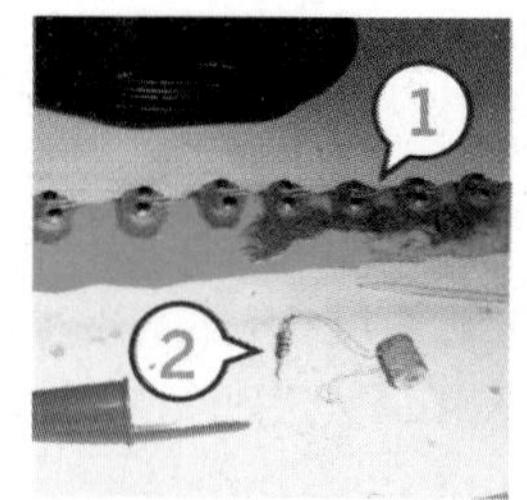

Figure S: 1. Partially assembled 2.4 volt LEDs; the resistors get added after the epoxy hardens. I added a resistor to the batteries once I started assembling the 2.4 volt LEDs.

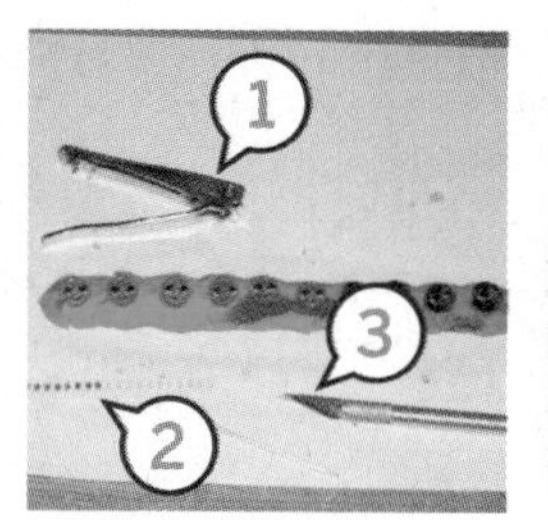

Figure T: 1. I used the nail clippers to cut the looped-back wire so I could add the resistors 2. 330Ω surface-mount resistors 3. Double sided tape on the tip of the blade makes it easy to pick up and position the resistors

Figure U: Just before soldering. The resistors are in place, clamped under the LED wires.

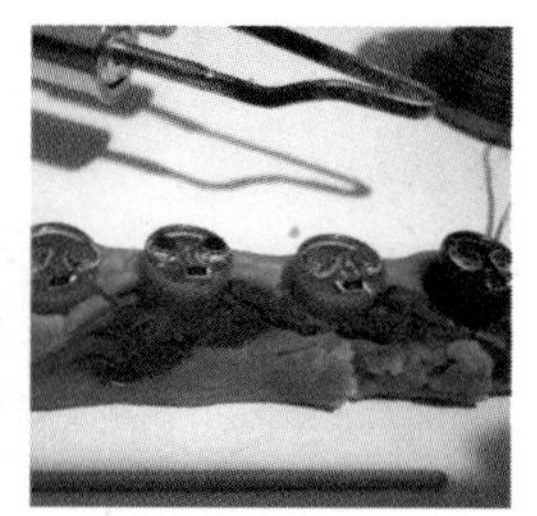

Figure V: After soldering

7. Add resistors to the 2.4 volt LEDs (red and yellow)

Once the epoxy on the 2.4 volt LEDs has hardened, use nail clippers or small wire cutters to cut the overhanging wire loop off the LEDs (see Figure T). Don't add too much stress to the wires as this might pull off the newly epoxied wire loop and magnet. If this happens you can use Krazy Glue to reattach it.

Cut and bend the wires to make room for a surface mount resistor. The resistor should rest about halfway between the edge of the magnets and the edge of the LED itself. Use an X-Acto blade or toothpick to bend the wires down a bit so that they will clamp down on the resistor once it is wedged underneath them.

Attach a small piece of double-sided tape to the end of your blade. This will make it easy to pick up and position the tiny resistors. Figure U shows the LEDs with the resistors in place.

Note: You can use a toothpick to position the resistors, bend the LED pins into place, and also occasionally to displace accidental solder bridges between the two pins.

Now it's time to solder in the resistors. This takes a delicate touch. Make sure to clean the soldering iron's tip often and lightly touch the points you want to solder with both solder and pre-heated soldering iron. If the solder doesn't "take" the first time, wait until the wires cool and try again. Make sure not to touch the magnets with the soldering iron. The

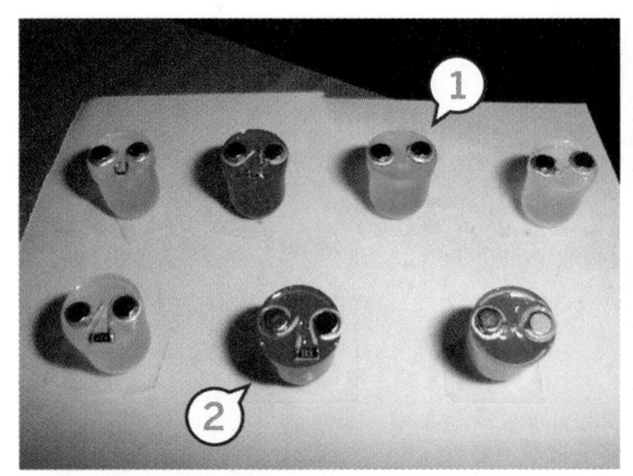

Figure W: 1. I used even smaller surface mount resistors at first before moving onto the larger ones below 2. Examples of the finished LEDs: 2.4 volt LEDs have resistors, and 3.6 volt LEDs don't.

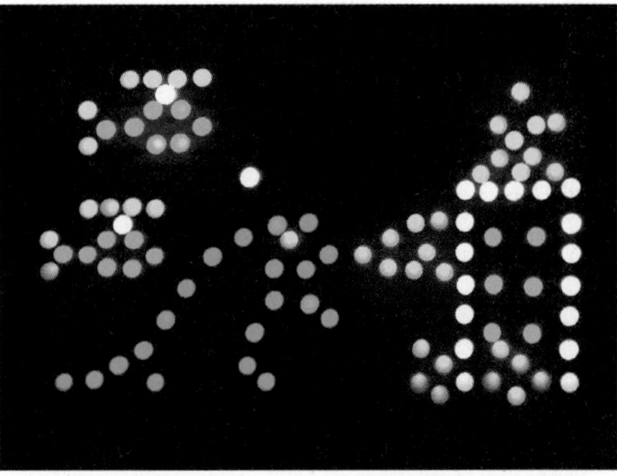

Figure X: An action scene with the magnetic lights

magnets will break down if exposed to excessive heat. If you accidentally solder a bridge over both wires, use the toothpick to push apart the solder as you reheat it.

For those who are unsure of their soldering ability, it might be possible to omit the soldering step altogether, if you make sure the wires firmly clamp down on the resistors. Test the LEDs to make sure they're working and then put a dab of epoxy over the resistor to keep it from moving out of its current (working) position. After the epoxy hardens a bit, be sure to test the LEDs.

Repeat steps 5, 6, and 7 until all the LEDs are assembled. Figure V shows some assembled 2.4 volt LEDs.

8. The finished project

Figure W shows a variety of assembled LEDs. Gather all the LEDs and place them on the fridge. Now it's time to start making pictures on your fridge as shown in Figure X.

When not used in the grid, the magnetic LEDs can also be used as regular refrigerator magnets—although they won't light up unless they're placed in the grid.

John Kowalski is a software engineer who has a passion for electronics, art, photography, and anything retro. Oh, and MacGyver is his hero.

User Notes

John, author of this Instructable, in reply to a question from bitchmobile, Instructables member: 900 [LEDs are] quite a bit more than this project will easily handle. You will have to use copper tape leading up to the grid instead of conductive paint to better handle the current and also stop the dimming that would happen if that many LEDs were put on the grid.

The change to copper tape leading to the grid would reduce the resistance of the electric circuit, so the power supply would also have to change to 3.5 VAC to keep the LEDs from burning out.

Because the LEDs are lighting on AC and not at their full brightness, they won't actually consume anywhere near 20mA each, but that will still require a hefty power supply... 5 Amps?

Mad Scientist's Light

A simple, adjustable tube-lamp to light your lair

By Timothy Johns

Figure A: Warm, but also Mad Scientist-y

Figure B: The interior of an unfinished pine box used for the base

Figure C: Wiring the sockets and dimmer switch

Figure D: Display case light bulbs

Here's a captivating light source suitable for normal usage and able to be dimmed down like a nice relaxing night light. This project is a Maker-friendly version of The Tube Lamp by Nik Willmore (www.e-dot.com).

1. Intro

Note: This project involves live electrical current and wiring that can be potentially dangerous. Take all proper safety precautions when working with AC electricity.

Note: Please do not try to reproduce this project for the purpose of making money—to do so would hurt the original designer.

A few years ago, I saw this little beauty called The Tube Lamp pop up online, designed by Nik Willmore. I decided it was exactly what I needed for my little Mad Scientist Laboratory. Unfortunately, I couldn't spare (or justify) the disposable cash needed to purchase it, though I would like to some day as his lamp still has a captivating hold on me. :-)

This project shows you how to make a much more cost-effective version of the tube lamp design. Rough estimate of costs involved is around $20 depending on the supplies you choose.

2. The base

A simple unfinished pine box with sliding lid (lid not shown in Figure B) that I picked up at the local arts and crafts store (Michael's) for about $2. It was the perfect size to fit the four light sockets. I sanded it all down and painted it after cutting the holes for the sockets in the bottom.

To make the lamp, the box was flipped over and the bottom became the top, so if I have to make any adjustments or replacements, I can turn the box on its side and slide open the lid.

3. Wiring

In Figure C you can see the inside of the box/base showing the back side of the light bulb sockets as well as the back end of the dimmer switch box and all the interconnecting wires. The light sockets were wired in parallel (one linking to another like a daisy chain) with one end of the power cord connected to the daisy chain of sockets, and the other end of the power cord connected to the dimmer switch.

The most expensive part of this whole project was the standard dimmer wall switch (push in to turn on and off, rotate to dim or brighten). It's the black box on the left in Figure C. It ran me about $7 at a home supply store.

To wire up everything, take all the black wires coming out of the sockets and bunch them together; I used wire ties to keep them all together. Do the same with the white wires coming from the sockets. Use a wire nut to connect all the white wires together and have them connect to one of the

Figure E: Light sockets set into the base

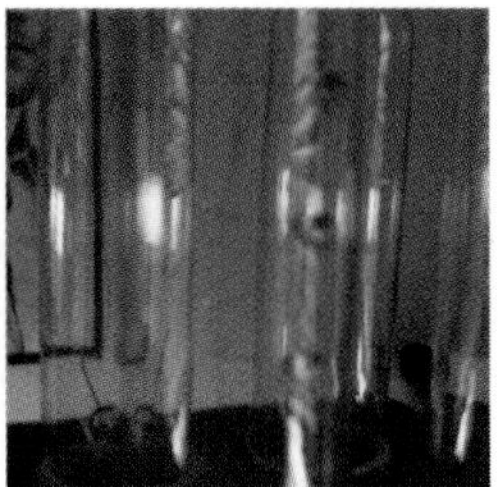

Figure F: Light bulbs before life-giving electricity is applied

wires from the power cord. Connect all the black wires from the sockets together as well and wire-nut them together with one of the two black wires coming from the dimmer switch. Then connect the remaining black wire from the dimmer switch to the other wire on the power cord.

4. The bulbs

I used simple 40W tube display case light bulbs, available at most home supply stores (I purchased mine at Lowe's). You can also look around online and find other display case bulbs with different filament patterns inside. Some are quite bland, others look spectacular when dimmed down, and you can trace the path of the filament in them.

5. Setting the sockets in the box

The light bulb sockets set in the cut-out holes. I made sure to cut the holes a bit small, and then lightly hand-sanded them till the sockets fit in properly with a few millimeters of the white socket exposed. I used a little bit of clear glue around the entire socket and the inside of the hole to attach it, as well as a thicker ring of glue around the sockets on the inside of the box.

The light sockets themselves were purchased at Lowe's and are fairly cheap as well ($2 or so), and are designed for ceiling lamp repair and replacement.

6. Light bulb innards

In Figure F, I tried to get a close-up of the bulb's insides. Didn't turn out perfectly, but you get the idea.

7. Finish

The finished product (Figure A) lets off a nice glow and is certainly an eye catcher. Everyone wants to know what it is and where I got it; not bad for a lamp I made for less than $20 (not including the sweat and blood you may or may not donate to this project).

I used a nice black radio knob to replace the boring beige knob that comes with the dimmer switch.

If you liked this project, be sure to check out the smoke stack lamp Instructable that I'll be posting, where I used some of the things I learned from this project. Right now, I'm still putting the finishing touches on that project, but you can see some shots of it in my Flickr photo stream (www.flickr.com/photos/timmyj/sets/72057594103809737).

Thanks for checking out my first Instructable!

Timothy Johns currently resides in south Jersey, working as a graphic designer, while striving to perfect his art, photography, and design projects.

User Notes

Trevor Myers built one with an old treasure chest

Dave Britt used halogens inside the lamps.

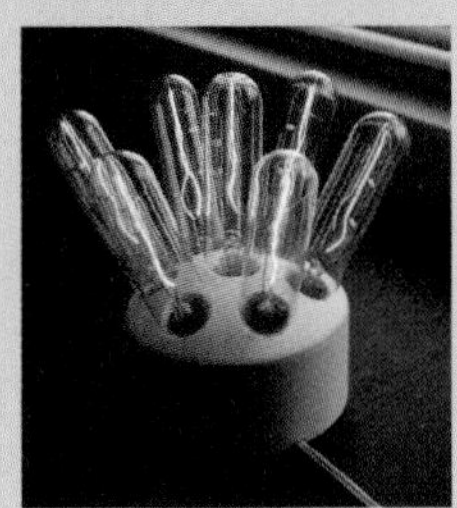

Jason Hull used a PVC end cap for the base

Ten-Green Modular Shelving

These great bottle-and-board shelves are made from recycled materials

By Roy Mohan Shearer, Zero-waste Design

Ten-Green is a modular shelving system, constructed simply from local recyclables.

This prototype was built with the Coach House Trust in Glasgow, using bottles from their recycling center, and wood from the maintenance of their own sites. You can use any found timber, and adapt the dimensions to suit the wood you are working with. Some examples are for sale at our shop here (www.etsy.com/shop.php?user_id=5250348).

The design featured in the online version of this Instructable is released under a share alike/attribution CC license by Zero-waste Design (www.zero-waste.co.uk).

Figure B: Shelf detail, with bottles and turnbuckles

1. Get your bits 'n' bobs

For one module you will need:

- Bottles (4), preferably all the same brand/height/shape. Wash them thoroughly in hot soapy water.
- Hook and eye strainers (2), sometimes called turnbuckles. I used the smallest I could find—14mm OD at the buckle, 5mm OD at the threads.
- Planks of wood (2) with four holes drilled in each (Figure C). The diameter and location of the holes depend on the sizes of your planks and bottles. The holes should be large enough for the ends of the bottle necks to fit through, and the plank to rest partway down the necks. Don't drill the holes too close to the edges of the planks.
- Sturdy cup hooks (2 per shelf)—4mm diameter ones are best

Revision: I would now NOT drill the holes in the lower shelf all the way through, but drill them from below to half depth, using a flat bit. They will then rest on the tops of the bottles of the module below, rather than allowing the necks to pass straight through as was the case in this prototype.

Screw in the cup hooks halfway between the holes on either side, as shown. Depending on the height of your bottles, you may need some "S" hooks to help the strainer reach between the upper and lower hooks.

2. Place the bottles and top shelf

Place the bottles concentrically over the holes of the lower shelf. Then carefully lower on top the upper shelf, allowing the necks of the bottles to pass through the 46mm holes in the upper shelf. Allow the shelf to rest on the necks of the bottles. Then adjust the bottoms of the bottles so that they sit flush with the lower shelf.

3. Tense up!

Attach the hook and eye strainers and "S" hooks (if needed) between the opposing hooks. Tighten up thoroughly.

4. Assemble into a larger unit

Make as many modules as you need of whatever sizes you require. As long as the bottles are of the same height, and your holes are drilled precisely, you can stack the modules on top of each other.

5. Experiment!

Other Ten-Green-type furniture I have worked on since: a kitchen shelving unit, that is more of a kit really, and two coffee tables that were made to commission/to suit specific bits of wood!

The tables are just the same as the shelving modules but flipped upside down, and without holes in the shelf at the base of the bottles (the table top).

For more information on this project, visit our website at www.zero-waste.co.uk.

I would be up for a trial of the new "Collaborate" feature of Instructables with someone on this project. Message me if you're interested.

Roy Mohan Shearer is a self-titled "post-industrial" designer based in Glasgow, UK. He started Zero-waste Design in 2005 to experiment with and practice sustainable, appropriate, and open design. When he is not making something he is likely to be found playing the drums.

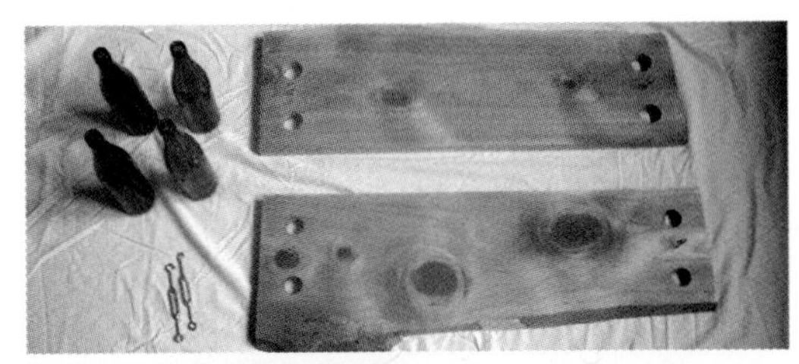

Figure C: Materials—simple!

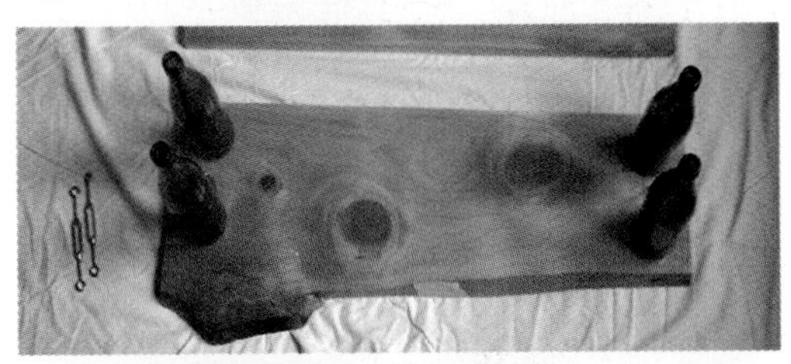

Figure D: Placing bottles on lower shelf

Figure E: Resting upper shelf on bottle necks

Figure F: Tensioning the turnbuckles to stabilize the shelves

Figure G: Multiple shelf modules—they're stackable!

Figure H: Experiment! Make a coffee table.

Figure I: Which are the "real" shelf supports?

Figure J: Make a tri-leg table!

User Notes

Scott made a stepped version

Ronnie Hinton made some without cutting the holes all the way through

Invisible Bookshelf

Amaze your friends with your blatant defiance of the law of gravity! By Troy Broadrick

Figure A: Books behaving strangely...

Cast levitation level 7 on your books!

1. What you'll need

- Book
- Good-sized "L" bracket
- Carpet knife
- Pen
- Small wood screws (of the flat headed variety)
- Large wood screws
- Tape measure or ruler
- Glue
- A stack of books for weight

To make your shelf, use a book you don't mind never reading again. A friend at work gave me a stack of old books that were collecting dust in his garage, and my previous two shelves were made out of free books that my local public library was trying to get rid of.

2. Measure twice, cut once

Measure the halfway point, make a mark, set down your bracket, draw an outline, measure again, cut out a hole deep enough that the bracket will set flush. You are doing this to the back or bottom of the book, the part that will be visible in the finished product.

3. Don't put away the carpet knife yet!

Use your knife to make a notch for the "L" bracket so the book can sit flush against the wall.

4. Screw old books!

...or at least place your small wood screws to secure the "L" bracket in place, one on either side toward the edges of the book to secure the book's pages together. Hanging the bracket over the edge of a desk, chair, or counter is helpful. The pages will try to rise up the screw as you insert it, so make sure you put some pressure on it to keep it in line. If this part is not done correctly the end result will show the pages as wavy and will tip off the viewer that something is amiss. The picture shows me using a cordless drill, but I found that doing it by hand was much more effective.

5. Glue and apply pressure

The glue will hold the bottom cover of the book in place, and the screws hold the pages together. Put the stack of books on top and wait overnight.

6. No, seriously—wait overnight

Did you think I was kidding?

7. Attach it to the wall

Sorry to switch books, but I wanted to finish the Instructable while the glue was drying. Use a large wood screw to attach the whole mess to the wall. Find a stud first.

8. Load it up!

Put some books on it to cover your handiwork. Make sure you put on enough books to cover the bracket.

9. Forget that it's there...

...and wait for the crazy looks on peoples' faces when they notice it and can't figure it out.

Troy Broadrick is a single father and is raising his son Jace in Louisville, Kentucky. Troy is a musician and would like to be a worship leader some day.

For additional information, discussion, and more, please visit the Instructables project page:

Figure B: Materials

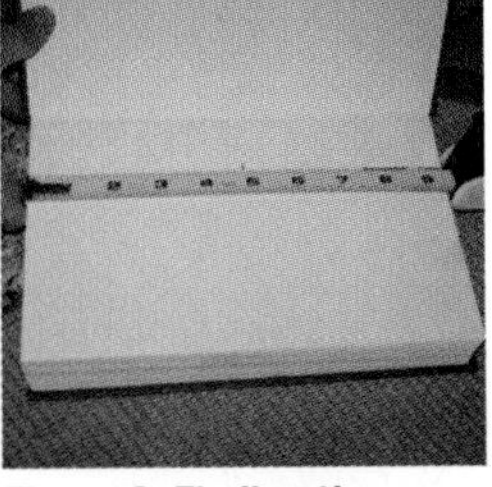

Figure C: Finding the halfway point

Figure D: Marking the bracket outline

Figure E: Cutting out pages for bracket placement

Figure F: Bracket sitting flush

Figure G: Book closed on bracket

Figure H: Notching cover

Figure I: Book will sit flush against the wall

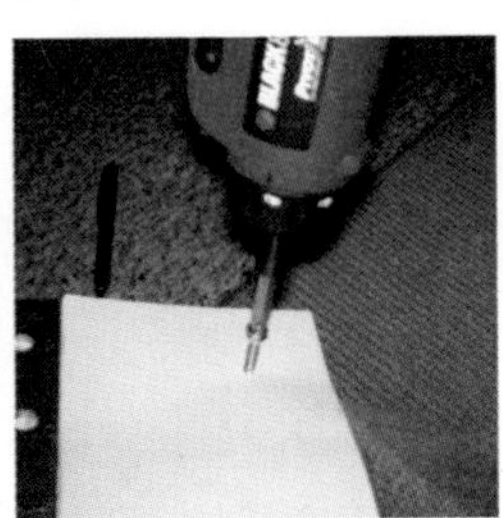

Figure J: Drilling holes for screws

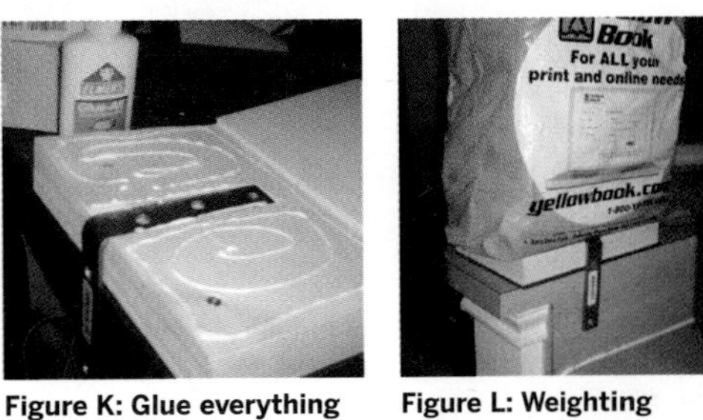

Figure K: Glue everything

Figure L: Weighting while the glue dries...

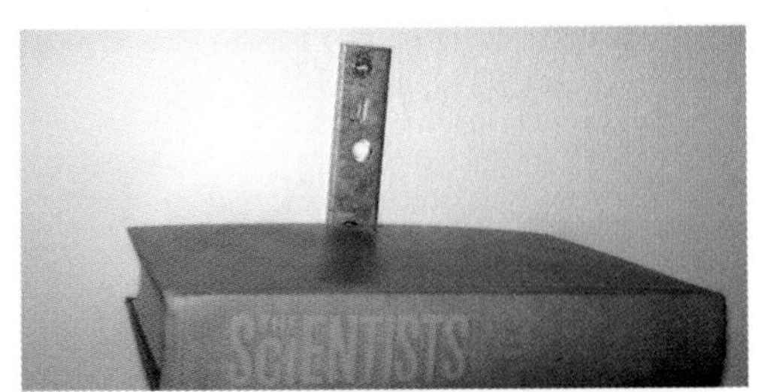

Figure M: Attaching bracket to wall

Figure N: Add books

Geometric Cut Paper Table Lamp

Add a warm glow to your life with this kirigami lamp

By Andrew Barnes

Figure A: Finished lamp

I've been making cut and folded paper lamps for a while now and get a lot of pleasure from designing, building, and enjoying them. I was thinking that I've gotten so much from Instructables I would give a little back.

You can probably get this lamp made in an afternoon (depending on paint drying time) and it should cost about $15-$20. It uses simple techniques of symmetric and asymmetric pop folds that can be the basis for more complex designs. My goal when designing is to make things with a single sheet of paper that look like they couldn't be made with a single sheet of paper.

1. Gather materials

Supplies:

- Paper—24-3/4" x 18" (see step 4)
- Long metal ruler or metal straight edge and ruler
- Craft knife with new blade
- Something to indent the paper—embosser, blunted nail, empty ballpoint pen, etc.
- Tool for pressing folds—bone folder, hard thing, etc.
- Double-stick tape, 3/4" wide
- Puck light with cord, switch, and plug
- 3/4" wood cut to 5-15/16" square
- Bumper feet about 1/4" tall
- Cord holder clips
- Heat shrink tubing or liquid electrical tape
- Drill with 1/4" and 3/32" bits
- Soldering equipment
- White paint
- Possibly some other stuff

2. Building the base

Most pucks have a place to wrap the wire on the underside. Use this to secure the wire so a tug on the cord won't cause damage.

Take the wood piece and choose a side to be the back (any side will do). Center the puck light on the base with the cord toward the back. Mark the screw holes through the puck light and mark where the cord comes out from under the light.

With the 3/32" bit, drill pilot holes for the puck. Use the 1/4" bit to drill the hole for the cord at a slight angle from the center of the top towards the back of the bottom.

Turn the wood over and drill pilot holes for the feet near the corners. Clean up the edges of the holes with sandpaper.

Now paint the entire base. You will probably lose sight of the pilot holes doing this, but when the paint dries you should see divots where they are.

Figure B: Marking the base around the puck

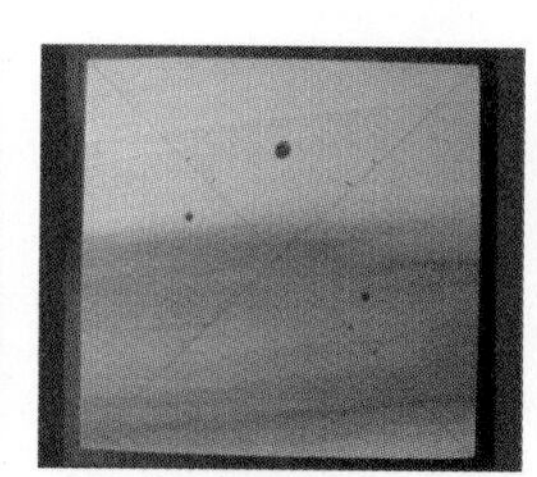

Figure C: Marked base top

Figure D: Feet go here!

Figure E: Threading the cord through the painted base

Figure F: Re-attaching the wires and wrapping with heat shrink

Figure G: Securing the wire to the base

Figure H: Attach the feet

3. Assemble the base

Cut the wire about 3" from the puck and feed the wire from the puck through the base (from the top). Now screw the puck into position and re-cut the wire so it stops about half way between the hole and the edge of the wood. Strip the puck wire and the power wire about 1/4" back.

Solder the wire back together using heat shrink tubing or liquid electrical tape to isolate and wrap the junction.

Now is a good time to test the lamp. If it's working, use the wire clips to secure the wire to the wood. The clips I had used nails to secure, so I drilled a small pilot hole in the base. Make sure that the cord is in snug enough that a tug on the cord won't strain the junction. Finally, attach the feet.

4. Laying out the lamp design

I like to use Strathmore 500 Bristol 3-ply for my lamps but anything good and stiff should work. For reference, the Bristol is 375 grams/sq.meter (gsm).

On the lampshade design drawings (Figures I and J), the solid red lines are cut lines and the broken blue lines are fold lines. The little piece on the right edge will be referred to as the flap.

Draw the pattern on the backside of the paper with a pencil. Using a very sharp blade and the straight edge, cut where appropriate. Use the indenting tool to dent the paper along the fold lines. Make sure all the cuts go right to the corners and the dents are well defined. If you do not have deep enough dents in the paper it can be difficult to get a clean fold. Now carefully erase the pencil lines.

(The downloadable files, found on this project's Instructable page, are .jpg versions of the lampshade design drawings in both inch and centimeter scale. Please note that the two designs are slightly different, and I built mine in inches. If you go metric you'll have to adjust other parts on your own.)

5. Folding the lamp

Note: I was folding and shooting the photos at the same time during the building so my technique is poor with a lot of pressure and curved paper and the like. You should almost always use two hands, and usually with one on either side of the paper.

Using a straight edge, table edge, or other method, fold the three corners that don't have the cut out design, and crease well. Unfold enough so the paper is relatively flat.

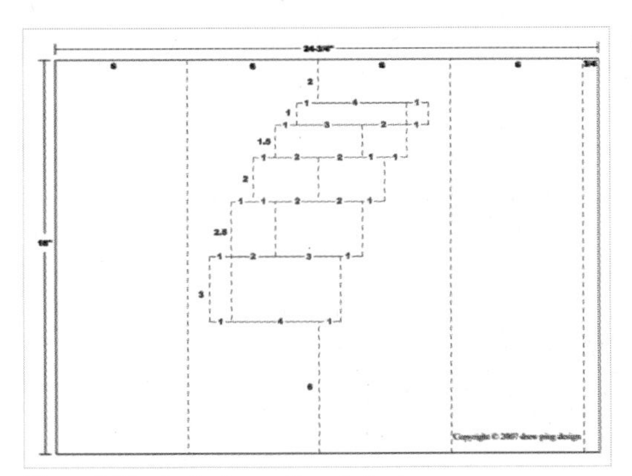

Figure I: Shade design pattern in inches

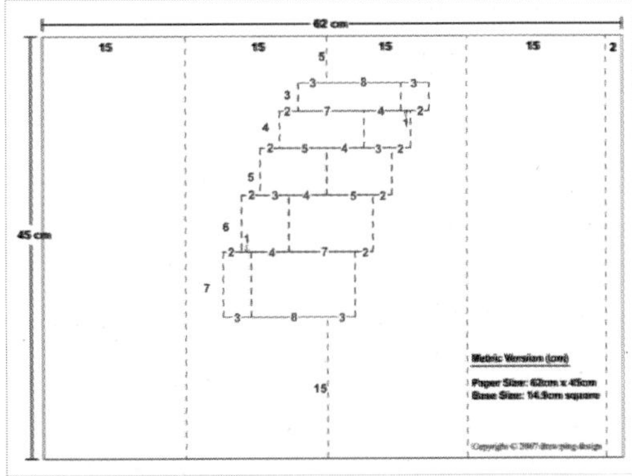

Figure J: Shade design pattern in metric

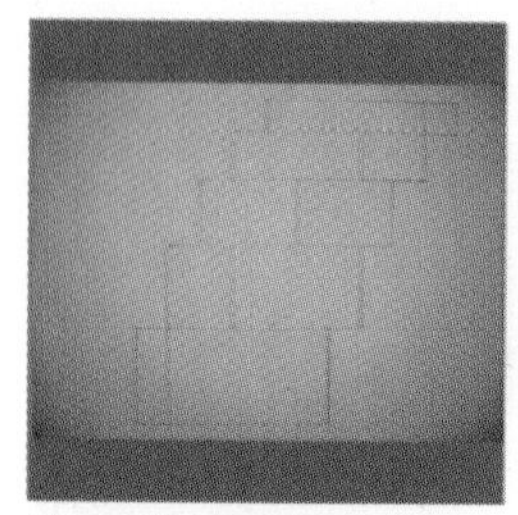

Figure K: Design penciled on Bristol

Figure L: Folding on the edge of a table

Figure M: Folding with a straight-edge

Figure N: Using a bone folder to press the fold flat

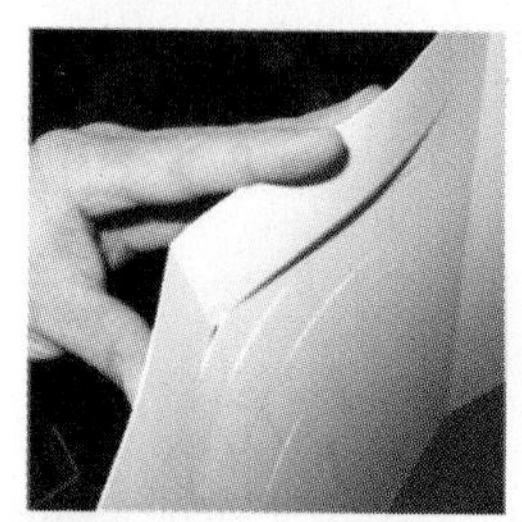

Figure O: Start folding from the top of the shade

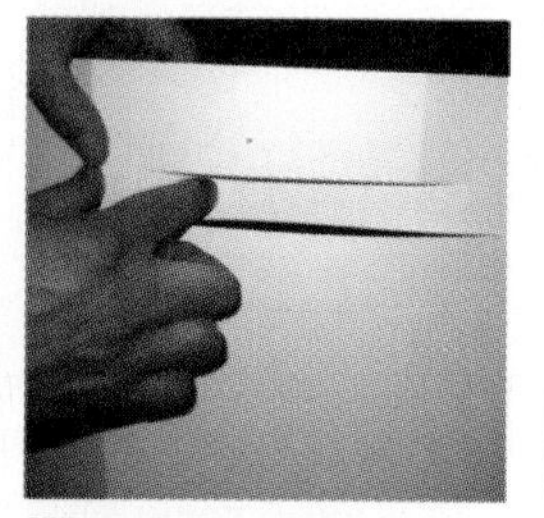

Figure P: Pressing the crease in from the outside and backing up the crease from the inside

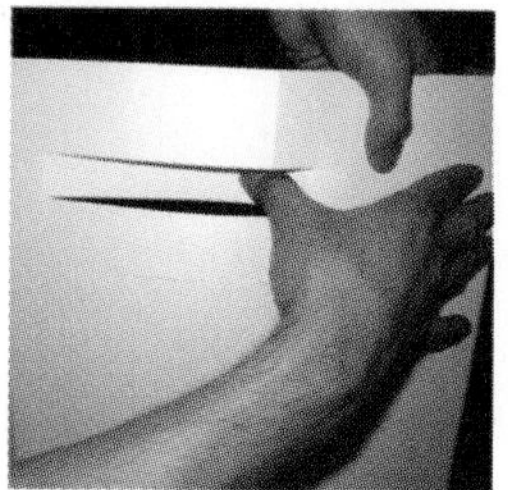

Figure Q: Pressing in with thumb

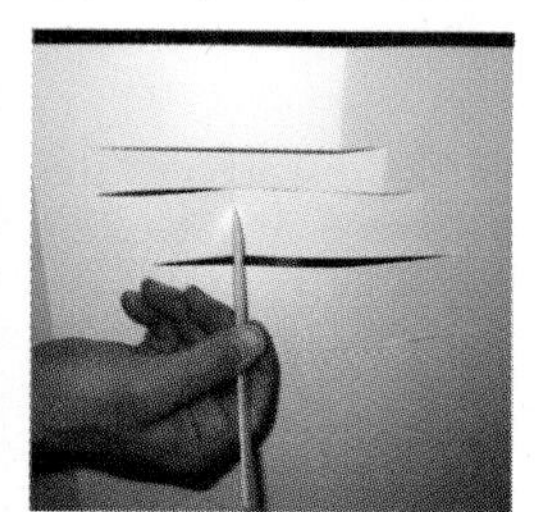

Figure R: Using the folding tool to push in the crease in the middle—work down each of the strips

Figure S: Starting to look like a lampshade!

Figure T: Squash fold

Now comes the tricky part. Starting with the top and bottom creases, fold the paper as far as it will go without putting undo strain on the paper. With this accomplished you should be able to start pushing the design into the body of the lamp.

Standing the shade upright at this point can help. Doing one at a time, use the indenting tool or your fingers to push and press the paper into position. Don't do any one all at once; in order to avoid unwanted creases, tearing, or curved paper you have to do a little bit on each until the folds are established.

Once everything is going the way it should go, you can lay the paper flat on the table and carefully push it down (all the while pushing and pressing the individual folds) until the whole thing is flat. Now crease every fold very well, if you don't the lamp will tend to be out of square.

With the lamp folded over, place a piece of double stick tape on the outside of the flap. If the tape has a protective strip on it, leave it in place for later. If you put a big flat thing on top of the lamp to press it down the tape will be easier to apply.

6. Final assembly—attach the paper

Put a strip of double stick tape all along the edge of the base (remove protective strip if present).

On a flat surface, place several risers at least as tall as the feet under the base. I used DVD cases and they worked very well. What you are trying to do is create a surface that is level to the bottom of the base so when the paper is applied it will be flush with the bottom.

Open up the lamp and place the flap on the back left corner of the base. Press into the tape to adhere. Aligning this corner well is essential to ease construction. Continue around the base adhering the paper to the tape.

If all of your components were accurately measured, it should fit easily. If it's a bit tight you can *gently* tug the paper and hopefully it will give enough to fit. If the paper is too big you should try to put an equal amount of gap on both sides.

Figure U: Putting double-stick tape on the flap—note weight on top of lamp

Figure V: Putting double-stick tape on the edge of the base

Figure W: Propping up the prepared base before attaching the shade

Figure X: Starting to attach the shade—note the wire coming out the back, with the flap at the back left corner

Figure Y: Continuing to wrap the shade around the base

Figure Z: Around the third corner of the base—it's getting exciting!

Figure ZA: Securing the vertical seam

Figure ZB: A nice, even seam

Figure ZC: Hang the wire off the edge of the table

Figure ZD: Pressing the seam down from the inside

Figure ZE: Pressing the shade onto the base

Figure ZF: Pinching the shade to square it up

7. Final assembly—vertical seam

Now that the paper is attached all around the base, begin to secure the vertical seam. Put one hand inside the lamp and one outside. Starting at the bottom, attach the two sides. The flap will be on the inside and the edge of the other piece should be just shy of the corner. Continue adjusting, attaching, and pressing the pieces together until you reach the top. Hopefully you will have a nice straight seam.

Place the lamp on its back with the flap towards the ground and the wire just off the edge of the table. Now press all along the seam to get really good contact. Press around the base to attach the paper firmly there as well as doing the top edge, then rotating 90 degrees all the way around.

Stand the lamp up and look at it from above. If it is not square, gently adjust the folds until it is.

8. Light up your life

Now plug it in and flip the switch. Of course, like so many things on Instructables, this idea can be expanded upon and be made significantly cooler. These are great with better lighting (high power LEDs—thank you, Dan: www.instructables.com/member/dan), modular bases, and much more complex designs.

Andrew Barnes is a paper art enthusiast and lighting fanatic. He is the founder of drew ping design (www.drewpingdesign.com), a company devoted to legitimizing creative whims.

Plank Chair

Make this sturdy, easy-to-store chair out of one piece of wood By Jesse Hensel

Figure A: Finished chair

Figure B: Saw set-up

Figure C: Plank cut in two

Figure D: Back leg

Figure E: Cut slot to fit back leg through

Figure F: Checking for fit

Figure G: Marking angles for fit

Figure H: Trimming angles

A simple and satisfying chair made from one plank of wood.

1. Get a large plank of wood

The piece of wood that I used for this chair was 7' long, 11" wide, and 1-3/4" thick. I originally used driftwood planks to make these chairs because they look nice and are free. If you choose to use found wood be careful that it's not pressure treated.

2. Cut the plank in two

Cut the plank into two pieces. For this chair I made one piece 40" long and the other 51". The dimensions can be varied to fit your preferences.

3. Make the back leg

Cut two inches of wood off of the sides of the board starting 15" from one end of the plank.

4. Cut a hole in the second plank

Place the tapered end of the first board 15-1/2" from the bottom of the second board and trace around the edge. Remove enough wood to slide the first board through the hole in the second at an angle.

5. Try it out!

Fit the two pieces together and try it. This is your chance to readjust anything you're unsatisfied with.

6. Fine tuning

The planks will not intersect at right angles, so some wood will need to be trimmed. Figure G shows the marking process. Use a saw to redefine the angles.

7. Finish and relax

I wanted to remove the glitter on the boards so I resurfaced the plank with a chisel. However, the only necessary step here is to relax in your new chair.

Jesse Hensel is a contemporary artist whose work is informed by the Yup'ik woodcarving tradition. His rugged natural sculptures use an Alaskan frontier perspective to critique contemporary society. San Francisco exhibitions have included *Man Cradle* at the Diego Rivera Gallery and *Keggtuli (Pike)* at the de Young Museum.

Instructables project page: www.instructables.com/id/Plank-Chair

Stool Made from Bike Parts and Crutches

Make this adjustable stool from found materials By Ryan "Zieak" McFarland

Figure A: The finished stool

Figure B: The raw materials

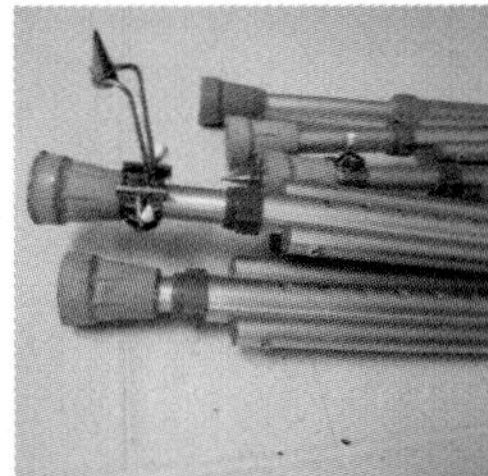

Figure C: No need for these pointy tips for use during icy conditions!

Figure D: Cut the crutches to length at the hole

Take a few metal crutches, a couple of bicycle wheels and inner tubes, and a few bolts, along with some insulation and a piece of plywood, and you have the makings for an adjustable height stool.

1. Supplies and materials

You will need:

- Crutches (3–4)
- Bicycle wheels with a 16" diameter (2); BMX bike size
- Inner tubes from a 26" tire (4); mountain bike type
- Sheet metal screws
- Carriage bolts, washers, and nuts
- Rigid insulation
- Plywood

Tools:

- Drill and drill bits
- Hacksaw
- Ratchet and sockets
- Sabre saw
- Marker
- Scissors
- Rags
- WD-40

I purchased the metal crutches from a thrift store. They were $2 per pair. Try to get two pairs that are as close to identical as possible. (You might consider buying an extra pair that fit you in case you need them some day. A few years ago I tore a muscle in my leg and was charged $80 for a pair.) Remove the handles from the crutches. The bicycle rims don't need to be perfectly straight—the ones I used were in horrible shape. I didn't bother trying to true the wheel; I wanted to see if this idea would work first. You can probably get the wheels and inner tubes at any bike shop for free. You'll want to remove the axles. Wash the grease off of the hubs.

2. Cut the crutches to length

Use the hacksaw to cut the crutches off. Set the crutch to the shortest height and then measure 22" up. Mark that spot with the marker and cut at the nearest hole. That will give the leg strength where the rim attaches. I used 22" because that was the height of a stool that I have. I wanted this to start at that height and go higher.

3. Calculate the leg locations

I used three legs, but in hindsight, four would have been better. Three legs is a little tippy when the stool is high and a person is getting on or off of it.

Figure E: Calculate the leg locations

Figure F: Drill pilot holes in the crutches

Figure G: Drill pilot holes through the rims

Figure H: Assemble the top. Starting to look like a stool.

Figure I: Marking the bottom wheel rim for notching

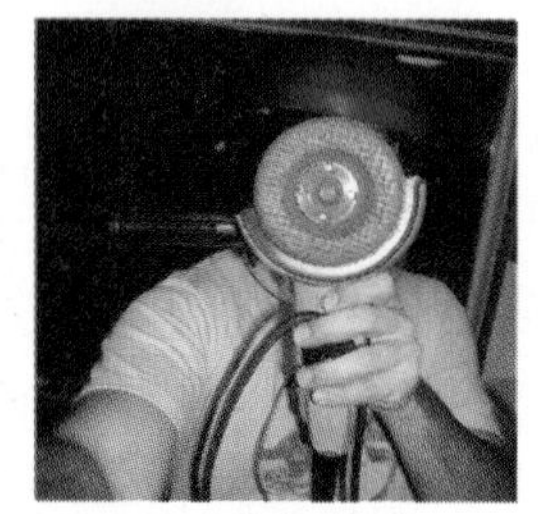

Figure J: Safety first!

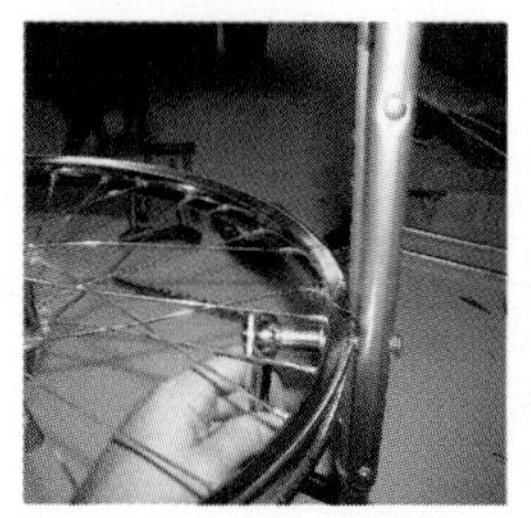

Figure K: Bolting on the lower rim. Supervisory cat in the background.

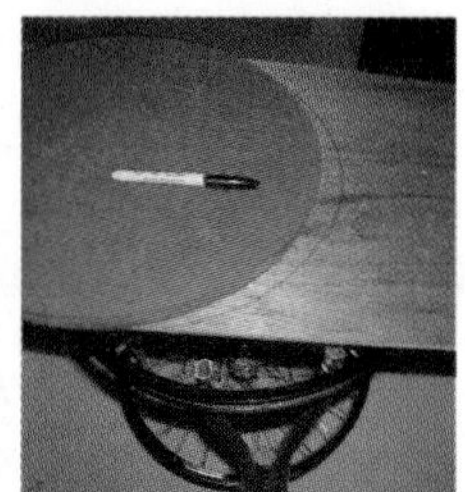

Figure L: Marking the plywood shape for the seat, using the foam for a template

Figure M: Getting ready to put the seat on

Figure N: Cutting the plywood with the sabre saw

Figure O: Cutting a hole in the center of the plywood to fit over the hub

Figure P: Tie the end of the inner tube and tuck it under a spoke to anchor it

Figure Q: Don't forget to cut off the valve stem

Figure R: Cutting down the length of the inner tube

Figure S: Wrap the lengths of tube across each other until they cover the seat

Starting where the opening for the inner tube valve stem is, count the spokes on the rim you plan on being the seat. Mine was 36. I divided that by 3 and used that number to space each of the crutches out 12 spokes. Note that, since each leg has two attachment points, the spacing will be from one leg "Y" to the corresponding "Y" of the next leg. Mark the locations of the 6 attachment points with the marker (Figure E).

4. Drill pilot holes

I drilled a hole slightly smaller than the threads of the screws I used through the steel rim and used an even smaller bit for the pilot holes in the aluminum of the crutch leg. This is a safe place to experiment because it will be covered up unless you upholster the stool differently than I did. I also used the scrap pieces from the crutches to get a feel of how far I could put the screw into the aluminum before the screw stripped the hole. If you have problems with this then just bore the holes out and bolt through like the bottom rim will be (Figures F-G).

5. Assemble the top

Put the screws in and admire the progress! Now measure 16" down from that rim and mark the inside of each of the legs. This is a comfortable distance (for me) to have my feet resting on the bottom rim. If you have a favorite stool, measure the distance to find what works well for your leg length. By attaching the rim here, the distance will stay where we like it no matter what height the stool is adjusted to (Figure H).

6. Ready the bottom wheel

After a bit of trial and error, I found that it was important to notch out a section of at least the top flange of the bottom rim to allow the crutch to adjust in height while bolting the legs on securely. I drilled through either side of the adjustable leg portion and through the corresponding part of the rim. Then, by sliding a few of the carriage bolts in, I could mark the sections of the rim that needed to be trimmed back. I used the marker to blacken the end of the bolt that was closest to the valve stem opening of the rim. That made sure I didn't have to fiddle with the rim and the legs trying to get bolts through misaligned holes. Cut the areas out for the center section of the leg. This could be done with a hacksaw and some frustration, or you can use an angle grinder. Making a smaller cut on the bottom flange of the rim isn't a bad idea; if you tighten the rim and legs together too far, the rim might press into the crutch leg and restrict the adjustment of the height (Figures I-K).

7. Cut the insulation and plywood to fit the rim

Use the rim as a template to cut a piece of rigid insulation or a thick foam pad. Then use that to mark plywood and cut the shape with a sabre saw. Cut out a 3" hole in the center of the board. This will let the hub pass through and allow the wood to lay a bit flatter. Using a sabre saw would work, but a 3" hole saw worked fantastically (Figures L-O).

8. Use inner tubes to cover the seat

Cut the valve stems off of the inner tube and then cut along the full length. If there are lines on the rubber to follow, use them to help cut a straight line. Tie a knot in one end of the inner tube and tuck it under one of the spokes. Stretch the inner tube over the top and wrap under until you have gone across the top three times—rotating around the seat as you go. Tie another knot in the end and tuck under another spoke. Repeat three more times but switch the direction of the wraps on each inner tube to prevent the wraps from forming a fan shape (Figures P-S).

9. Clean the seat

I used WD-40 to clean the inner tubes, but just about any cleaner certainly would work. Especially something designed for rubber.

For more of my projects and activities, visit my website at zieak.com.

Ryan "Zieak" McFarland lives in Alaska and spends his days as the parks and recreation director for a small town. Nights and weekends he does computer stuff, tinkers with projects, and spends quality time with his friends.

Wireless Home Router with Analog Utilization Meter

Using an old analog gauge to display network information in a more human readable form

Pauric O'Callaghan

Figure A: The completed router

Figure B: The router in action

I grew up in and around boats making wiring looms and control panels, and have a collection of gauges and dials that would normally be connected to small marine diesel engines.

Note: This Instructable was among many that were featured in *The New York Times* article "In a Highly Complex World, Innovation from the Top Down," by G. Pascal Zachary, published July 29, 2007. See www.instructables.com/forum/TW6BU3YF4R3E4Y9 for more details.

Today I work as an interaction designer creating interfaces that simplify complex data. As such, I like re-using the old analog gauges to display network information in a more human readable form... tying my past to the present to some degree.

I used a 3" rev counter—a simple clean design—that came off one of the boats my dad owned when I was a kid. I wired it into a wireless router I had lying around at work.

The rev counter is a rough approximation of the traffic between my home network and the Internet.

1. How it works—an overview

There are many ways to find out how much bandwidth is being used. This being the first pass at visualizing the usage, I opted to simply use the uplink LED as an indication of the amount of traffic passing between my home and the Internet.

This has some serious limitations. I do not know whether the hardware (Broadcom chipset) or firmware (DD-WRT) contains the sampling algorithm that drives the LED (it's probably the chipset).

Here's the first issue: an LED must be on for around 30ms for the human eye to register it properly. Networking packets are much much shorter than this. So the router must do a little math and translate real network traffic into slower LED blinking. So there is a sampling loss, and the LED is a rough approximation of the actual traffic.

I must boost the 3.3v that drives the LED up to 14v required for the rev counter (most automotive dials and meters like this are linear 0-12v or 14v). For this I used a basic op-amp circuit (Figure C shows a rough sketch of it). Without some swanky digital-to-analog conversion I again lose a lot of resolution.

The result is not a very good representation of the traffic bandwidth being used, but the further I got into the project, the more it became an interesting object of art and less a solution to the original problem.

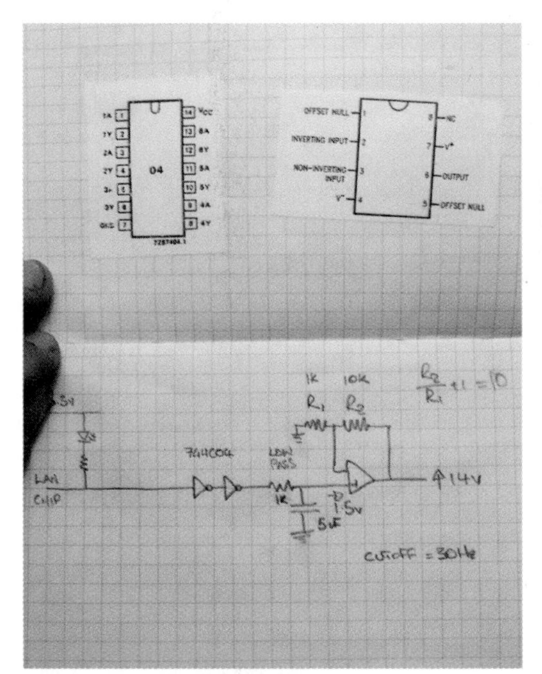

Figure C: A sketch of the circuitry for this project

Note: I've been working with the guys from DD-WRT (http://dd-wrt.com/dd-wrtv2/index.php). If your router supports it, I highly recommend you upgrade your current software to this feature rich open source firmware.

2. Front panel

I desoldered the LEDs from the router's PCB and routed them out to the front with a ribbon cable and header connector.

Then I designed an overlay in Omnigraffle on a Mac, printed it to an overhead projector transparency with a laser printer (you can get transparencies for ink jet printers as well). I also chiseled out a channel for the overlay so it had a nice inset look.

I fell in love with a power switch from a piece of development hardware at work, unfortunately it was a momentary switch so I

Figure F: The back of the panel face with the meter and indicator LEDs attached

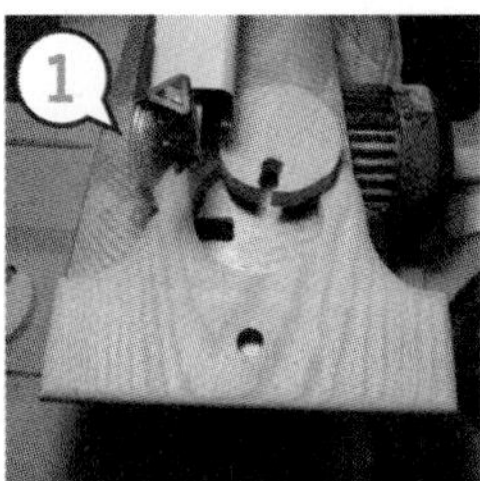

Figure D: 1. A fish tank pump that blows away the dust as you cut

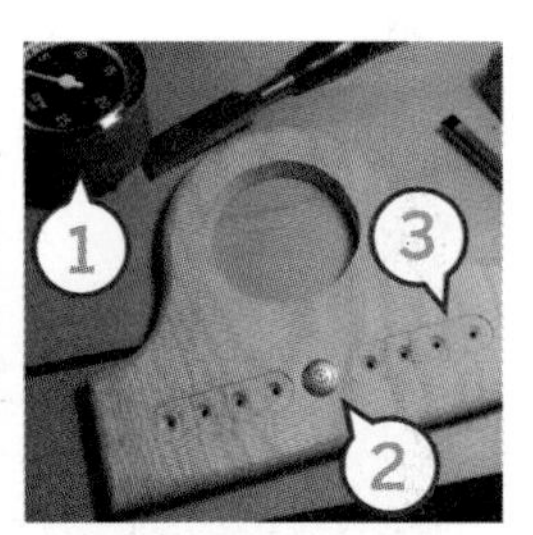

Figure E: 1. Rev counter from a Lister Perkins 30 HP marine diesel engine 2. A damn good looking front power switch 3. Inset for the front decals

spent quite a lot of time retrofitting a NO (normally open) microswitch with a blue LED glued to the tip.

I routed the edge of the large front panel section (see Figures D and E). Figure F shows the rev counter attached to the front panel. You can see the bezel in Figures G and H.

3. Op-amp circuit

There are two stages between the LED on the main router board and the rev counter:

A. Isolate the op-amp circuit from the router board. This is done with a buffer in the form of a 74HC04, a hex inverter with gates that will not draw any current from the router and will output a signal based on the inverse of its input. This guy comes with 6 gates, so if you want to get the same output signal as the input signal you tie the gates back to back.

Note: I had an intermediary stage that was designed to smooth the square wave signal driving the LED to a nice analog rising/falling charge to the rev counter. However, the mechanics of the counter provided the smoothing I wanted. So, in some of the diagrams you'll see an RC Low pass filter.

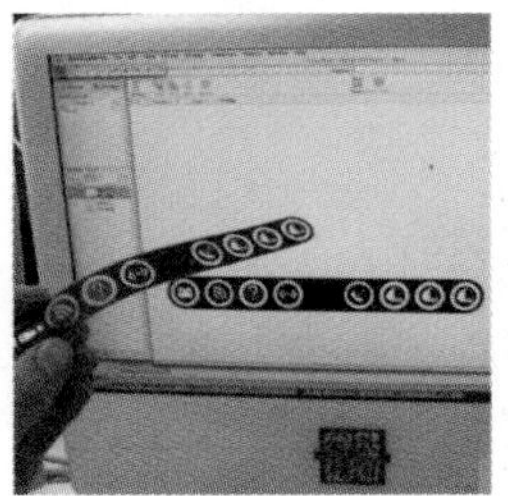

Figure G: Front decal design

Figure H: Paper prototype of the front decals

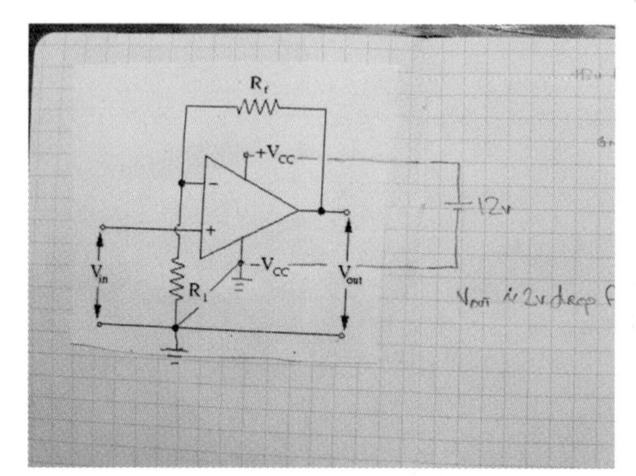

Figure I: The op-amp circuit for converting the network activity LED into an analog signal

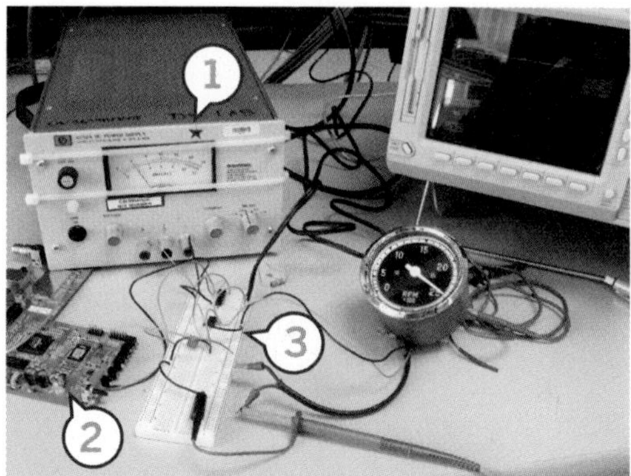

Figure J: 1. I'm very fortunate to have access to a lab equipped with every imaginable tool you could possibly need for designing a project like this 2. Linksys router board. Although I ended up using a different one from Buffalo, I used the Linksys router to get a flashing LED for testing the circuit 3. Solderless breadboard

Figure K: 1. Trimpot for tuning the gain on the amp 2. Input pins 3. Op-amp 4. Hex inverter/buffer 5. Output 6. Twelve volts in; for testing purposes I had to mess around with the Vcc for the hex inverter. I ended up with its rail being around four volts 7. Twelve volts in

Figure L: The rev counter and button

Figure M: 1. I drilled right through this section and in about a 1/4" to the back of the front panel, then set in some dowels. I then glued this section to the rest on the left hand side in this picture. This gave me two nubs that aligned when putting the case together.

Figure N: Drilling with a Forstener Bit

Figure O: 1. I sealed the wood with a semi-gloss finish 2. I used this to rub the stain in and wipe off the excess

B. The op-amp. I chose a very old chip, the LM 741, that worked but came with a lot of limitations that drastically affected the design. The rev counter never goes to zero, and the range seems to hover around the center of the dial. Limitations of the op-amp. It's a lesson learned, and in future, I'm going to improve this circuit to have a wider range of output.

Figures I, J, and K show the design of the circuit progressing from paper, to solderless breadboard, onto a prototyping board.

4. Parts & tools

Parts list

- Rev counter (Figure L): It's a VDO meter that still had some fishing boat gunk on it. Testing showed its characteristics were that it liked to be fed a linear input of 0-12v.
- Router: The smallest PCB form factor in the market that I found was from Buffalo
- Software: I installed the DD-WRT open source router firmware (www.dd-wrt.com/dd-wrtv2/index.php).This isn't strictly necessary but I can't prop this software enough.
- Wood: Oak plank from local hardware supply
- Solderless breadboard for testing
- Protoboard/perfboard for the final op-amp circuit
- Op-amp: LM 741
- Buffer: 74HC04 hex inverter
- Decal: transparent stock for laser/inkjet

Tools

- Electronics: Soldering iron, multimeter (if you have an oscilloscope, use it for fault testing)
- Carpentry: scroll saw, table saw, carving chisel, mortising tools, glue, and dowels
- Decal: any drawing application such as OmniGraffle, Adobe Illustrator, or Microsoft Paint
- Most of your digits still attached

5. The case

This is a very organic way to make a case, the right way would be to use some cheap chipboard or pine and veneer it with a nicer wood. In solidarity with the unskilled, I layer-caked the case up from individual sections cut from a plank.

I milled out the areas needed for the rev counter, router PCB board, op-amp protoboard, and added a channel for the aerial coax to the rear.

I marked each section from a template, cut with the scroll saw, glued, and sanded. Then I finished it with a dark stain and semi gloss.

Figures M, N, and O show various steps in assembly. The following video shows more detail: video.google.com/videoplay?docid=3821193407635705452.

Pauric O'Callaghan says: "Design is the art of deleting the non-essential. Making and hacking skills allow us to re-design products to meet our needs and desires while removing unwanted or superfluous functionality."

User Notes

Pauric, author of this Instructable, in reply to a question from Instructables member Mourtegoul as to whether the op-amp could be replaced by some components tied to the square wave output that drives the activity LED: Very true, there is any number of solutions to the problem. Some suggested a microcontroller. The op-amp works, and I will look at a simple transistor. The thing to think about is smoothing; with a flat out square wave driving the meter you will notice every little blip, I wanted some form of cumulative gain. Even with the op-amp giving me the desired output, it still swings a lot.

Universal Lamp Shade Polygon Building Kit

One simple cut-out shape lets you build all sorts of different designer-looking lamp shades By Dan Goldwater

You can make dozens of different geometric forms using various numbers of this cut-out shape made from paper or plastic. The pieces just fold together by hand into rigid forms, and you can take them apart and build them into new shapes any time! This is fun and educational for kids and adults alike, and you get a really nice lamp shade when you are done (as seen in *ReadyMade* magazine, Dec. 2007/Jan. 2008 issue).

Figure A: This one is made from 30 pieces of 0.5mm acetal (Delrin)—the photo looks amber but the lamp is actually white

1. What you need

I saw a lamp shade made from a simple repeating shape at a friend's house, so I traced the shape and made my own. The lamp I saw used thin plastic for the pieces. I believe the original design for this lamp was done over 30 years ago by the firm Iqlight (www.iqlight.com); they sell pre-cut parts in case you do not want to make the parts yourself.

The basic working shape is shown in the illustration in Figure F; the DXF file can be found on the Instructable page.

You will need sheets of paper or plastic that allow light to get through. The stiffness of your material determines how large your pieces can be—stiffer material for larger pieces and larger lamps, thinner material for smaller pieces and smaller lamps.

You will need a lamp fixture—just a raw socket on a cord. I found some nice ones at IKEA for $4, and some fluorescent bulbs. Use a compact fluorescent bulb so you can get more light without melting the plastic.

I experimented with a number of different plastics and sizes for the parts, here are my results:

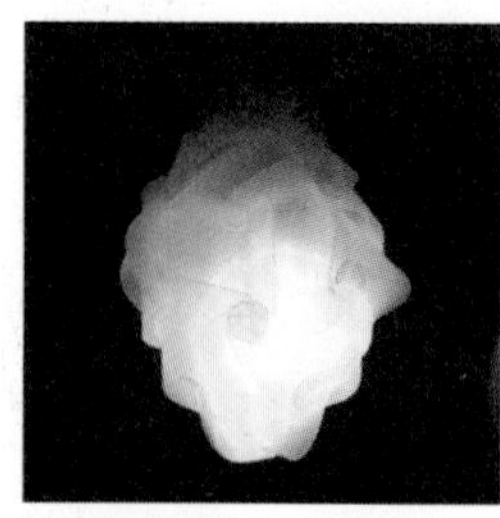

Figure B: 24 pieces of 0.8mm nylon—a bit annoyingly yellow-ish

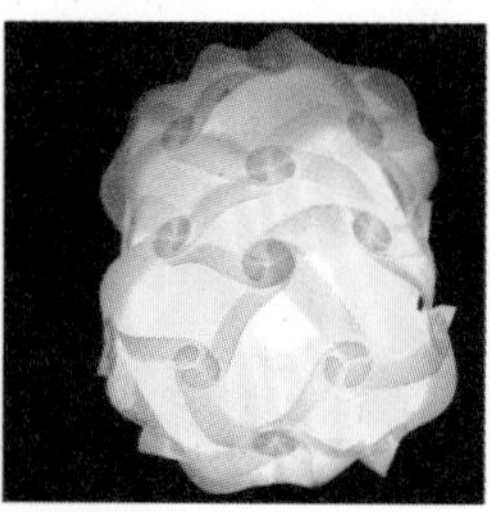

Figure C: 35 pieces, basically a tube with 2 end-caps, of 0.8mm HDPE

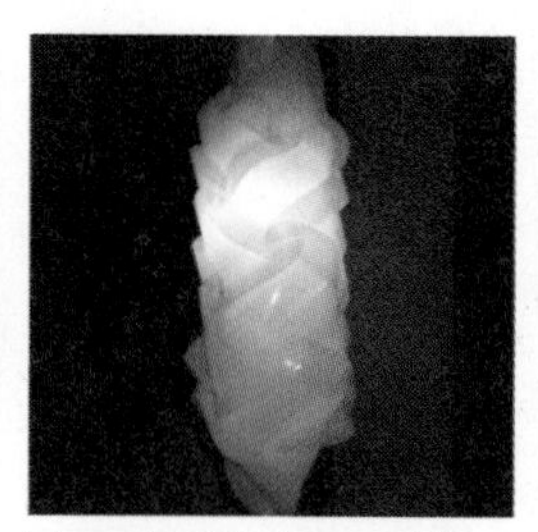

Figure D: 24 pieces, with a cool-white bulb—the others have warm-white bulbs

Figure E: 80 pieces

For additional information, discussion, and more, please visit the Instructables project page:

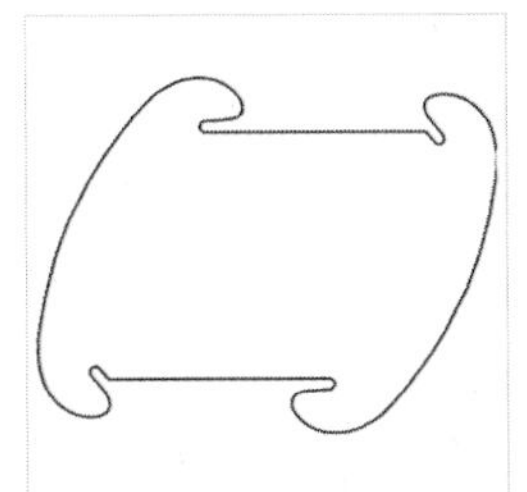

Figure F: The basic shape—surprisingly simple!

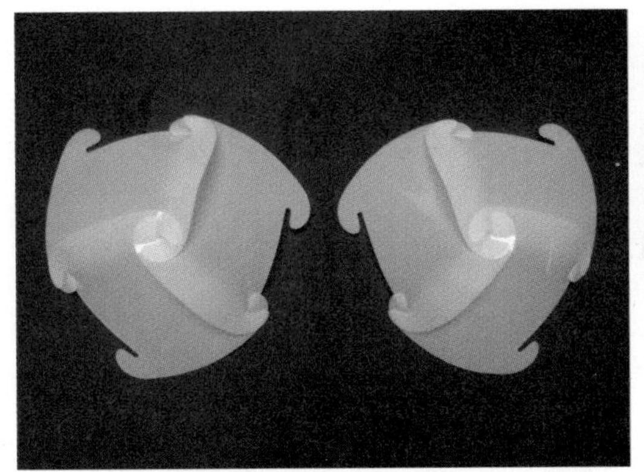

Figure G: Right-handed vertex and left-handed vertex—mirror images of each other

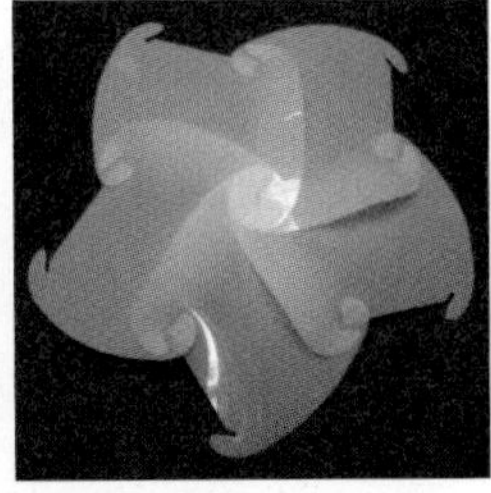

Figure H: Five narrow angles meet at the vertex (corner)—this is a top view

- **HDPE:** works well, looks good, cheap and easy to get. I used 0.8mm thick stock for 8cm pieces (measured flat-side to flat-side). That's about the smallest you'd want to go with that thickness, you could go up to 12 or 15cm with that thickness. This is the least expensive plastic by far. One of the commenters suggested using plastic milk-bottles, which are made of HDPE, this is a good idea!
- **Acetal (Delrin):** this seems to be the best choice for looks, it has the purest white color and best light dispersion (basically, looks just like acrylic except it doesn't crack as easily). I used 0.35mm stock for 6cm and 8cm pieces, and 0.5mm stock for 8-12cm pieces, and 0.65mm stock for 12-15cm pieces. It is still a bit brittle and harder to work with than HDPE or paper, and more expensive—I don't recommend it for your first lamp. If you cut this with scissors it will be somewhat tricky due to the brittleness, but possible.
- **Vinyl:** I didn't try it, probably want to use thicker pieces as it's not very stiff. But you can use colors!
- **Paper:** I did not try paper but it should be good for smaller constructions.
- **Laminated paper**: this is a great idea suggested in the comments section. Try laminating color tissue paper, then cut the pieces from that. Very unique and colorful!
- **Acrylic:** too brittle, it will crack when you try to assemble. Too bad!
- **Nylon:** has a yellow-ish look when lit up which makes it undesirable.
- For a large construction use stiffer material

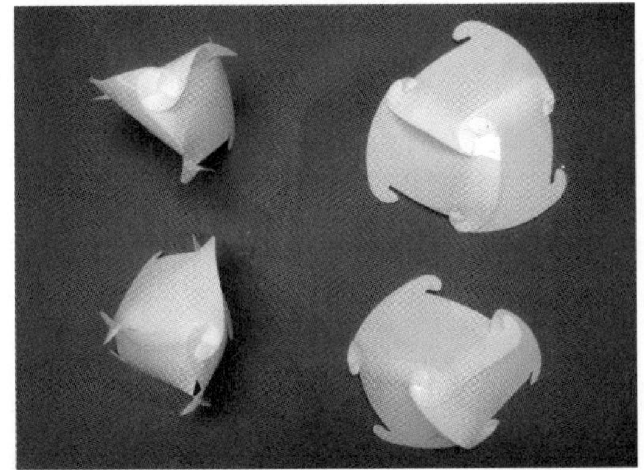

Figure I: (clockwise from top left) 3 narrow angles at the vertex; 3 wide angles at the vertex; 1 narrow angle and 2 wide angles at the vertex; 2 narrow angles and 1 wide angle at the vertex

Figure J: Bottom view of the same 4 assemblies as seen in Figure I

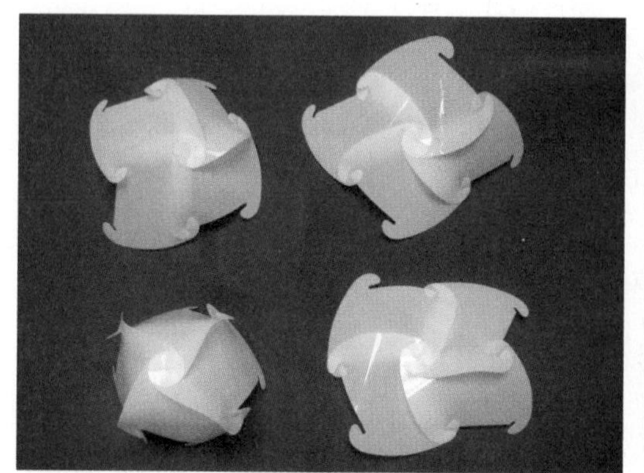

Figure K: (clockwise from top left) 3 narrow angles and 1 wide angle at the vertex; 2 narrow and 2 wide angles meet at the vertex in a checkerboard pattern—the 2 narrow angles are across from each other; 2 narrow angles and 2 wide angles, with the narrow angles adjacent to each other; 4 narrow angles at the vertex

Figure L: Bottom view of the same 4 assemblies as seen in Figure K

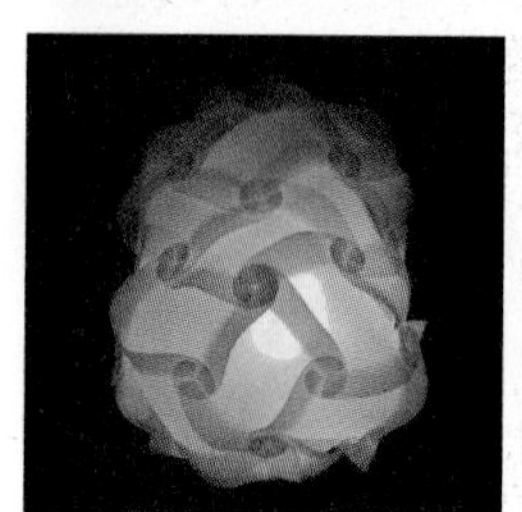

Figure M: This lamp has 3-piece, 4-piece, and 5-piece corners

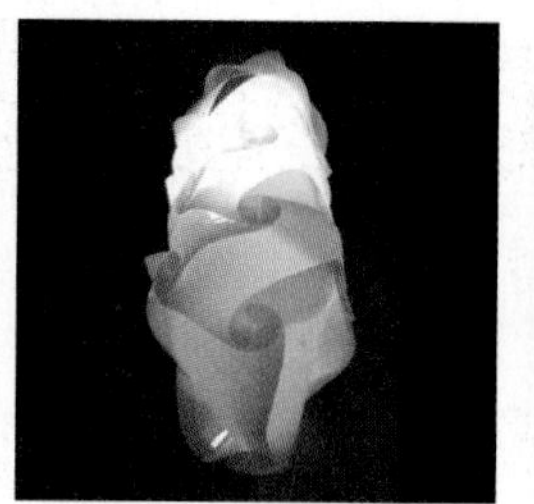

Figure N: Another view of the 24-piece lamp shown above in Figure D

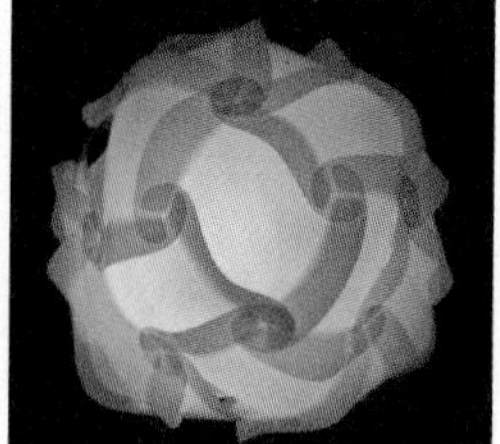

Figure O: 30 pieces, in a less-rounded configuration

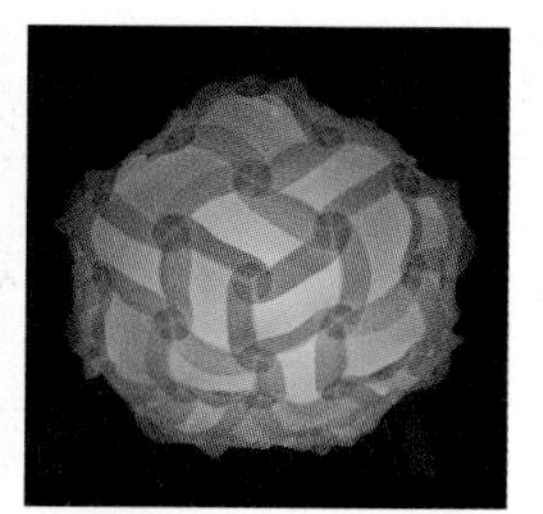

Figure P: Another view of the 80-piece lamp

for the same size piece. For example, if you are making a lamp shade with 12 pieces use thinner material than if you are making a shade with 100 pieces, assuming same size pieces.

Where to get it:

- HDPE 1/32" sheet: www.usplastic.com, item number 42584
- Acetal sheet, 0.015" and 0.020": www.mcmaster.com, item numbers 8738K52 and 8738K53

2. Cut out your pieces

You can cut your pieces in a couple ways:

- Trace them out with a marker, then use scissors or a knife. This is slow, but can be done easily.
- Make a "cookie cutter" out of sheet metal in the shape of the part, then heat the cutter with a torch and use it to stamp out the parts (only works for plastic).
- Use a laser cutter

3. Assemble

Just try fitting the parts together! They go together in many different ways. You can make a variety of corners with three, four or five adjoining pieces.

There are several ways to think about and categorize the different types of geometric shapes that can be constructed. I've shown top and bottom views of every different type of vertex (corner) that can be built. All larger assemblies are made up of a combination of the types of corners shown, so think of them as your building blocks.

Corners can also be "left-handed" and "right-handed"—mirror images of each other.

4. Assemble (part two)

You can make forms with anything from 8 to 100 or more pieces each. The only drawback seems to be that there is no way to make concave corners, only convex corners are possible.

You can see my examples, and you can see more

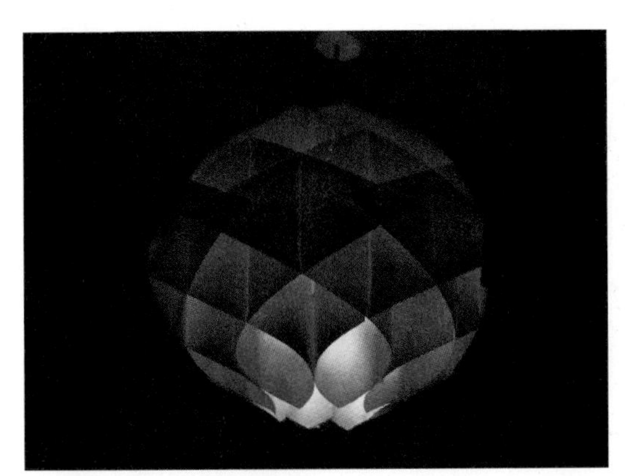

Figure Q: Lamp seen in China

at this site: www.iqlight.com. That site also sells the parts if you don't want to make them yourself.

5. Other ideas

I think the design I've presented is my favorite because it is reconfigurable into lots of shapes, but I thought I'd put some links to other styles which can also be made from sheets of paper or plastic.

I spotted the lamp shade shown in Figure Q hanging from the eaves of a hotel in China. It is made from (I think) six identical pieces (possibly seven or eight). Each piece is a large diamond shape with slots cut in it to allow sliding the pieces together. It looks like each diamond is slid into itself to form one of the central vertical tubes; its free tips are then slid into the other pieces to hold the whole thing together. I have not actually made one yet, so please comment if you have.

Also see: www.instructables.com/id/Flower-Calender; www.yasutomo.com/project/paperlampshade.htm.

You can also usually find one or two polygon-inspired lamp shade designs at your local designer lighting store, and at IKEA.

Dan Goldwater loves beams of light and the devices that make them. Dan is a co-founder of instructables.com and SQUID Labs and now makes practical digital light art for bicycles at Monkeylectric.

User Notes

Benito Cavazos made several different sizes!

Travis Swaim's variations include one with squared-off pieces

Thomas Tonino made a rice bag lamp

Steve "the gibbon" Gennrich shows one hung from a cord

Paul "Manggo" made a standing lamp

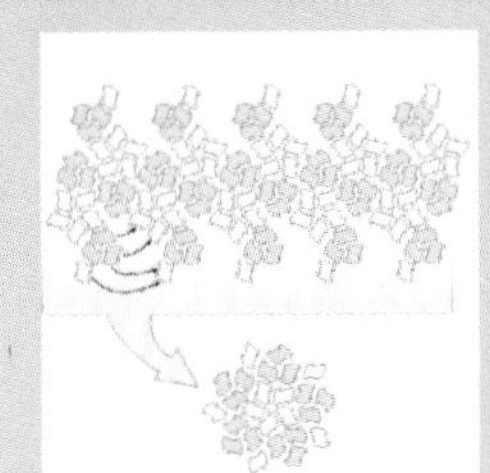

Nicolás De Francesco, aka nevermore78, created a construction roadmap

Travis Swaim also made one with CDs!

Pipe Dream Bed

Have sweet industrial dreams on this sturdy, adjustable bed made from iron pipe

By Limor Fried

Figure A: The finished bed

A bed made from iron pipe. The design is simple yet sturdy. Not too pricey, simple to make, and easy to adjust/disassemble, this bed will give your room an industrial feel.

1. Acquire Materials

Here are the required materials:

- 1-1/2" Schedule 40 iron pipe: available at plumbing/hardware stores; it's rather heavy. You can buy 10' lengths and then cut it yourself. Note that 1-1/2" pipe is not actually 1-1/2" in diameter. It's really closer to 1-7/8"!
- Hacksaw/pipecutter/chopsaw: for cutting the pipe down
- Couplers: you can buy these from McMaster Carr; they are used to build railings. They are sometimes called "speed rail" or "speedrail." Search the online mcmaster.com for "speedrail" and order the couplers you want. I got mine for free and they were all of the "4 outlet crossover" kind so I used only those, but there are more suitable configurations.
- Allen wrench: for opening and closing couplers
- Plastic pipe-end covers: get these at the place with the pipe; they protect your floor
- 1 x 8 wood panels for making the slats, and 2 x 4s for holding the slats together.

2. Prep

Measure your mattress and couplers to determine how long your pipe pieces have to be. For my queen size mattress and couplers I needed:

- 57" (x4) for the head and tail of the bed, including the headboard and tailboard

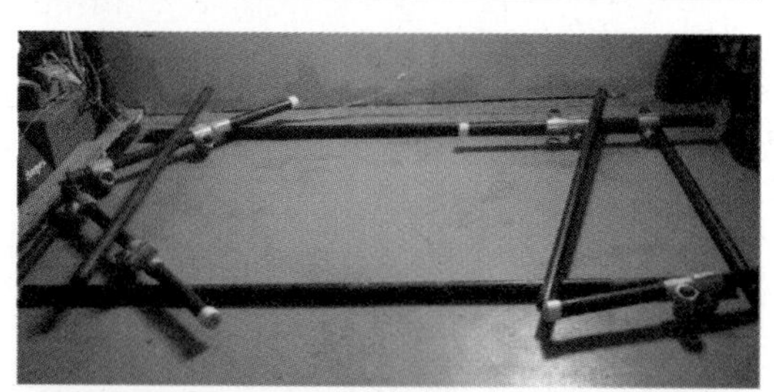

Figure B: Measured and cut pieces

Figure C: Laying out the headboard

- 42" (x4) for the four legs
- 87" (x2) for the lengthwise pieces

Cut and deburr your pipe.

3. Headboard

Slide the couplers on the pipe, and make up your headboard. It doesn't have to be in its finished shape, but keeping it square as you work will make it easier to assemble the whole thing. Use the Allen wrench to tighten down the couplers.

4. Tailboard

Do the same assembly for the tailboard.

For additional information, discussion, and more, please visit the Instructables project page:

Figure D: The assembled frame

Figure E: Slat assembly

Figure F: Another view of the slat assembly

User Notes

Stephen T says: You can adjust the feel of the bed by changing the size/spacing of the support slats. Wider boards as shown will give a quite firm feel. Smaller (1" x 4") slats will give the bed a softer, more springy feel... just have to use more of them. For people w/o tools... no problem. The lumber yard/home center will chop the boards to size for you... same with cutting the pipe.

Dennis Vogel has an Instructable for an Aluminum Pipe Bed: www.instructables.com/id/Aluminum-Pipe-Bed

Mark Rehorst built a loft bed with PVC: Yes, PVC can be used for this sort of thing. I built a loft bed for my son about 6 years ago using 2" PVC pipe. You can see it and DL plans here: http://mark.rehorst.com/PVC_Loft_Bed/index.htm

Jon Rutlen built it! This is mine. It took me about an hour. I used 1" pipe and I actually got Home Depot to cut the pipes for me. I used different fittings that made the cross bars equal with the long bars.

Jon Rutlen's build

5. Frame assembly

Have a friend or two help you hold up the head and tailboards while you slide the lengthwise pieces in. Now you should square the design, make sure it's level, and get it into a final configuration.

6. Slats

Use the 1 x 8 wood to make slats that will form the support for your mattress. Use three 2 x 4 pieces to hold the slats together, using wood screws to hold the whole thing together.

7. Done!

Add your mattress and sheets. Now take a nice nap.

Limor Fried designs and sells open source hardware and kits at her company, Adafruit Industries, located in New York City.

Beanbag Sofa/Bed

Build the biggest, baddest beanbag ever!

By Dan Goldwater

Figure A: Our finished big, bad beanbag sofa/bed

I started out thinking about building a beanbag, but being the prototypically exuberant man's man that I am, I thought: "I'm going to go huge! Why make a typical beanbag just like everyone else? I'll make the biggest, baddest beanbag evah!!!"

Well, in the end, I did make the biggest, baddest beanbag ever, but I think any practical assessment would say that it's too large to be convenient (e.g., hard to get through doorways and it tends to absorb all available floor space in room). But I learned enough through this effort that I can advise you, dear reader, how to build a more practically sized beanbag sofa.

1. Materials and tools

- Bean bag beans (see next steps)
- Spool of sturdy cloth (see next steps)
- About 3 feet of Velcro
- A sewing machine
- A 4–6" diameter cardboard tube about 2–3' long
- Optional: a shop-vac

2. Choose your sofa size

The size of your sofa determines how much cloth you'll need. Where a standard beanbag chair is a sphere (ball), the beanbag sofa is a cylinder shape (tube with end caps). There are several possible sizes of sofa you can make, with convenient names (see Table 1).

Table 1

Sofa Model	Diameter	Length
Pea	3 feet	3 feet (standard beanbag chair)
Garden slug	3 feet	8 feet (recommended)
Banana slug	4 feet	10 feet (bad! diameter is too big!)
Centipede	3 feet	30 feet (around-the-room sofa)
Earthworm	1.5 feet	8 feet (back rest only)
Millipede	3 feet	1000 feet (block party sofa)

It turns out there is also a jellyfish sofa. The jellyfish is what you get when you don't have enough beans in the bag to keep its shape and it just squooshes out all over the floor. The problem with the banana slug that I built—and the reason I don't recommend it—is that when it's filled enough to not be a jellyfish, it will no longer fit through a standard doorway. And it's just way too huge for a normal room; it tends to envelop everything else in the room. Months later, you'll discover your little dog Toto buried in the folds of your jellyfish.

Oh yes, your spool of cloth must be as wide as the diameter of the sofa. For the recommended "garden slug" sofa, you'll need a spool 3' wide and 32' long. Use something sturdy! I made mine out of corduroy. There's a good selection of fabrics for $4–8 per yard at your local fabric store.

3. What about the beans?

You'll want to fill the beanbag at least 80% full of

For additional information, discussion, and more, please visit the Instructables project page:

beans. A normal beanbag chair is not this full, but the sofa will not hold its shape unless it is nearly full. For the earthworm type used as a backrest, you may want to pack it 100%. Calculate the volume of the sofa (a cylinder) like this: pi * radius * radius * length. So the garden slug is 56 cubic feet (3.14 * 1.5 * 1.5 * 8). Beanbag beans are sold by the cubic foot, so you'll need about 50 cubic feet of them.

Where to get the beans: you will need a lot of beans, so the best thing is to find a local styrofoam (EPS) products manufacturer. Try epsmolders.org to find a list of EPS manufacturers around the country. Getting the beans will be a fun excursion unto itself, as you can see below. The technical term for what you want is "Expanded Polystyrene Beads," or "EPS beads" for short. Usually they will have both new beans ("virgin beads"), and used beans ("regrind"). Virgin costs about 10x as much as regrind, usually more than you'll want to spend on some lunatic project you read about on Instructables. My local foam manufacturer (Bay Foam in Hayward, CA) sells 35 cubic foot bags of regrind for $10. They had a huge mountain of them behind their factory. The fancy places online that sell beanbags will tell you that virgin beads have a better feel in the beanbag, but my sofa seems perfectly nice with the regrind. Remember: regrind = recycling!

How to find a styrofoam products maker: they do not seem to like to list themselves under "foam" in the phone book, that seems to be reserved for the lowly urethane-foam folks only! Instead, you'll want to look under "packaging & shipping materials," or whatever else they might be making out of styrofoam, such as architectural trim.

No airbag needed! On your drive back from the foam factory, you'll be about as safe as you've ever been in your motor vehicle. Now is the time to enter a demolition derby. It took some effort, but I was able to wedge about 120 cubic feet of beans into my minivan after I took most of the seats out.

4. The plans

We're making a basic cylinder out of cloth. Of course, I made the banana slug model (I would have gone bigger, but I had not heard of the centipede or millipede until after I'd finished):

- Cut two circles of fabric 3' in diameter—these will be the ends of the cylinder
- Cut three rectangles 3' wide and as long as you want the sofa (8' or 10' long). These three will form the tube of the cylinder.
- Cut a 2' square, this will cover the fill-hole

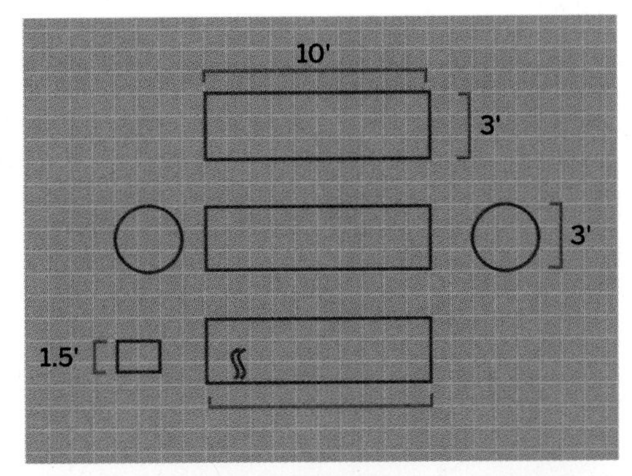

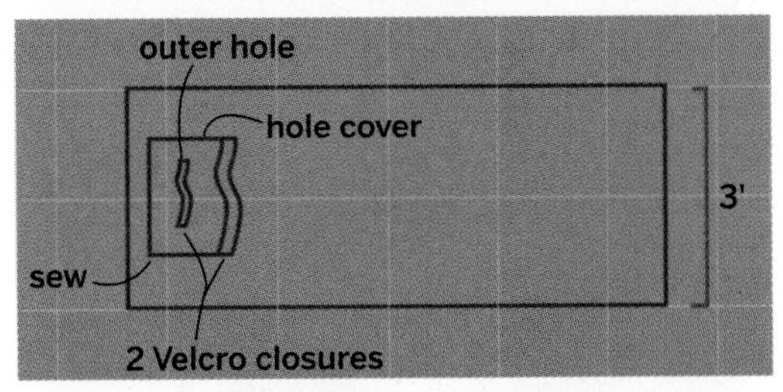

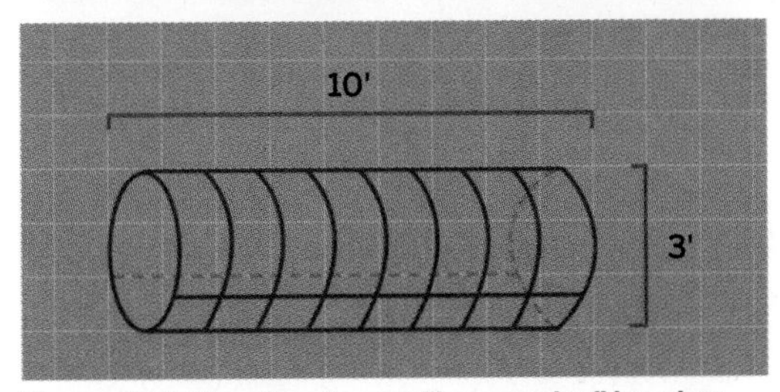

Figures B-D: The plans for our "banana slug" beanbag. I recommend doing the smaller "garden slug" design.

5. Make the double-sealed filling hole

Whether you're making a slug or a centipede, surely nothing can be more important than a HUGE GAPING MAW TO SWALLOW UP YOUR CHILLUN AND NEVER LET THEM OUT AGAIN!

The filling hole is how you'll get the beans into the sofa. I made a double-seal to ensure that no beans can escape by accident. The double-seal also reinforces the Velcro closures so that you'll be able to jump all over the sofa with no fear of it opening up.

- Cut a one-foot slit in one of the 3' x 8' fabric rectangles, about one foot from one end of it. Put Velcro on each side of the slit so that you can close it up. This is the outer seal.

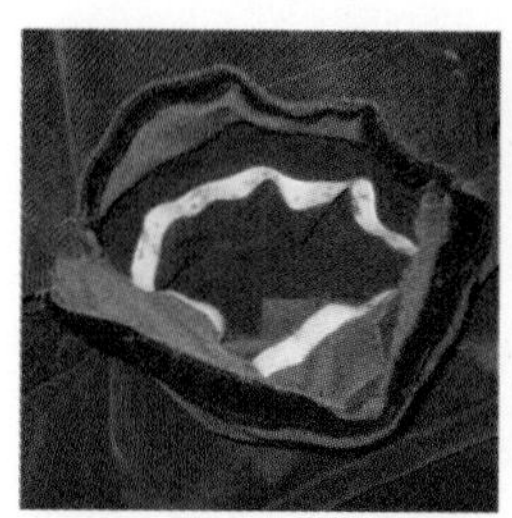

Figure E: The outer and inner Velcro seals. The outer seen here in black Velcro, the inner, in white.

Figure F: This is my 2' x 2' rectangle, with Velcro, before attaching to the bag

Figure G: Finished, empty bag, waiting for some beans

Figure H: The fill assembly, with a cardboard tube placed in the bag's fill hole (left) and duct tape sealing the tube to the bean supply (right)

- Sew the 2' fabric rectangle onto the 3 'x 8' piece so the middle of it covers over the one-foot slit. Sew it only on three edges; on the fourth edge, attach Velcro. This is the inner seal. Of course, do this on the "inside" side of the 3' x 8' fabric.

6. Sew the sofa!

- Sew the three 3' x 8' rectangles together along their long edges, i.e., make a 9' long x 9' wide sheet, then fold the sheet in half and sew its 8' edges to make a tube (do it inside-out). When you're done, you'll have a 8' long tube with open ends.
- Go over all seams twice for strength
- Keep the tube inside-out
- Sew on one of the end-cap circles (also inside-out). Remember the circle circumference is a teeny bit larger (9.4') than the tube circumference (9'). If you're a perfectionist, that means you'd actually want your circle to be about 2.9' diameter, not 3' in diameter.
- Sew on the other end-cap circle, inside-out.

7. Flip it! Fill it!

Like an unhappy sea cucumber, it's time to pull the entire sofa out through its mouth, to make it right-side-out.

Once it is righted, the fun starts! How are you going to get all those beans into it? The most important thing to remember is: no matter how you do it, you'll be finding beans all over your house (and probably your neighborhood) for the rest of your life. So you might as well have fun trying:

Figure I: Bag partially filled, using the Boring (aka staircase) Method

Figure J: Our finished banana slug, too large to fit through doorways!

The Shop-Vac Method 1:

- Put mouth of sofa over head of Shop-Vac
- Suck beans out of bag from whence they came

The Shop-Vac Method 2 (the Bean Cannon):

- Turn the Shop-Vac into blower-mode
- Put the Shop-Vac head into bean supply (careful not to clog it)
- Point nozzle at unsuspecting passersby. When bored, point nozzle into sofa mouth to fill it.

The Leaf Blower Method:

- Just like Shop-Vac Method 2, but with a leaf blower.

The Boring Method:

- Hold the mouth of the sofa with the rest hanging down a staircase
- Attach 6" diameter cardboard tube to your bag of beans
- Insert cardboard tube into the sofa mouth
- Pour in your beans

Dan Goldwater loves beams of light and the devices that make them. Dan is a co-founder of Instructables.com and SQUID Labs and now makes practical digital light art for bicycles at Monkeylectric.

Concrete Light Bulb Wall Hook

This is how to make an excellent excuse for driving a lag bolt into your wall

By Ray Alderman
aka Whamodyne

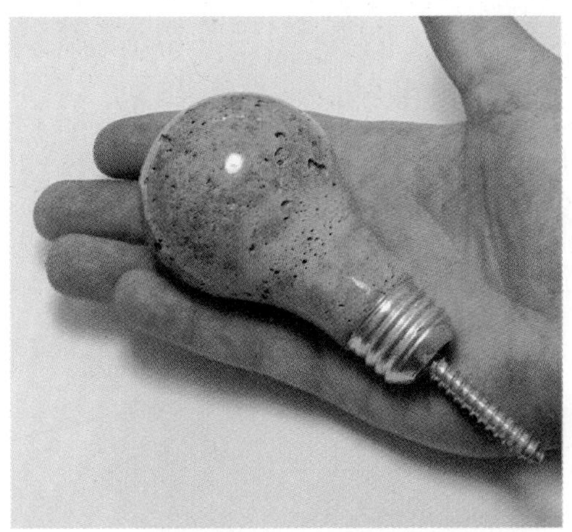

Figure A: The finished Concrete Light Bulb Wall Hook

The Concrete Light Bulb Wall Hook is an excellent excuse for driving a lag bolt into your wall. Functional yet stylish, it gives a nice industrial design feel wherever you mount it.

Last winter after breaking out the serious cold weather gear, I found myself fighting the coat rack next to the front door. It was, to put it bluntly, failing miserably. Tipping over, breaking off, it was a mess. I swore before the next winter I would drive some serious hooks into the wall that would handle all my heavy overcoat needs. I just haven't seen any kick-ass hooks yet that I liked enough to justify making serious holes in my walls.

Cut to the last few months. As mentioned in my blog (www.whamodyne.com) I've been playing around, trying to make a concrete light bulb. Why? Because I find the contrast of blending a new material like concrete into an everyday shape like a light bulb to be a great design element. So while messing around with these guys, I realized this would be a great excuse to drive lag bolts into my wall for hooks. By embedding a lag bolt into the concrete light bulb, I could make a wall hook that was useful enough to handle anything I wanted to hang off it. Thus this project was born.

This was originally an entry in the Etsy/Instructables SewUseful Contest, because I'm the kind of guy who enters a sewing contest with concrete and no sewing. If you are not familiar with Etsy.com, it's a place where people sell their handmade goods, and I set up a store to sell my bulbs there. So if you love this idea but don't want to make it yourself, not to worry! I have them for sale in my Etsy shop at http://whamodyne.etsy.com.

1. Tools and materials

You will need a work area where a little sand, concrete mix, and even glass shards won't be a problem. Make sure you have a small brush and dustpan available at all times. Normally you wait till the end to shatter the light bulb, but it can happen at any point in this process so be ready for cleanup from the very start.

Caution: Do not attempt this Instructable with Compact Fluorescent Bulbs (CFLs), which contain highly toxic materials. CFLs require special handling whenever they break or are disposed of.

Tools (see Figure B)

- Small pair of pliers
- Small pair of wirecutters
- Small screwdriver
- Carbide scribe. You can use something like an awl or even a long skinny nail, but I found my trusty old scribe to be invaluable for this.
- Plastic tub to mix the concrete in. I used an empty 5-pound tub of spreadable margarine.
- A scrap of wood to mix the concrete with. You could use an old wooden spoon or something like that if needed.
- Plastic spoon to put the concrete mix into the light bulb.

Figure B: 1. Cheap light bulbs 2. Tub from 5 pounds of spreadable margarine 3. Measuring cup and spoon 4. Rags are always handy 5. Small wirecutters 6. The carbide scribe 7. Small screwdriver 8. Pliers 9. Plastic spoon 10. One 5/16" lag bolt, 3.5" long 11. Scrap wood to mix the concrete

- A measuring cup and measuring spoons for adding the correct amount of concrete mix and water.
- A toothbrush you won't be using for your teeth anymore. I picked up a twelve pack of cheap toothbrushes from the dollar store. Every workbench should have a set of these.
- Coffee stirrer and plastic cups
- Gloves and safety glasses. These are a must because the glass bulb often breaks and little shards go flying in all directions, including straight at your eyes.
- Misc. items like Sharpies, some rags, etc.

Materials:

- Quikrete Mortar mix. I got the 10-pound bag at the local home improvement store for $2. This is enough to do over a dozen light bulbs. I could have purchased the 60-pound bag for $7 at a much lower cost/volume, but this project really doesn't need that much.
- Light bulbs. Just the cheapest standard sized incandescent light bulbs you can find. I got mine at Wal-Mart. A pack of four for 77 cents. Can't beat that with a stick.
- Water. You'll need about 4 tablespoons worth. I kept a bottle of water nearby on the bench and refilled it from the tap when needed.
- Lag Bolt. I'm using a 5/16" lag bolt, 3.5" long. 5/16" was the largest sized lag bolt I could fit into the light bulb without cutting off the head. I didn't want to do that because the head gives the bolt a lot of grip when embedded in the concrete. With a lag bolt 5/16" in diameter, I can drill a 1/4" hole in the wall to get a good balance between grip and ease of installing. In other words, it turns easily into the wall, yet holds really well.

Note: Concrete is a mix of cement, water, and aggregates. My research showed that a sand mix, aka mortar mix, is good when using a smooth-surfaced mold like the inside of a light bulb. It gives a very high shine when cured. A sand mix is different from your generic concrete in that the aggregates doesn't have any gravel, just various sizes of sand. I decided to do it with mortar mix instead of your standard bag of generic concrete.

2. Hollow out the light bulb

In the beginning I found the directions at TeamDroid (www.teamdroid.com/diy-hollow-out-a-light-bulb) to be a great help on how to do this. Now that I've done it over a dozen times it's routine. With practice this becomes quick and simple. There is another Instructable describing how to hollow out a light bulb here (www.instructables.com/id/EH8FGYFOB9EQZJIN8P) to get another perspective on it. TeamDroid linked to a GE tech spec on a standard bulb at www.gelighting.com/na/business_lighting/education_resources/environmental/downloads/msd/msds_incandescent_lamps.pdf.

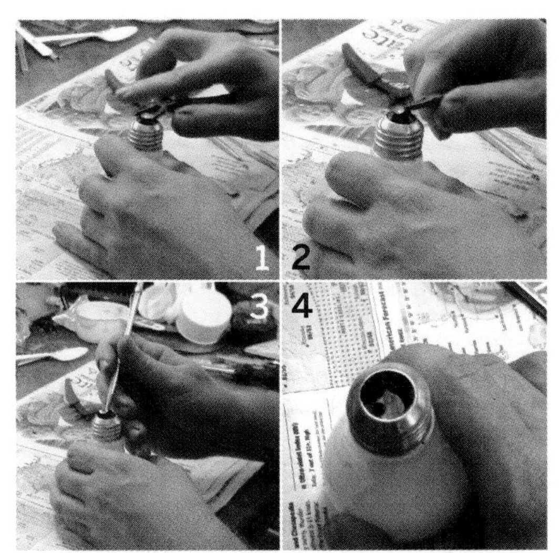

Figure C: 1. Prying up one edge of the metal circle with the blob of solder in it at the end of the bulb 2. Pulling up the edge (just rip that sucker right off) 3. Prying up the purple glass insulator pieces 4. Some of the purple glass insulator has been removed—take the rest out with the screwdriver

Figure D: 1. Folding over those metal tabs on the inside of the light bulb. This removes a "lip" that broken glass bits can hide in 2. Breaking the last glass cylinder inside the bulb by pushing down with the scribe until something breaks 3. Cleaning up the last bits of glass that could block the hole on the inside of the bulb 4. All the internal parts have been removed and the glass bits shaken out.

For much of this step, you will be poking around inside the light bulb trying to break off the internal glass bits. Do this over a trash can and often shake the light bulb out over the trash can to get rid of the glass shards. Wear safety glasses at all times. More than once some glass flew up towards my face when I was doing this. Figures C and D illustrate the procedure described here.

First, grip the metal circle with a blob of solder in the middle at the bottom of the light bulb with your pliers and gently pry it up from the dark purple glass insulator. This is pulling a wire in the middle that you want to break, so just pull it off.

Once that is done, take your carbide scribe and over a trash can, pry into the hole you just made in the purple glass insulator and break up that purple glass. You want to remove all the purple glass insulator from the light bulb body. I use the scribe to start some cracks and lift off a section of it, then I follow up with the screwdriver to get the rest. Turn the light bulb upside down and shake out all the glass bits that have fallen inside.

Inside there is a small glass tube that pokes up into the glass insulator. You might or might not have already broken that off by now. If not, just use the screwdriver as a lever against it until it snaps loose. Empty into the trash can.

Now you have a hole in the bottom of the light bulb. At this point I take my pliers and gently bend over the metal tabs on the inside of the hole so there is no "lip" on the inside. Later when you have broken up the rest of the inside pieces, there isn't anywhere for the bits to catch and stay in the light bulb when you shake it out.

There should be a wire visible inside that's soldered to the side of the metal screw piece. Take your wire cutters and cut the wire as close to the side of the bulb as possible.

Now, the inside has a glass cylinder you need to break off and clean up the edges to finish the job. Take your scribe or screwdriver and put it down into the light bulb until it meets resistance. Tap it gently until something breaks. Then using the screwdriver, lever against the side of the light bulb to clean out whatever remaining glass bits are left. You want the neck of the light bulb to be clear from the hole all the way down the body. Turn the bulb over and shake it out one more time to get rid of the last of the internal glass pieces floating loose.

Take a toothbrush and while dry, push it into the light bulb and start loosening up the dry white powder. Pay special attention to the neck of the bulb.

Figure E: Doing the final cleaning. All the white powder residue should be removed by now.

Figure F: A hollowed out, cleaned and rinsed bulb. Set it aside to dry.

Figure G: A cup and a quarter of the mortar mix ready to be mixed up, with the tablespoon at the ready.

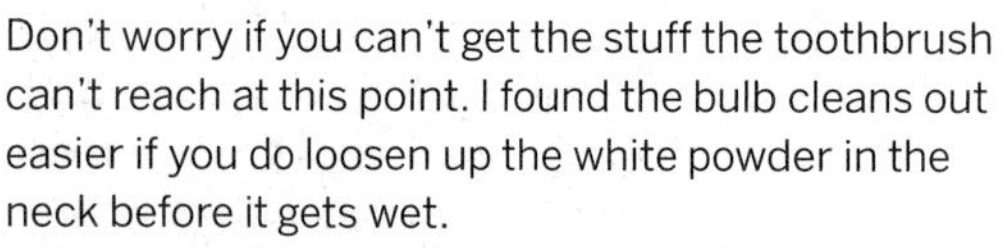

Don't worry if you can't get the stuff the toothbrush can't reach at this point. I found the bulb cleans out easier if you do loosen up the white powder in the neck before it gets wet.

Now take it over to the sink and add a little soap and water. Scrub around with the toothbrush and shake the bulb to get the water everywhere, as shown in Figure E. Pull out the toothbrush and then wash out the soapy water from the bulb. At this point it's all nice and clean inside without any soapy residue. Set it aside to dry out (Figure F). Now it's time to mix up the concrete.

3. Mix up the concrete

This is the part that's more art than science. I've found that in the small batches this project calls for, it's very easy to add too much or too little water to the mix. When you are mixing an entire 60-pound bag of concrete, being off a teaspoon on the water doesn't matter that much. When you are mixing up just a cup of concrete, that teaspoon starts to matter.

Mortar mix, when cured in a glass mold like we are using, gives a very nice gloss surface. The lower the amount of water you use, the smoother and more glass-like the surface is, and the stronger the resulting cured concrete is. However, the lower the amount of water you use, the harder it is to have it fill in the gaps on the sides and it leaves lots of holes and divots. Getting the mix liquid enough to spread out but not too liquid that it loses its strength is one of the issues on the water-to-dry mix ratio. Finding the correct compromise between these two issues is really a matter of practice and personal taste. I would suggest you play around with it in multiple bulbs if you are interested in getting the best result you can.

The ratio I've found works well is about 1.25 cups of the mortar mix and a hair under four tablespoons of water. This is more mortar mix than you need to fill a light bulb, but there is always some spillage and trying to reduce the amount means even more accuracy on the water measurement. This is a good place to start (Figure G). So measure out a little under four tablespoons of water and put that into your plastic tub.

Slowly mix in the mortar mix a little at a time. Let a little bit get wet, then a little more, then a little more while stirring the whole thing. It's a bit like making biscuit dough at this level, but you're pouring the dry into the wet instead of the other way around. The consistency should be good enough that the mortar mix wants to stick together in one large clump, but it isn't sopping wet. If you feel you need to add more water or mortar mix to get it correct, then go for it. Just do it a little bit at a time. A small amount of either material makes a large impact at this point.

Once it's at a consistency you like, keep stirring nice and slow for a few minutes. You want everything to be thoroughly wetted as much as possible. See Figure H.

Caution: At this moment the clock starts and you have 30 to 45 minutes to finish the light bulb before the mortar mix starts to harden up.

Once you have thoroughly mixed up the mortar mix, bring out the light bulb and start filling.

4. Start filling up the light bulb with the mortar mix

You are working on a time limit at this point as the mortar mix starts to set. If you can get it all done in thirty minutes or so it should be fine.

Put your light bulb into a small plastic tub with the hole pointing upwards. When you are adding the mix there is always spillage and you don't want that all over your work area. I put a little bit of sand in the bottom so it will stay straight early on; once you add a few spoonfuls of the mortar mix it stands up straight on its own.

Figure H: 1. Add a bit under four tablespoons of water to your tub 2. Slowly pour in the dry mortar mix to the water while stirring 3. Starting to get there—it's just reaching the point where it wants to clump up in one large ball 4. Yum. Bake at 375 degrees for 12 minutes... wait, sorry, wrong Instructable!

Figure I: 1. A scoop of the wet mix in the plastic spoon, use your other hand to feed it into the light bulb 2. This is before any shaking—the mix needs to be vibrated and tapped to spread out and coat the inside of the bulb 3. Whole lot of shaking going on 4. After shaking and tapping, it's starting to liquefy and spread out.

Take a plastic spoon and scoop up a level amount of the wet mortar mix. Holding the end of the spoon over the hole in the light bulb, use your finger on the other hand to push it down into the hole. Some will spill over the edges of the hole; that's ok. You made almost twice the volume of mortar mix as will go into the light bulb.

After four or five spoonfuls into the light bulb, you want to vibrate it so the mortar mix liquefies and spreads out evenly on the inside. Shake it back and forth to make it liquefy. If there is an air bubble visible on the side that just won't fill in no matter how you shake, tap it repeatedly with your fingertip. This moves the bubble up and the liquid towards the tapping.

Repeat these (a handful of scoops, much shaking and tapping to fill in the gaps and make it all liquid, all shown in Figure I) until you reach half way up the neck of the bulb. At that point (Figure J) it's time to add the lag bolt.

5. Put in the lag bolt and finish filling the bulb

Before putting in the lag bolt, mark off with a Sharpie where 1.5" is from the pointy end so you know how far to push it down.

Put the lag bolt into the mortar mix (Figure K). Because the light bulb shape has some undercutting with the mold, the mortar mix wants to clump up in the middle of the bulb and not grip the sides. You can use the head of the lag bolt to tap down the pile in the middle of the bulb so it starts filling in all the gaps.

Keep adding mortar mix around the lag bolt to fill it up. Holding the lag bolt firmly to make sure it does not rattle, keep shaking the bulb and tapping the sides to fill in gaps and liquefy the mortar mix.

Take a coffee stirrer (Figure L) and use that to tap down the mortar mix. You want to keep adding the mix, tamp and fill, tamp and fill. Finally the entire bulb is all done and you are tamping on the top. You

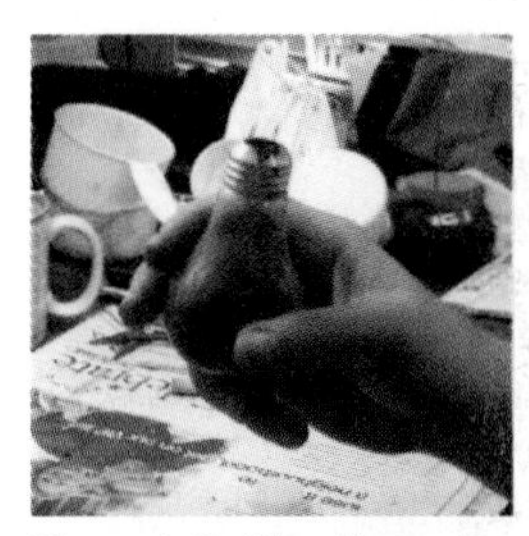

Figure J: Getting there. At this point it's time to add the lag bolt to the light bulb.

Figure K: Adding wet mortar mix around the lag bolt.

Figure L: Tap it down with a coffee stirrer; you could borrow one from Starbucks.

Figure M: Full and ready to go, the metal threads have been cleaned up of concrete gunk

Figure N: Tapping the light bulb over a trash can to create cracks and get some of the glass to fall off the bulb.

Figure O: The bulb after much whacking

want the mortar mix to be level with the top of the hole but not sticking out in a bulge.

When you are all done, put the light bulb with the lag bolt pointing up in a spare plastic cup. Like my coffee stirrer, the ones I used were also borrowed from Starbucks. Take a rag and clean up the metal threaded part of the light bulb as well as you can—you don't want the mortar mix drying on it.

Set this to the side for at least five days to let the mortar mix cure as shown in Figure M. I normally write the date it was made on the glass with a Sharpie so I can keep track.

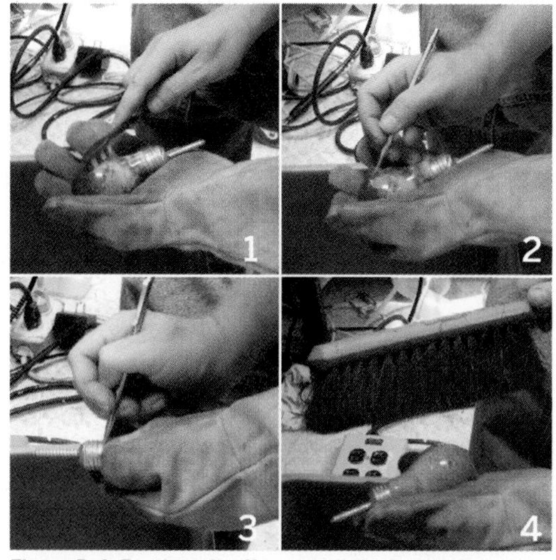

Figure P: 1. Brushing it off to get all the loose glass and especially glass grit off the bulb and into the trash can 2. Prying off the glass on the bulb piece by piece with the carbide scribe—it likes to fly off sometimes so be careful 3. Cleaning up the neck of the bulb where the metal meets the glass, making sure there is no glass showing or that could come loose later 4. Brushing the whole thing down with the shop brush to get out the last of the debris

6. Crack the light bulb

Let the mortar mix cure for at least five days. A few days more is a good thing.

This is the part where you are breaking up the light bulb glass into little bits and pieces **on purpose**.

Glass will be flying in all directions, so wear safety glasses and at least one glove (like in the pictures) at all times!

Over a trash can, hold the bulb in one hand that has a glove on it. I've tried to use gloves on both hands at this stage and didn't have the fine control I wanted. You might be able to do it.

Taking a hard metal object, start striking the side of the light bulb. In Figure N I'm using my carbide scribe. The glass will start to crack and form spider web fractures. Keep hitting. Eventually small pieces will fall and/or fly off of the bulb.

When you have formed a good number of cracks around the bulb (Figure O), take a toothbrush and scrub it vigorously over the entire bulb. This will brush into the trash can any glass grit or loose pieces. Then take your carbide scribe (or awl or sharp nail) and start to pry up the edges of the glass left on the bulb. Some of it will come off in large chunks, some of it will come off a little piece at a time. Try to aim for the trash can but know that it won't all go in there, some of it will pop up and go in any direction. If a large piece of glass doesn't want to come up and doesn't have any cracks in it, beat it with the metal object some more to create the spider web. Every so often take the toothbrush and rub down the light bulb to get rid of any ground-in glass or loose bits.

When all the glass has been removed from the light bulb, take your scribe and carefully go around the neck of the bulb where the metal met the glass. Make sure there are no loose glass shards under the

Figure Q: It's done! Look at the high gloss smooth finish on that sucker. The holes give it character.

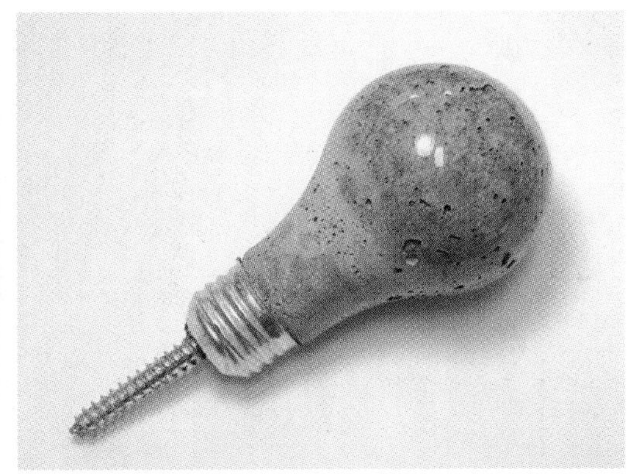

Figure R: All set with a very nice surface

Figure S: There be a stud here somewhere

Figure T: Drilling the hole. The blue tape is the depth stop at 1.5".

Figure U: Just hand screw it in

Figure V: Hang your hat. Showing off my pride for my beloved Oakland A's

lip of the metal. Shake and tap the bulb to see what floats loose. When you are done, take a shop brush and run it over the entire bulb one more time to get any glass grit or loose pieces out of the holes in the concrete. Figure P shows all the steps needed to get the glass off.

It's done! The mortar mix often leaves pits and holes even with all that shaking and tapping you did, but I think it adds a bit of character to the whole thing and makes each one you do unique. Feel the glossy smooth glass finish on the bulb and how it reflects the light (see Figures Q and R). Now it's time to mount it!

7. Mount it on the wall

Now that you have this killer Concrete Light Bulb Wall Hook, it's time to mount it into the wall. The steps here are fairly simple.

Find the stud: This hook needs a wood stud in the wall. There are multiple ways to find a wood stud. In Figure S, I'm using a cheapie stud finder.

Drill the hole: Using a 1/4" drill bit (Figure T), drill a hole 1.5" deep into the wall.

Screw it in: The hole is big enough that you can screw it in (Figure U), but the bite is good enough that it can hold just about anything you want to hang off it.

Heavy overcoats, small children, whatever strikes your fancy.

Hang your... Or you could use it as a hat hook (Figure V). It's all good.

Ray Alderman putters around and makes stuff in his workshop in Central Virginia.

Home & Garden | Food | Photography | Science | Computers | Electronics

The Best of Instructables
Food

Anyone can cook, and everyone should.

Don't know how? That's where Instructables comes in. If you can follow simple instructions, operate a knife or stove without injury, or even just push microwave buttons, all you need are some basic tools, a bit of practice, and you'll soon be making better food than you can buy.

But the real fun begins when you start to *play* with your food. I've never been one for religiously adhering to recipes, preferring instead to understand the underlying theory so I can modify the dish to my tastes, selection of ingredients, or weird whim. You learn the traditional skills, then devise custom tools and techniques. That's the basis of "food hacking"—play with the permutations, learn what's going on at the molecular level, and start tinkering from there. So what if that's not the way your mom cooked? Your great-grandparents certainly experimented—they called it "learning to make do with what you've got."

So jump in and try something! Make a better burger, build a pit oven, mod your toaster, carbonate some fruit, make a 3D dinosaur cake, test variants of your grandma's chicken salad. Document everything and post the results to Instructables for feedback, and you'll learn even more from the community of cooks gathered there. It's an easy way to start making awesome things you can share with others.

But the best part about food hacking? At the end, you get to eat it.

—Christy Canida (canida)

Figures A–H: An assortment of different styles of bento lunches

Crafting a Bento

How to make beautiful and delicious Japanese box lunches By Clamoring

Bentos, or boxed lunches, have a long history rooted deep in ancient Japan. They originally began as simple meals requiring little or no effort to assemble. Today they are a vibrant art form that is popular worldwide.

This project will attempt to provide the basic design principles, resources for obtaining the necessary tools, and some of the traditional rules for making beautiful and delicious bentos.

1. Know the rules (then break them!)

Like many other Japanese art forms, bento-making has its own set of guidelines. Traditional bentos follow a couple of basic rules:

- The 4-3-2-1 rule: 4 parts rice, 3 parts protein, 2 parts vegetable, and 1 part "treat" (usually either pickled vegetables or something sweet)
- Sushi should be prepared with more wasabi than usual
- Pack foods with flavors that might run or stick together with a divider. Separate wet foods from dry using a nested or altogether separate container such as a cupcake form. Sauces and dressings go in their own bottles (usually with a lid or cap).
- Oily foods (like gyoza) should be packaged on top of an absorbent material
- Bentos should not require any refrigeration or heating
- Above all else, your bento should be equally as nice to look at as to eat! (Note that this is the only rule that is *not* optional! :-)

2. Assemble hardware

If you're into kitchen gadgets, making bento boxes can be a very fulfilling pastime. There are tons of super-cute accessories with which to decorate your lunches. Many of these items can be found online (check the end of this article for links). If you're lucky enough to have a large Asian market in your town, you'll probably be able to find everything you need there. However, if you don't have one nearby, don't fret. We'll talk about options using readily available items you probably already have in your kitchen.

The first thing you need is a bento box. This will influence your portions, your shapes, and even what types of food you use. There are several types (see Figure K). Cute shapes like the bullet train and Pandapple boxes are most popular for kids. Tiered boxes, like the shamrock bento, are more often used for adult lunches. Lock & Lock boxes are fantastic for two reasons. One, they come with individual, removable dividers. Two, they lock completely air tight. I've started seeing Fit & Fresh brand in stores. The orange one pictured here has a separate ice ring you can freeze as well as a folding spoon. If you want to get started right away and don't have any of these types of boxes, you can also use a standard container. The actual shape of your box will have a lot to do with the final design of your box.

Figure I:
1. **One part treat**
2. **Packed separate from wet tomatoes**
3. **Two parts vegetables (heirloom tomatoes)**
4. **Three parts protein (egg and tofu)**
5. **Pickled daikon (a type of radish), red bell pepper, parsley, wasabi (hot mustard)**
6. **Four parts rice**
7. **Bento sushi is made with more wasabi than usual**
8. **Fresh ginger**
9. **Wasabi, blackberry, pickled ginger**
10. **Soy sauce**
11. **Bell peppers (used mostly as a space filler)**

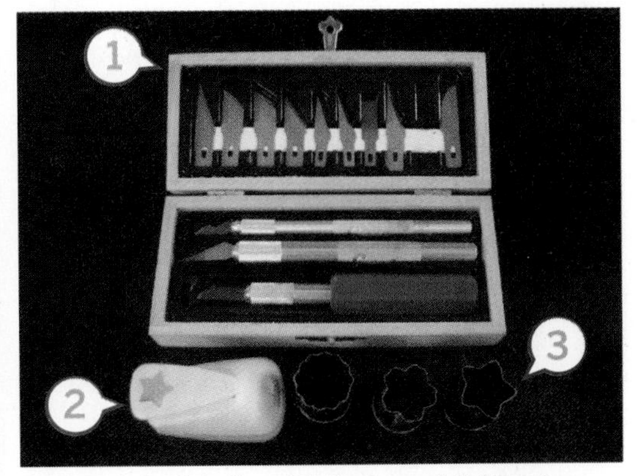

Figure J: 1. Craft knife set 2. Craft punch 3. Miniature cookie cutters

Cupcake forms and dividers are very handy when keeping flavors from mingling. Mini forms fit well in bentos. There are also silicone forms available that are great if you have something really wet or messy (like spaghetti). The most common divider is green plastic grass but there are lots of other specialty designs.

Many colors, shapes, and sizes of specialty forks designed especially for bento boxes are available. Regular shrimp forks are small in size, easy to find in stores, and fit in many boxes. Skewers or toothpicks can be cut to size and decorated should you be so inclined. A nice pair of chopsticks will round off your bento set. I like the ones that come in a matching box.

Many of the fancy patterns you see in bentos are made with some form of cutter. A cutter can be a cookie cutter, craft punch, or craft blade. I use my craft blade more than any other bento tool! Cutters are especially handy for cutting nori (seaweed/sushi paper), vegetables, or sliced tofu or meat. Who doesn't want little carrot stars on their salad?

Figure K: Bento boxes, both child and adult styles

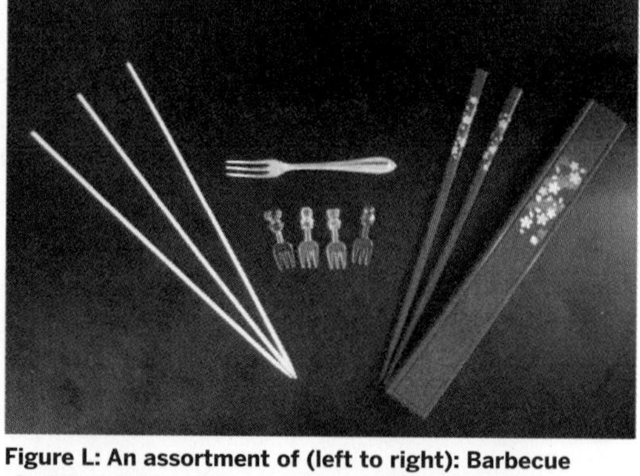

Figure L: An assortment of (left to right): Barbecue skewers, shrimp fork, mini bento forks, chopsticks with box

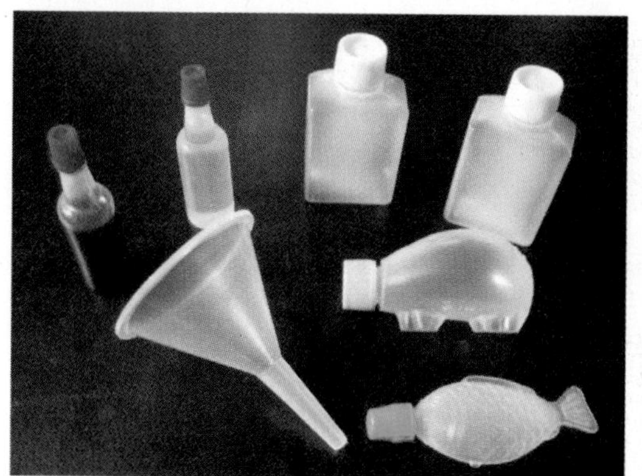

Figure M: An assortment of sauce bottles and a funnel

Figure N: Bento supplies (left to right)—Sushi rice, tonkatsu sauce, furikake (rice seasoning), wasabi powder (just add water), mirin, Japanese-style bread crumbs, tempura mix (no egg required)

Probably one of the more difficult specialty items to substitute is a sauce bottle. Barring proper bottling, you can also put sauce into a ziplock bag (towards one corner) and secure the sauce with a rubber band. This would be something like a pastry frosting bag only very small. At lunch, you can clip the tip off and squeeze the sauce out. I have also folded tinfoil into a little cup shape. If you go that route, be sure you use it for a thicker substance (like peanut butter) rather than something liquidy (like soy sauce) as it will probably leak.

Once you have your supplies together you're ready for some serious bento making!

3. Gather specialty food items

A trip to your local Asian market is important in creating a traditional bento. But if you live in an area where you do not have access to such items don't fret. There are lots of creative and healthy ways to make a bento from seasonal local ingredients.

- **Calrose rice** is your first choice for sushi rice. Minute rice isn't going to cut it.
- **Tonkatsu sauce** is simply good on everything. It's mostly used on tonkatsu (fried cutlet, usually pork) but it is also very delicious on steamed veggies. It's similar to Worcestershire sauce.
- **Furikake** is used as a rice seasoning to spice up bland rice. It is usually a mix of dried seaweed bits, sesame seeds, dried shrimp, and various salts (vegetarian options such as the one below are available).
- **Japanese bread crumbs** are primarily used in making tonkatsu and fried shrimp. I think you

Figure O: Making a pocket in the sushi rice. Sprinkle salt on hands when forming a ball and keep hands moist to prevent rice from sticking.

Figure P: Form the pocket

Figure Q: Stuff the ball. Here hard boiled egg and seitan (wheat gluten) cooked in soy sauce is used. When stuffed, cover the hole with more rice.

Figure R: One option is to cover the entire onigiri in nori

could use regular bread crumbs in a pinch, but the ones marked Japanese seem to be lighter. (Maybe it's just my imagination!)

- **Mirin** is a sweet light syrup used in making sushi rice and tomago (egg) sushi.
- **Tempura mix** can be used to make tempura batter. You can make your own mix but if you use a premade one you can eliminate the egg.
- **Wasabi** can be purchased in powder or paste form. Don't let the pleasant light green color fool you—this stuff will clear your sinuses!

Now that you've done your shopping I suppose you're wondering what to put into your bento. Let's look at a few options.

4. Stuff to put in your bento—onigiri

Onigiri, rice balls with filling, are a wonderful comfort food. They are fun to make, fun to look at, and fun to eat. They also serve as a nice parcel to decorate as they have a large surface area. The simplest onigiri, and maybe the most traditional, is simply a rice ball with an umeboshi (pickled plum) in the middle. Umeboshi are extremely popular in bento boxes and especially onigiri.

First you must decide on your filling. Just like sushi, you could put anything you like in an onigiri. Something with a little body is best as anything too fluid will tend to seep. Some commercially packaged onigiris pack the nori separately so that it stays crispy. Some common fillings are tuna, chicken, curry, boiled spinach, umeboshi, or tofu. It is also common to flavor the rice.

After you have decided on your filling take some rice and form a ball. You can make it as large or as small as you like. Using your fingers or a utensil to make a pocket. Add your filling and top with some rice. The triangle is probably the most common onigiri shape. Just form with your hands (see Figures O-S). It's so easy!

5. Stuff to put in your bento—tempura

Tempura is a crispy batter coating used on vegetables or shrimp (although you could use it on anything that will hold together in hot oil). It's fantastically cheap and easy to make.

The batter consists of: 1 egg, 1 cup ice water (it is important that the water is ice cold), and 1 cup all-purpose flour. Mix gently until blended but still lumpy. Use immediately. While you can certainly fry this at this point, it is extra delicious if you also bread your food. Japanese bread crumbs are light and give a big crunch.

You should have three containers: one with the batter, one with the breading, and a pan with hot vegetable oil. Dip your item in the batter, roll in the breading, shake off excess. In Figure T we're doing string beans. Fry the beans in the oil, drain off excess on a paper towel. That's it!

Often served with tonkatsu sauce (see step 3 for more info).

6. Stuff to put in your bento—sushi

Sushi is probably the most versatile food you can put in a bento box. Believe it or not, it is quick and easy to make. Aside from cooking the rice, making a sushi roll can take as little as five minutes.

Sushi could certainly be its own Instructable. Instead of trying to cover every type of sushi you could find in a really fantastic bento box, I'm only going to cover one just to get you started. Maki sushi is the round, nori-wrapped (seaweed) sushi

Figure S: The finished onigiri. The one on the left is wrapped in nori, the one on the right has been lightly fried to give the rice a crispy texture and then covered with a nori star band (with the stars cut using a craft punch)

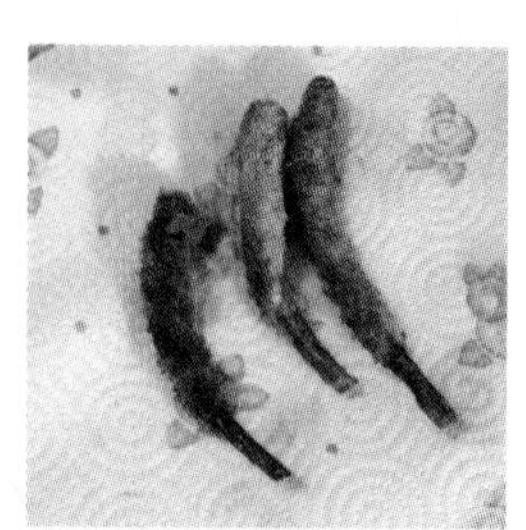

Figure T: Tempura, battered in tempura mix and Japanese bread crumbs and fried in a hot vegetable oil

Figure U: Sushi rice, cooked and ready to roll

Figure V: With wet hands, squish the rice gently into the nori. When covered, leave an inch at the top and add wasabi.

roll with any number of tasty fillings. Let's use the tempura beans from the last step to fill this one.

Prepare your sushi rice and let it cool. Calrose rice with a dash of rice vinegar and a dash of mirin is what I like to use. The kind of rice you use is important—use sushi, calrose, or sticky rice.

Now you have your rice, nori, wasabi, and your fried beans. It's time to roll!

Place your nori paper (seaweed) shiny side down on your table or bamboo mat. Prepare a shallow dish of water to dip your fingertips in to keep the sticky rice from sticking to your fingers. Cover the entire sheet with rice except for a strip about an inch wide at the top. Place the wasabi and beans about two inches up from the bottom. Roll tightly, bottom to top, and stop just short of the bare nori strip at the top. Wet the top strip with water from your fingertips and complete the roll. Cut in half and in half again (see Figures U-Z).

You now have sushi to add to your bento boxes! It's small, doesn't require refrigeration, and it's fun to make!

7. Design your bento!

When I start a bento, the first thing I look at is my main item (e.g., onigiri or sushi). I almost always have this in mind before I start cooking. I then choose my bento box and go from there.

As we previously discussed, your bento box will have a great deal to do with your overall design. For example, if you have a traditional lacquerware bento box with rounded corners you are going to need to fill in some odd spaces. Likewise, boxes with dividers built into them may or may not be the right size for what you want to do.

After space filling, you'll want to consider color. By far, fresh fruits and vegetables are going to have the best colors. Fresh, slightly steamed veggies will yield a brilliant spectrum. It is considered unappetizing to have a bento that is uniform in color.

Texture is also very important. Smooth shiny surfaces next to spiky shapes next to billowy veggies treat the eye to a visual array of excitement. Noodles look great furled up into a "bird's nest."

Giving your bento a name or a theme can pull the whole piece together and inspire details. Although you would think it might be the first step in the design process it is often the last. It is surprisingly easy to put together a bento that is full but not finished. Many times just a small pair of eyes cut out of nori, a few carrot hearts, or a tiny little fork can make an ordinary lunch into a bento box.

8. My romp into bento madness

From 1998 to 2000, I lived in Narita, Japan, where I worked at a summer camp teaching English to little kids. On my very first day, I was blown away by what the kids brought for lunch. There was not a single PB&J or bag of potato chips to be found. Instead, vegetables were lovingly cut into animal shapes, anime characters graced lunch box lids, and personalized messages were crafted right into the food. I could tell right away that the Japanese bento is more than a school lunch, it is an art form.

Bento fun isn't exclusive to children. A lunch prepared for a husband might not include a sausage cut lovingly into the shape of an octopus but might instead include a handmade onigiri (rice ball) with a curry filling. A lunch purchased on the bullet train

Figure W: Add your filling, in this case, our tempura string beans

Figure X: Roll the roll. Two hands are required (unless taking photos with one).

Figure Y: Wet the unriced 1" strip with water and finish the rolling to seal

Figure Z: The sushi roll is ready to be cut to the desired size

might not include a tomato with a face, but could well include hand-dipped fried shrimp.

While modesty prevented them from saying so, I learned that my students' mothers woke up very early every morning to ensure their husbands and children left the house with bentos they could be proud of. Making a nutritious and cheerful bento is an investment of time. It takes longer than microwaving a pizza or throwing together a bologna sandwich. It is also considered an extension of the preparer's love for the recipient. Originally simple meals that required little preparation, bentos have blossomed into an exciting new trend. The intricate designs and unusual foods were intimidating at first but I have learned and compiled ways around these challenges. Like any gadget enthusiast, I have collected a pile of tools with which to play with. And like any busy crafter, I have collected just as many time-saving techniques.

Figure ZA: The female figure was cut out of nori with a craft knife. Make sure the rice is cool before applying or the nori will shrivel. Note the use of contrasting colors and textures.

9. Inspirational links

Bento ideas are commonly shared among bento-makers worldwide. Below are some of the places I regularly go to for supplies and new ideas.

Jbox (jbox.com)
Cooking Cute (cookingcute.com)
Bento TV (bentotv.com)
Obento, My Bento (Flickr group) (flickr.com/groups/367772@N22)
Bento Boxes (Flickr group) (flickr.com/groups/bentoboxes)

Note: The online Instructable for this project also includes a 3-part tutorial on making gyoza (Asian dumplings).

Besides making beautiful and delicious bentos, Kayobi Tierney (aka Clamoring on Instructables) likes to dabble in jewelry making and electronics. You can see more of her tinkerings at clamoring.com.

contest WINNER!

"1UP Mushroom" Mushroom Burger!

Your nerdy veggie friends will gawk in awe at this dee-lish burger with a sense of humor

By Karen Chu

Figure A: Our finished, delectable 1UP Mushroom burger

Figure B: Gathering our ingredients together

Remember all those childhood summers spent playing the NES indoors when it got too hot outside? Summer vacation, BBQs, and the NES? Ah... those were the days.

When I cook dinner for my friends, I tend to put as much attention into the vegetarian dishes as the meat ones. Why not dress up the veggie burgers to make your guests feel special and have a spiffy time at your next barbecue?

This Instructable was inspired by Barry Rosenstein's The Red Fez Burger from *Build a Better Burger Challenge* from the Food Network a few years ago. To reflect a Moroccan theme, he colored his top bun red to look like a fez. I thought that was the coolest frakkin' thing ever!

For this project, I made my own veggie mushroom patties. They taste like okonomiyaki, with a soft fritter/pancake/omelet texture. Okonomiyaki is a Japanese household savory pancake dish where vegetables, meat, and noodles are all mixed together in a pancake batter and placed on a griddle, then topped with a special soy-based sauce. Feel free to change according to your taste: Boca burger, Gardenburger, meat patties, or just a slab of meat... unless you're Bowser, who desperately needs to cut down his meat intake if he wants to be in good princess-kidnapper shape.

Ingredients for the patty

(Makes 4 burgers):

- Half of one white onion
- 2 large portabello mushroom caps (or 6-8 ounces of whatever mushroom you like)
- 1 tsp. of kosher salt
- 1 egg
- 1 tbs. of pancake mix
- 1 tsp. of minced garlic
- Black ground pepper to taste

Ingredients for the fixins:

- 4 large slices of mozzarella cheese
- 1 tomato, sliced at least 1/4" thick at room temperature
- 4 spoonfuls of pesto
- **Optional:** mayonnaise (I personally hate mayo, but it's like crack to some people)

For additional information, discussion, and more, please visit the Instructables project page:

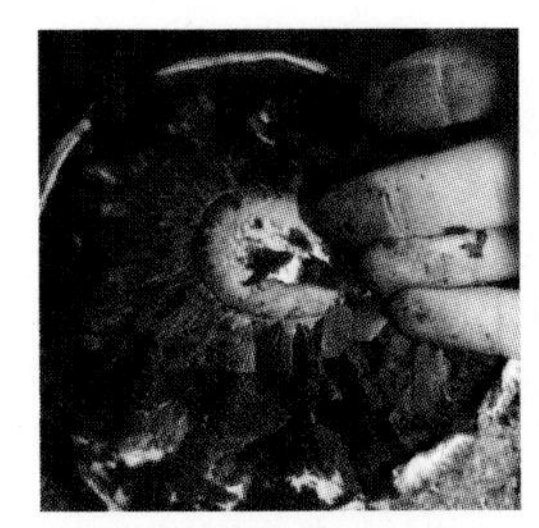

Figure C: Washing off the portobello mushroom gills

Figure D: The diced onions and mushroom caps

Ingredients for the buns:

- 4 *very round* rustic, crusty buns. The type I decided to go with are my favorite whole wheat kind.
- 1/2 tsp. green food coloring
- 1 tsp. of water
- 4 large slices of mozzarella cheese for the circle decoration
- **Optional:** I also tried this recipe with basil juice instead of the food coloring in an attempt to be more "natural." The green looked good but it made the burger WAY TOO PUNGENT. Whew!

1. Prepping the patties

First, wash your hands! Then wash your produce! (Washing is important!)

If you're deciding to use the portobello mushroom caps, here's an extra step. As most foodies know, portobello mushrooms, when cooked, turn black and muddy. This is because of the dark gills on the underside of the mushroom caps. They also tend to burn easily and make the whole thing taste a bit burnt and bitter (a little like gasoline). To avoid this, use your fingers and rub off the gills. Do this under a running faucet and the gills will come off fairly easily.

Dice! Dice! Dice! Chop the onions into itty-bitty pieces. To be exact, I would say into less-than-1/4" rough square pieces. Now, chop the mushroom caps into the same rough size. Add the chopped mushrooms, chopped onion, minced garlic, salt, black pepper, and the olive oil into a bowl. In a small bowl, whisk the egg and the pancake mix together until the lumps are gone. Add it to the mushroom mix.

I know that the pancake mix sounds weird—usually people use breadcrumbs, flour, or cornstarch as a binding agent. I personally like the buttery taste of pancake mix.

Figure E: Grilling the patties

Mix the whole thing well until the pancake mix and egg are well incorporated into the mushroom bowl bonanza.

2. Grilling

The ideal grill temperature for cooking the patties should be medium-low.

When it's time to put on the grill, with your bare hands take enough of the mushroom mix to form into the size of a Ping-Pong ball. Then flatten the ball with your hands into a patty shape about 1" thick. Place onto a plate and spray the surface of each patty with cooking spray. Now it's fire time!

If you're using an outdoor grill, I suggest placing the patties on the outer rim for 6 minutes on each side and 3 minutes hanging out on the top rack or on the outer rim. This will give it some steam action too.

If you would like to save time, especially at a popular BBQ function, make the patties the night before using a skillet. Cook both sides for 6 minutes over medium heat and stick it under the oven for about 3 minutes. Then carefully store the patties in the fridge.

Of course, this whole patty thing is technically optional. Feel free to substitute any patty you like for this step. If you're a meat eater and would still like to partake in this festivity, you should! This particular mushroom patty reminds me of a Japanese household dish Okonomiyake and that's why I like it.

3. You won't get none unless you got buns

While the patties are cooking, it's the ideal time for prepping the buns. Mix the water and food coloring together in a small bowl. Brush the outer part of the

Figure F: Food-coloring the buns

Figure G: Stacking the pesto and tomato on top of the patty

Figure H: Melting the cheese over the stack to create a stem effect

Figure I: Power up with this little token. Yummy!

top halves with the green solution. This is time for personal judgment. Food dyes differ, and buns differ. I suggest having a throwaway bun to use as your trial bun for coloring. If the color is too strong, dilute with water and try to reach the optimal solution. This is why using crusty, hearty breads is better than using airy, porous soft buns—they won't turn into soggy mush.

The green will look really vibrant at first, but it will gradually mute as the buns dry.

Using a shot glass or a small bottle cap, punch holes into the slices of cheese to make small circles. These are for the mushroom spots on top of the buns. Each slice yields about five small circles.

After coloring the buns, place them on the grill (color face up) to toast and give them some heat. Remove from grill after a light toast. Place the cheese circles onto the colored surface to resemble the mushroom spots. The warmth of the bread will make the cheese stick.

4. The fixins

Don't take the patties off the grill after they're done cooking. They should be resting on the top rack, and it's time for the works. The goal of the next few steps is to make the burgers look like the white 1UP mushroom stem.

First, take a dollop of pesto and place it on each patty (still on the grill!). Align a tomato slice onto the patty, over the pesto. Now you should have a nice little mini tower.

Note: I'm kind of a tomato snob and I'm really picky about them. The tomato I used is a local black heirloom tomato that has less gooey innards and is nicely tart. Since the mushroom, onion, and bun all have a nice sweetness to them, I wanted something more acidic in this burger, and that's why I opted to put in such a thick slice.

Take a large slice of the mozzarella cheese, aim and center, and place it on the tower. The goal is to have the cheese melt and drape over the tower without ripping. This is why the tomato slices should be at room temperature so the cheese won't have a difficult time melting evenly. After the cheese has achieved the ideal drape, remove from the grill and place it on the center of the bottom burger bun. Place the green bun on top, and there you have it, a "1UP Mushroom" Mushroom Burger!

It will jazz you up with enough energy to jump on goombahs for a while in World 1-2 so you can get even more 1UPs! (Doesn't everyone know that trick?)

Hello. My name is Karen. I like dinosaurs. I like games. I like games with dinosaurs in them.

How to Make Carbonated Fruit

Add sparkle (and burp factor) to your fruity snacks with dry ice By Noah Weinstein

Using dry ice, cut-up fruit, and a strong plastic bottle you can make carbonated fruit. It's refreshing, bubbly, and totally unique.

Many thanks to Instructables user Argon (www.instructables.com/member/argon) for coming up with this idea and giving me tips on how to make it.

Caution: Dry ice must never come in contact with the skin or any other living tissue.

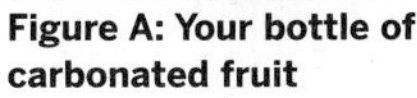

Figure A: Your bottle of carbonated fruit

Figure B: Gather your materials

1. Materials

To make carbonated fruit you only need to gather a few things:

- **Fruit—**When making carbonated fruit it's best to use firm fruits, like oranges, apples, and pears. I tried doing it with softer fruits like kiwis, strawberries, and bananas and it just doesn't work as well. Apples in particular seem to work the best.
- **Bottle or container—**You will need a plastic bottle or a container to put the fruit into. I have found that a wide mouth Nalgene works best. You can use an empty 2-liter soda bottle however, just be careful not to add in too much dry ice, more on that later. DO NOT use a glass jar. The bottle will be under pressure and broken plastic is safer than broken glass. If you have a vessel that is designed to take pressure, like a beer keg for example, then by all means, try using that.
- **Dry ice—**The final thing you will need is a block of dry ice. You will only need a tiny, tiny amount of dry ice to make the carbonated fruit, but it's hard to buy less than a large block of the stuff. Now, chances are that you have never seen dry ice for sale. You can't make it on your own and you might not be able to find it easily.

I used the Dry Ice Directory (www.dryicedirectory.com) to find out where it was being sold locally—they have listings for all over the world. I live in the east bay of California. I was surprised that in all of Oakland there was only one distributor—the ampm gas station on Market and Grand in West Oakland. They oddly enough had a ton of the stuff for sale, and they are open 24/7! I was very impressed that I could buy dry ice anytime I wanted even if it was only for sale at that one place.

Note: Before you obtain dry ice, you should review the Dry Ice Safety Info website at www.dryiceinfo.com/safe.htm, which includes detailed information about the safe handling, transportation, storage, and disposal of dry ice. This informational site is maintained by a group of manufacturers and sellers of dry ice.

2. Cut the fruit and put it into the bottles

The first step is to cut up the fruit and put it into the bottle(s). Cut the fruit as if you were making fruit salad—no seeds or orange peels are wanted here.

I cut smaller pieces to fit through the narrow neck of the soda bottle and bigger ones for the wide mouth of the Nalgene. I highly recommend using a Nalgene to make carbonated fruit.

3. Add the dry ice

The next step is to cut off a small chunk of dry ice from the block. You only need about two grams, or a piece about half the size of your thumb. There is no harm to putting in too little dry ice—you will simply end up with only slightly fizzy fruit. However, putting in too much dry ice *is* dangerous and could make a really big mess.

Figure C: Placing fruit in bottles

Figure D: Cutting up the dry ice. Don't touch it!

Figure E: Add dry ice to fruit and put on the caps

Figure F: Refrigerate the fruit for a day or two.

Dry ice is constantly sublimating, not melting, from its solid form of CO_2 to CO_2 gas. Unlike regular ice made from water, it goes directly from its solid phase to its gaseous phase with no liquid phase in between.

As a result, the dry ice block will produce gaseous CO_2 until there is nothing left of the solid block. The bottles are going to be sealed tightly with their caps, so if too much CO_2 gas is built up inside the bottle, they might explode (the soda bottle bursts at around 115 psi). We are looking for only 30 psi, so you must not use a big hunk of dry ice.

The dry ice in the picture (Figure D) was enough for both of my bottles of fruit; each one got about half of the small chips you see below.

4. Wait a day or two

As soon as I put the dry ice into the bottles and sealed the top, I could see it turning into its gaseous phase. Most of the dry ice will sublimate in an hour, so that's all the time it will take for the bottles to fully pressurize. Waiting overnight is a good idea to let the CO_2 gas work its way into the fruit.

I put the bottles into an empty drawer and closed it for the first hour; I have to be honest, it was the first time I was doing this and I didn't know what would happen. After an hour I could see that the bottles were under pressure, but not in any danger of exploding, so I transferred them to the refrigerator for the night.

You can only carbonate things that have water in them. I thought about doing fizzy meat, but I don't think it has enough water to dissolve the CO_2 into.

I went to bed, and brought the bottles with me to Instructables HQ the next morning.

Figure G: Bleed the pressure and cut the top off

Figure H: Serve and burp!

5. Open, eat, and burp

Once the bottles have sat overnight you are ready to open, eat, and burp.

Bleed the pressure from the bottle by opening the cap like you would open a shaken soda bottle.

I cut the top of the plastic soda bottle off with a sharp knife and poured it out into a bowl. You can simply pour the fruit out of the Nalgene bottle through the wide mouth of the bottle.

Now that the fruit is out of the bottles it's ready to eat! It loses its fizzyness pretty quickly, so make sure you chow down in the first 15 minutes after opening the bottles.

Carbonated fruit tastes like regular fruit, but it tingles on your tongue. It's a totally unique eating experience, and makes you burp a whole lot if you have done it right.

Noah Weinstein enjoys making all kinds of things, from tree houses to lasagna. If he's making something that he can't eat, chances are, it's for Instructables HQ where he works and plays.

How To Build a Pit Oven

Have fun doing a little "backwoods cooking" in your yard with a pit oven

By John Lock

Figure A: Digging the hole for the pit oven

We had a BBQ on the weekend and thought it might be fun to do a little backwoods cooking with a pit oven. The meal was to be a whole salmon that had been taking up valuable real estate in our freezer for about a year, so we thought this would be the perfect opportunity to get rid of the beast. After a bit of research on the Net, we set to the task. It was a fun little project and we all had a good laugh pretending to be cavemen!

1. Dig the hole!!!

No great surprises here. Grab yourself a spade and start digging. You'll need to find an open piece of earth with no fire hazards overhead (overhanging trees, etc.).

Our hole measured roughly 2' (L), 1' (W), 1' (D). We tried to keep everything pretty square, but that's only because we're slightly anal and I don't think it's that important in the great scheme of things. We made sure we kept the pile of excavated earth near the pit so that it was close at hand when we needed to bury our feast!

At this point you may want to make sure that any nosy neighbors, etc., haven't gotten the wrong idea about this hole in the ground you're digging; police searches and BBQs rarely mix!

2. Line the pit

We searched around the garden and found a load of medium-sized stones (like those you might find on a river bed) and some flat slabs of rock from a collapsed wall. We lugged these back to the pit in a wheelbarrow and began to line the pit.

The earth at the bottom of the pit was loosened up to give the rocks something to bed into. Starting with the walls of the oven, we used the slabs to line each face. We then used the remaining slabs to form a base for the oven. With a bit of trial and error, we managed to find bits that were just wide enough to push against the wall slabs, giving them some support.

The medium stones were then dumped onto the base and spread out so that they formed a fairly flat and even cooking surface.

And there you have it, one pit oven, ready for action!

3. Build the fire

Now for the fun bit. Fire! We gathered together plenty of small, dry twigs for the kindling and some larger branches for the main fuel of the fire.

We started by loosely scrunching up some newspaper to form the base of the fire, added the kindling, followed by the larger branches. Next we lit the paper and danced around the fire like madmen, celebrating our power over nature. (Not really!)

It goes without saying, but take great care when building and lighting fires; even though we were careful there were still a few singed hairs!

We kept the fire fueled with larger and thicker branches, making sure that the fire burned evenly over the whole pit. What you're aiming for is a nice even distribution of glowing embers over the bed of the pit. This will heat the stones, and it is this heat, retained by the stones, that does the cooking.

We planned on burning the fire for about two hours to give the stones a thorough baking but after an hour we'd run out of wood. The solution? We just chucked a load of charcoal onto the embers and retired to watch England vs. Israel football on the TV.

4. Prepare the salmon

Just time before kick-off to prep Sammy (aka, the salmon). Traditional methods of pit cookery use large leaves to wrap the meat. This protects it from being contaminated with earth when buried. More modern methods tend to use kitchen foil as a substitute.

We rolled out a length of kitchen foil long enough to accommodate the fish and doubled it over for extra protection. It was then lightly greased with butter and on went Sammy.

The corners were pinched up to form a tray so that our baste wouldn't spill all over. Next we thinly sliced a lemon and put that inside Sammy along with herbs, seasoning, and a little butter.

The baste was made by melting some butter and adding lemon juice, honey, seasoning, and paprika. This was poured all over the fish.

We sealed everything up with another doubled up sheet of kitchen foil that was crimped together with the bottom sheet.

5. Bury the beast!

Half-time and the fire's hot. The charcoal's done its job well and it's time to put Sammy in the oven.

Guides found on the Internet suggested removing the ashes before placing the meat in (or at least scooping them up to one end of the oven) so that the meat is in direct contact with the hot stones. We didn't bother with this and just laid down a couple of layers of kitchen foil over the embers as an additional layer of protection.

In went the fish, another layer of foil, and then the earth. We shoveled carefully at first, progressing to great big spade-fulls, until Sammy was well and truly buried.

6. Let it cook

We reckoned about 1 hour and 30 minutes would be enough for salmon to be cooked through (purely guesswork). By the time the match had finished, and we'd had our burgers and hot dogs, it was time to get digging.

7. Digging for treasure

Obviously, you want to be careful at this stage, one false move with the spade and dinner's ruined!

Figure B: The finished pit oven, dug, fitted with stones, and ready for dinner

Figure C: You're going to need a LOT of wood. You can also use or supplement with charcoal.

Figure D: The fire in full flame

Figure E: The fire, down to cinders within an hour

Figure F: Sammy, our beloved dinner, about to be wrapped in aluminum foil

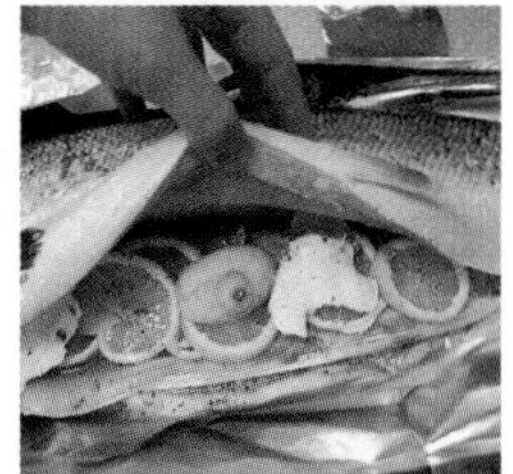

Figure G: The cavity, filled with lemon, herbs, seasoning, and butter

We knew Sammy was about 4" down and once we saw the silver glow of the foil, the rest of the extraction was done by hand. Again, be careful here. We were surprised by how hot the earth, the food, and the stones still were. After about five minutes, Sammy was out and ready for the table.

Figure H: Adding a kitchen foil bed on top of the coals

Figure I: Once the fish is in the oven, bury the evidence

Figure J: Out of the pit, ready for the plate

Figure K: Our pit-oven "Sammy," cooked to perfection

8. Enjoy your bounty

And now, the moment of truth. We gently peeled back the foil to a release of steam and a wonderful smell. It had worked! The fish was perfectly cooked; the skin peeled off easily, and the meat just fell off the bone.

Thanks, Sammy—you were one spectacular fish. There was more than enough to go round and everyone had their fill.

9. The clean up

Next day and time to clean up. After removing any non-biodegradables (like foil, etc.) we just filled the pit in, leaving a nice little mystery for archaeologists of the far future.

I hope this helps you to decide to make your own pit oven. After blundering through our first attempt, I'd recommend it; we'll definitely be building another one soon. Thanks for reading!

John Lock is a web designer and music producer from Birmingham, UK. He can be found at www.novawebs.co.uk and www.novatronic.co.uk.

User Notes

Kamakana Kaimuloa says: I am surprised you found success with river rock. It has been my experience that when you use river rock, or any other non-porous rock, there's a strong possibility that they will explode as a result of the heat. Shards of extremely hot rock can quickly ruin a pleasant evening.

Fatafehi Fonua says: In New Zealand, the indigenous Maoris call it "hangi," and in the Pacific Islands, they call it "umu" (earth oven). They've been cooking like this for a thousand years. You start a fire, pile on wood and stones, and wait for the stones to turn white-hot (the stones you have are ideal). Big embers are chucked out. Flat stones are settled at the bottom of the pit and the food is added (you have to work quickly). Criss-crossed sticks or branches are placed on top of the hole, supports are spread across pit (should be breathable). A thick blanket from the linen closet is ideal. You then cover with thin "veneer" of dirt to the point where you can't see steam rising from the pit. You can cook anything you'd cook in a conventional oven this way. My aunt bakes bread with this method. Cooking time depends on amount of food in your pit. In a pit about the size of the one in this Instructable, cooking time would be about 30 minutes.

Uncle Cy says: This has long been used for beans, called "bean hole beans." Families have their own recipes that they share with no one. These have been passed along for many generations. Another variation of this is, of course, the traditional lobster bake. Mainers place the lobsters over hot coals with clams and a few eggs smothered in seaweed. When the eggs are hard-boiled, the lobster and clams are done. Nothing tastes better. Seaweed provides the moisture for the steam, and also insulates the catch. Never better than on the seashore. Now I'm hungry. Thanks for sharing this ancient, and still used, cooking method.

Hobo Stove from Tin Can

Make the tin can stove that hoboes have used since time immemorial By Tim Anderson

Figure A: Salmon head and giblets soup on an olive oil can stove

Figure B: How NOT to leave a campsite. A hobo stove prevents this.

The tin can stove is quick to make, easy to light, and does a lot of cooking with mere handfuls of twigs for fuel. It doesn't make much smoke or shine much light, in case you don't want to be found. It also doesn't leave fire scars or start forest fires very easily, and that's good for both fugitives and environmentalists!

Another tin can would be the cooking pot for all you living-history hobo re-enactment enthusiasts. Preferably with a piece of wire through two holes in the lip to hang it like a little bucket.

In Figure A, I've used my favorite can for a hobo stove: a 3-liter olive oil can. I'm cooking salmon heads and giblets in a soup on a driftwood pile in the rain in British Columbia. I consumed the olive oil during the weeks it took me to learn to catch salmon. This is a new stove, the paint hasn't all burned off yet, and it needs more air intakes. With just one door, there will be charcoal left in the ashes. With three doors everything gets burned, and it's easier to feed it fuel.

1. Don't do this!

The hobo stove and some common sense will leave your campsite looking like wilderness.

Figure B shows what bad camping leaves behind. Not visible in Figure B is a pile of poo left by the users of this site. They thought they buried it, but rain washed the sand right off of it, and the bacteria from it washed down into the oyster beds and now the tribe won't be able to sell their oysters. In this area you're supposed to dispose of human waste in deep water in the current, or bury it a couple stones' throws from the water in an 8" or greater deep pit.

Strange to say, but kayakers are the bad guys in this case, much worse than power boaters even.

In case you're tuning in late and want the current eco-dogma: shellfish farming is usually good for the environment, whereas salmon farming is usually bad.

2. Pressure cooker on a hobo stove

I'm cooking a Plecostemus South American Armored Catfish (Figure C) I caught with my cast net west of Lake Okeefenokee, Florida, in the swamp along Fisheating Creek. I finished eating the olive oil in this can during the days it took to learn to throw the net. These are aquarium sucker fish that were dumped out and are breeding wild and growing large. The meat is very yellow and very tasty. The thick scales fit together in a very interesting way but

For additional information, discussion, and more, please visit the Instructables project page:

Figure C: Cast-net caught Plecostemus South American Armored Catfish going in the cooker

Figure D: Using a pressure cooker to prepare an entire day's meal on a hobo stove

I couldn't think of any use for them.

The pressure cooker is heavy but that doesn't matter so much in a canoe (Figure D).

It's a joy to use. I can cook all my food for a day at once. I would boil a dozen eggs with a stainless bowl of bread dough resting on top of that, all inside the cooker. I formed the dough into a bagel shape so the steam would cook it better, and it got steamed into a really good bagel. Especially when I used seawater in the dough. That was the perfect amount of salt. I just get a roaring fire going until the cooker is steamed up, then I forget about it, fire dies down and goes out. Everything gets cooked perfectly and I don't have to pay attention to it.

This cooker has two more bottom doors on the sides that you can't see, and burns the fuel completely. There's no charcoal left in the ash.

If you don't want soot on your pot, wipe soap all over it before putting it over the fire. Afterward, the soot will wash off easily.

3. Be your own hobo

We're going to make the hobo stove shown in Figure E. The license plate is optional, but something like that, even a folded piece of tinfoil, is nice so you won't harm the surface under your stove. This stove is tall and narrow, so you'll need to pound three sticks around it to support your pot, or put it between three rocks, or hang your pot over it.

Get the tallest can you can find. You need height for convection to give you good airflow. This coconut juice can is good. It's tall and the steel is pretty heavy for a drink can so it'll last a while.

Food cans had lead solder in the joints until 1993, so don't use old cans. Pineapple cans have zinc plating inside in case you think you need to breathe more zinc.

Figure E: A ready-to-use hobo stove on a license plate base to protect beneath the stove

Figure F: Carefully cutting an asterisk pattern in the top of the can

4. Make the initial incision

Cut an X or asterisk in the top of the can (Figure F).

Warning from my Granddad: "Don't cut toward yourself and you won't get cut."

These flaps are going to be the pot supports. The bottom on this can is heavier than the top, so I'm turning the can upside down to make them. The other end has the pry-tab open, and that makes it hard to make the pot supports turn out right. You'll get a better result by opening the can by cutting the X instead of whatever the vendor intended.

5. Pry up and crease the flaps

Pry up the flaps (Figure G). Don't cut yourself on the sharp corners. Crease each flap down the middle as shown in Figure H. That makes them a lot stronger. Figure I shows the top of the stove after all the flaps have been positioned correctly.

Figure G: Prying open the flaps

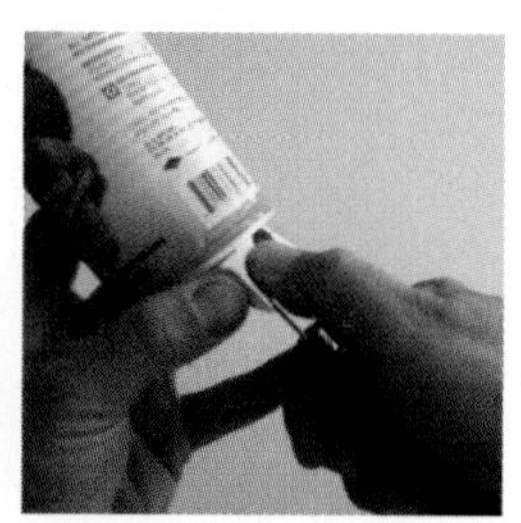

Figure H: Creasing each of the flaps

Figure I: The finished stove top

Figure J: Cutting the air-intake and stoking doors

6. Cut doors

Cut some drawbridge-style doors in the sides near the bottom. Those are the air intakes and stoking doors (Figure J).

7. First use

The first time you use it, wait for a while for the paint to burn off it. Stay back and don't breathe the fumes.

It's really easy to start a fire in one of these stoves. Start with wispy stuff or paper if you're still that close to civilization. Then work up to pencil-sized pieces. Thumb thickness is probably the most you'll want for cooking. Thicker than that tends to smoke, cuz you'll put in wet pieces by accident.

You can toss it in the top before you put the pot on, and then poke them in under the pot. You can feed longer sticks in gradually through the doors in Seminole star fire fashion.

It's so quick to start one of these stoves, sometimes I'll pull over and do my cooking at a rest stop or by the side of the road. I'll put an aluminum license plate under the stove so I won't leave a mark on the pavement. A folded piece of aluminum foil or flat rock is just as good.

I don't worry about trouble from authorities, because it's so easy to move the fire or put it out. But I've never been hassled. The rules about fires are usually about "open fires" and this isn't one.

Figure K: Using a three-stick pot stand and a palmetto plant as a windscreen

8. Three-stick pot stand and palmetto leaf windscreen

I'll sometimes pound three sticks into the ground to support my pot (Figure K). That's good if the pot is too wobbly or heavy on top the stove or if I want to get it a bit higher off the stove.

If it's windy, you'll want to put a windscreen around your stove. Otherwise the heat will all get blown away and cooking will take too long.

Here I've made a windscreen from palmetto leaves stuck in the ground and I'm cooking on a three-stick pot stand with a can of Mexican "Fuego" brand sterno. I'm in the jungle back from the beach on a biosphere reserve in the Yucatan Peninsula. This is oatmeal with olive oil, cocoa, and honey in it.

Tim Anderson co-founded zcorp.com, manufacturer of 3D printers that are computer-controlled machines that build sculptures. He travels looking for minimum-consumption technologies developed by poor people. He writes the "Heirloom Technology" column for MAKE magazine and has written 150+ Instructables.

How to Eat a Banana Like a Monkey

Chow down on "nature's fast food" like our simian siblings By Ed Lewis

The actual eating isn't too different from us humans. The real magic is in how monkeys peel their bananas to get at the goodness inside. Odds are, you peel your banana the way you've been taught and never thought much about it. So let's look at how the experts do it.

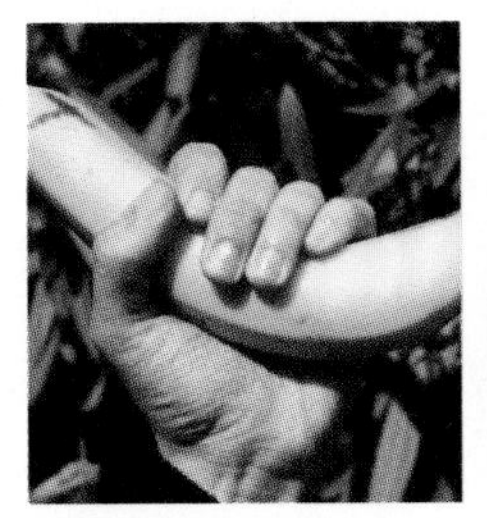

Get a banana I like mine to have more brown on the outside, others insist on a little green. Hey, it's your banana, I'm just here to tell you how to peel it.

Find the top Bananas grow in bunches pointing up with stems on the bottom. The other side is really the top.

Pinch the top Give a firm pinch on the top to break it. You might get some smooshed banana on your fingers, but with practice, you'll do it cleanly.

Peel your banana You pinched it and got a good grip on some part of the peel. Now pull!

Repeat Peel down two or three more times and you'll have a beautiful banana to behold and eat.

Eat it! This banana was a big one. It was ripe and delicious.

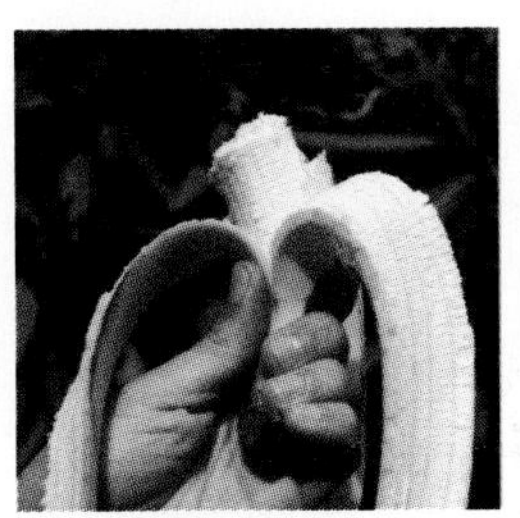

The beauty of it all With the stem still intact, you have a natural handle. There's no fumbling or trying to get the last bit out, it just happens, the way nature intended.

Dispose Toss the peel in the compost and get on with your day, a little happier knowing that you just ate a banana like a pro!

Gallery

Food Hacks

Weird and wonderful ways of playing with your food

Modular Pie-cosahedron

turkey+tek

/modular-pie-cosahedron

Who says pie needs to be round? Not turkey+tek, who used 20 triangular pie pans and magnets to make this geometrically delicious pecan pie.

The Butter Pen

j_l_larson

/butter-pen

Any food project that starts out with cleaning a used deodorant container sure sounds gross, but stick with us here. What you end up with (after some judicious sterilization) is a fun and utensil-free way of spreading butter on toast, corn, you name it. Thanks, Butter Pen!

8-bit Art Toast Maker

Alessandro Lambardi

/Mod-a-toaster-and-have-retro-art-toast-for-breakfa

Remember when they found the Virgin Mary in toast? Now you can emblazon your own breakfast with the deity or cultural icon of your choice using this ingenious toaster hack that creates a handsome silhouette in your Wonder Bread.

3D Dinosaur Birthday Cake Beth and Mike Simon

/3d-Dinosaur-Birthday-Cake

You might think making a three-dimensional cake this detailed would be for pros only, but it's really very easy, if you follow the directions here, and use the downloadable pattern. One baker even added wings and turned it into a dragon cake.

Mini S'Mores Grill

Jason and Ian Wilson

/Mini-Smores-Grill

Gather together some long bolts, cedar grilling plank, metal condiment cups, and toothpicks, and you have yourself an adorable little tabletop grill for toasting marshmallows. Quick and easy enough for you to make one for everybody at your table.

Carrot Caviar

Michael F. Zbyszynski

/Carrot-Caviar

"Molecular gastronomy" is all the rage these days (think: Julia Child cooking with a chemistry set). This Instructable shows how you can get in on the geeky fun, turning a bottle of carrot juice into tasty little caviar-like beads.

Chocolate Covered Squid

Noah Weinstein

/Chocolate-Covered-Squid---Valentine_s-Day-Candy-Fr/

Let's face it, nothing says "I love you" in the most unique way possible than the gift of chocolate-covered cephalopods. You can rest assured that, while others are giving the same ol' boring samplers and factory-grown roses for Valentine's Day, your gift will be truly one of a kind. And as long as your recipient is Wednesday Addams, your creativity should score big.

Making Kombucha

Arwen O'Reilly Griffith

/Making-Kombucha

People make all sorts of health claims about this currently popular fermented drink, made from tea, sugar, and a macroscopic solid mass of microorganisms. It's far more delicious than it sounds. And it's fun and easy to make.

To view these Instructables, and read comments and suggestions from other food hackers, type www.instructables.com/id followed by the rest of the address given. Example: www.instructables.com/id/Carrot-Caviar

Home & Garden | Food | Photography | Science | Computers | Electronics

The Best of Instructables Photography

Digital photography has ushered in a new era of experimentation. No longer tied to the time and expense of developing every roll of film, photographers are free to experiment and share what they've learned. Whether turning string into a tripod or using a magnifying glass to take better close ups, one need not let convention limit innovation. Transferring photographs onto blueprint paper, making an underwater housing for a video camera, or taking professional-level photos with a homemade light tent, these are just a few of the projects made available by the Instructables community. Feeling ambitious? Learn how to explore the world in infrared, convert a duck decoy into a spy cam, and even paint your photos in real time with light.

I remember building my first camera, back in the summer of 2003. The lens was the objective lens from an old telescope, the body was a Frankenstein combination of styrofoam, dowels, and curtains. Focusing was accomplished by sliding a piece of Plexiglas, frosted with sandpaper, back and forth. Once the image was in focus, a piece of blueprint paper was substituted for the Plexiglas. After setting up this monstrosity, the image could be exposed, a process that took up to eight hours in strong sunshine. Finally the image was developed with ammonia, and if everything went right a beautiful blue image appeared. Much more often an entire day was wasted. If I had only had Instructables at the time, you see, I'd have known I could get the same results in minutes by contact printing. What innovation will you be sharing?

—Robert H. Dutton (Tool Using Animal)

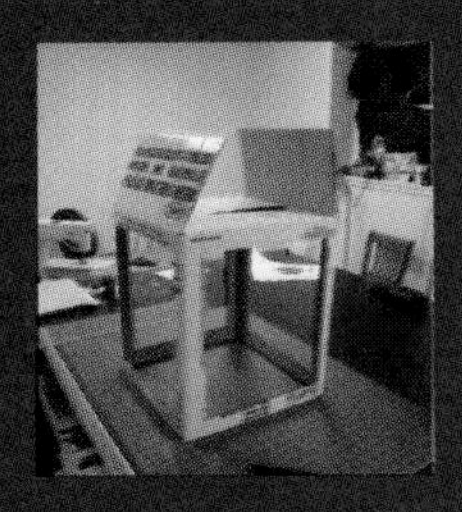

String Tripod

My take on an old photographer's trick

By Alexander Reben

Figure A: The String Tripod, attached to a camera

Figure B: Eye bolts and nuts

Figure C: Braided cord

The string tripod is an old photographers trick—here is my design. It's also sometimes referred to as a string bipod, string monopod, chain tripod, bipod, etc....

This device is used to stabilize a camera in order to get clearer pictures at a slow shutter speed. With more and more digital cameras coming out with vibration reduction (VR) or image stabilization (IS) systems, the string tripod has a new life.

Note: You can read more about vibration reduction here (http://tinyurl.com/5tmtdv).

Since image stabilization systems work best with rotational vibration, translational vibration can still create blurry pictures. By restraining the up-down left-right and back-front axis, you can lessen this vibration. Since with this design you still have rotational freedom, you can pan and follow something such as a bird or sports player. You can also recompose your shot with little trouble. Of course this technique also works well with non image stabilization systems.

Advantages:

- Cheap
- Easy to make
- No special tools needed (or really any at all)
- Hard to break
- Small
- You can use this where tripods are disallowed (such as museums)
- If the "tripod" gets dirty you can throw it in the wash
- If you misplace it, you can make another
- Rotational movement still available for panning and recomposition

Disadvantages:

- Does not hold as steady as a real tripod
- You can get some strange looks while using it

1. Parts

The parts you will need are as follows:

- A 1/4" x 20tpi stainless steel eye hook or eyebolt, 2" long or shorter
- A 1/4" x 20tpi nut (may come with the eye hook)
- 40' braided nylon and/or poly cord (parachute cord is recommended)
- Optional: a carabiner

All of these can be found at your local hardware store and should be had for about $5 total.

2. Prepare the cord

Cut your cord to about three times your height (more is better as you can cut off excess later). Make sure you melt the ends with a match or lighter to keep them from unraveling. Next create an overhand loop knot at one end as shown in Figure E. Pull the knot tight.

Figure D: Carabiner

Figure E: An overhand loop knot

Figure F: Making the ring hitch

Figure G: Pulling it snug

Figure H: Attaching the tripod to the camera

Figure I: Using it as a monopod

3. Connect cord to hook

Put the loop you just created through the ring on the hook. Then place the loop on the cord behind the threaded portion on the hook (Figure F). Pull tight and the cord will form a ring hitch.

It should look like Figure G.

4. You're done!

You're done with the construction. Now comes installation and use.

5. Camera installation

Now you're ready to install this on your camera. Simply screw in the hook into the tripod mount at the bottom of your camera. When the hook becomes snug, tighten down the nut to meet with the camera body. You do not need to screw this down with much torque, since it is only there to keep the hook from backing out. Figure H shows the tripod connected to a camera. You can cut the hook to exact length if you want.

Note: If your camera has plastic threads, be careful to not cross thread the hook. You should never have to force it in.

6. Using as a monopod

To use as a monopod, take your camera and put it just below eye level. Next take the cord dangling from the bottom and loop it under your shoe. Take the remaining cord and hold it tight in your hand while gripping the camera. Now pull up on the cord to camera eye level and take a picture. The cord should be taut. An advantage to this is that it prevents a rotation axis because you are holding the cord in your hand away from the attachment point. Figure I shows how it's done.

This is my preferred method when I need to move around a lot or need to set up quickly. If you have the hook already attached to the camera it will take less than a second to get into position.

You can also tie a large loop at the end where your foot would be using a overhand loop and put your foot through it. You would not have to hold the end in this configuration.

Note: For each of these methods it is important that you do not put excess force on your camera. As every camera is designed different, you need to determine the amount of force your camera can take. With that said, I have been using this method for years with many different cameras with no problems.

7. Using as a bipod

To use as a bipod, take a wide stance and loop cord under both feet. Bring the hook and end of cord to about where they would be when attached to the camera. Once you know this distance, secure the end of the cord into the hook as shown in Figures J and K.

You have now created a triangle in the cord. Attach the hook to the camera and pull up tight. If

Figure J: Attaching the other end of the loop for bipod use

Figure K: Pulling it snug

Figure L: Using it as a bipod

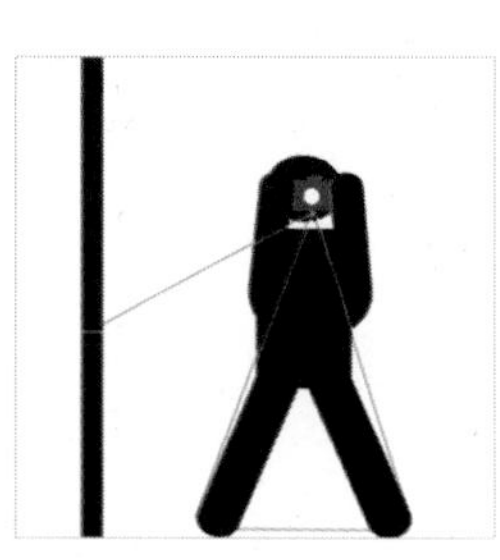

Figure M: Using it as a tripod

you miscalculated the distance, simply loosen your grip and pull the end of the cord tighter through the hook. Figure L shows how to use it.

This method is good if you have a little more time on your hands or need a more stable shot.

If you are in a crunch, you can also just skip the last step and hold the end of the cord like when using as a monopod.

Figure N: Attaching the carabiner

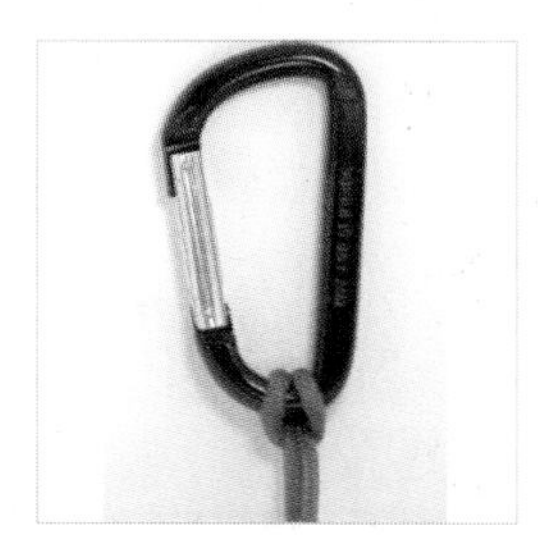

Figure O: Pulling it snug

8. Using as a tripod

To use as a tripod, follow the steps to use as a bipod. Then take the leftover end of the string and attach it to something (see Figure M). Optionally you can install a carabiner or other device at the end of the string as shown in Figures N and O. When using this method, it is imperative that you keep all parts of the string taut.

This method takes the most time to setup and requires something in the environment to attach to. Also, front to back movement is not as bad as left to right or up and down. Personally, I hardly use this method but it is always an option.

9. Tips & tricks

Shutter Speed:

Using the right shutter speed is essential, you should always use the maximum speed possible.

The rule of thumb for absolute minimum shutter speeds for a handheld camera is 1/focal length. For example, a 135mm lens with a 1.5x crop factor (on a SLR) makes it 202.5mm. So with a handheld you should expect somewhat clear pictures starting at 1/200 or 200 shutter speed. It also matters how far away the subject is: the farther away, the faster the shutter speed needed. The third factor is how far the tip of your lens is from the camera, the further it is—the more it will amplify the vibration.

The string tripod requires a bit of practice and getting used to. I did not notice much of a difference at first, but after a while it really helped. After some practice, it should improve the minimum shutter speed to 1/2 or 1/3 that of

Figure P: Tests—1/2-second exposure, 200mm at 10 feet

Figure P1: No string tripod; no image stabilization

Figure P2: No string tripod; image stabilization on

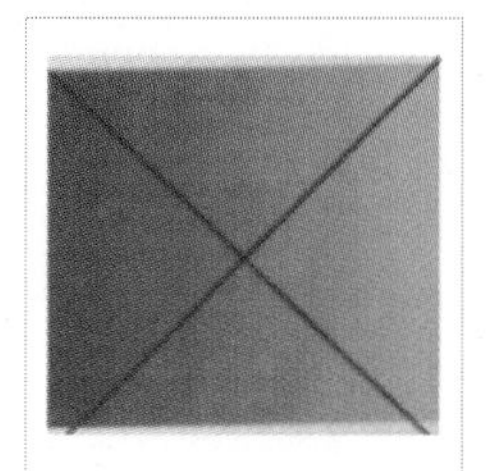

Figure P3: String tripod used; image stabilization on

handheld. With VR or IS and the string tripod, you should get 1/4 or 1/6 that of a handheld.

So, if you are first starting out I would not go any slower than 1/100 with a string tripod at that focal length. As you can see in Figure P, after practice and with VR I can get clear pictures at 1/2 sec shutter speed at 300mm equivalent at 10 feet.

With all of this said, it is extremely hard to generalize these numbers and each person is different. There are limitations on this design and those limitations will be different with each person. If you have not determined these limitations for you, leave the camera on P or Auto—this should optimize the shutter speed. Keep experimenting and don't give up, it will be worth it.

Technique:

To take steady pictures in general, use the following steps.

- Get a good footing
- Compose your shot
- Keep your elbows tucked in
- Take a breath and hold it in
- Slowly depress the shutter
- Keep the camera at your eye for a few seconds after the picture is taken

10. Conclusion

For $5 and a little work, you too can have a tripod in your pocket. Figure P showed some tests done (taken with a D200 and 18-200mm VRII).

Here are three pictures taken at 135MM (202MM equiv) at 1/15 sec:

- Figure Q is handheld, no string tripod or VR
- Figure R is with string tripod, no VR
- Figure S is with string tripod and VR

Each of these is directly from the camera with no alterations. Figure T was taken in real life with the string tripod. Tripods were not allowed in this aquarium.

Alexander Reben enjoys photography, simple elegant design, and building robots. He has worked on robots for photography at NASA and is currently a graduate student in The Media Lab at MIT.

Figure Q: No string tripod, no vibration reduction

Figure R: String tripod, no vibration reduction

Figure S: String tripod, with vibration reduction

Figure T: Aquarium photo

Super Simple Light Tent

How to Build an Indispensable Tool for Small-Item Photography

By Bill Wilson

Figure A: An apple photographed inside the tent

I was inspired to do this project after seeing the PVC light tent posted on the MAKE blog (www.makezine.com/go/PVCLightTent). This light tent uses a cardboard box, some white material (Tyvek), and lets you take reasonable photos of products such as bottles, watches, jewelry, small objects, etc.

1. Select Materials

First thing to do is find yourself a usable box. The box I used is a half of a resin plastic shelf. The dimensions are roughly 16" x 15" x 15". This size has handled most things I have put in it, however I think something a little wider would be easier to use.

Materials used:

- Masking or other heavy tape (duct, packing, etc.)
- X-Acto knife
- Ruler
- Glue stick
- Bristol board
- Semi-transparent white material (Tyvek, white suiting/Ripstop nylon, bed sheets, etc.)

2. Cut the Box

- Lay the box flat
- Using the ruler, add a 1" to 1-1/2" border to all sides of the box (top, bottom, left, and right)—essentially you want to cut a hole in all sides of the box. Tip: don't forget to add a line on either side of the center of the box as it lays flat
- Cut out the four panels of the box by using an X-Acto knife to cut on the lines.

3. Assemble the skeleton

- Open up the box and close the bottom of the box.
- Tape down the interior and exterior seam.

Figure B: Pick a good cardboard box

Figure C: Materials you will need

Figure D: Laying the box flat on a cutting mat

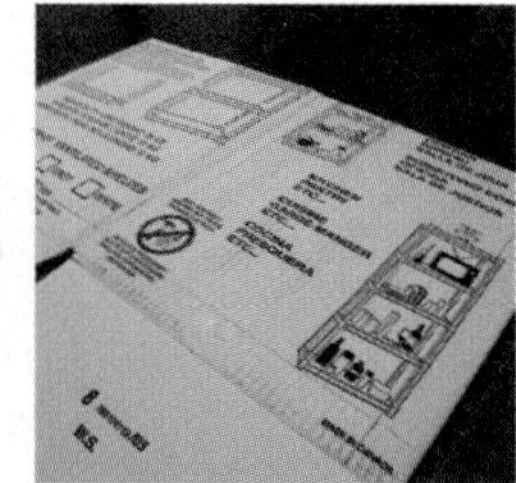

Figure E: Marking the borders before cutting

Figure F: Cutting the first hole

Figure G: Cut holes in four sides of the box

Figure H: Box upside down: close the base of the box

Figure I: Tape the interior seam

For additional information, discussion, and more, please visit the Instructables project page:

Figure J: Tape the exterior seam

Figure K: Measure and cut the Tyvek

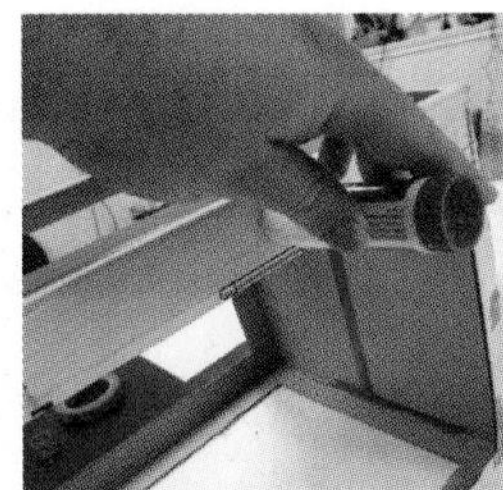

Figure L: Glue the material to the box

Figure M: Cutting the bristol board

Figure N: Installing the continuous background

Figure O: Setting up the light tent

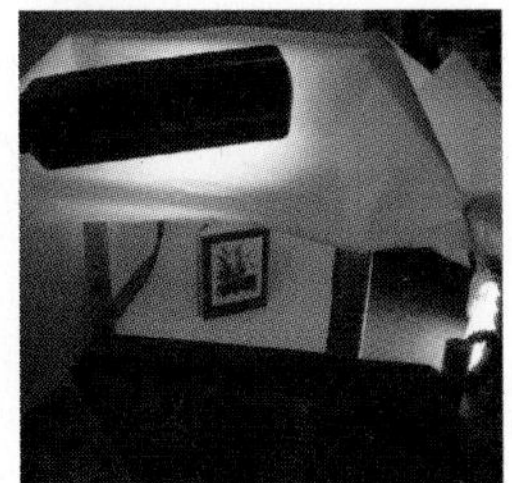

Figure P: Diffusing the light

Figure Q: Photograph taken in the tent

- The bottom of the box will serve as the platform for placing your objects.

4. Wrap the box

Using the semi-transparent material you have chosen, wrap it around the box so that it covers three of the four sides. I used sign printing grade Tyvek and attached it using a glue stick.

5. Add a Continuous Background

This is part of the magic of the light tent: creating a continuous background in your images. To do this we add a piece of bristol board cut to fit the box.

- Use the depth of the box + the height of the box as a rough measurement
- Cut out a piece of bristol board that matches the dimensions. This creates a nice white platform to shoot your images against. Try using other colors, blue, black, etc.
- Insert the bristol board into the box so the edge of the bristol board is placed against the front of the box and the board is allowed to curve like a wave, half-pipe... you get the picture... I hope :-)

6. Add light and enjoy

Now you're ready to take some photos. I used a desk lamp and a couple of Ott-Lites (13watt) for the apple shot in the beginning of this Instructable.

For better/different results I tried switching to the simple clamp-style fixture used in the PVC light tent with 100 watt bulbs.

Experiment with the light's location and try diffusing the light that shines through the top of the box with other semi-transparent material (such as nylon). The light entering the box will be diffused and the shadows will soften or disappear...

Note: At this stage (or perhaps before) you can probably cover the inside of the box with white as well, at least the frame. Or you could switch the white material from the outside to the inside. I bring this up because It was pointed out at DPReview.com (http://tinyurl.com/4r38y6) that there is a black reflection in the photos produced using this box.

7. End result

Figure Q shows a shot taken with this light box. To my untrained eye, it looks pretty good.

Bill Wilson is a long-time tinkerer, dismantler, and maker of stuff. His most recent obsession is kite building and flying. Taking this hobby to new heights, Bill's latest kite accessory is a camera mount that allows him to combine his passion for kites with his love of photography. Visit Bill's kiting blog at www.steadywinds.com or find his insightful comments on the www.kitebuilder.com forum, where he is a regular contributor.

LED Light Drawing Pens

Tools for drawing light doodles By Stuart Nafey

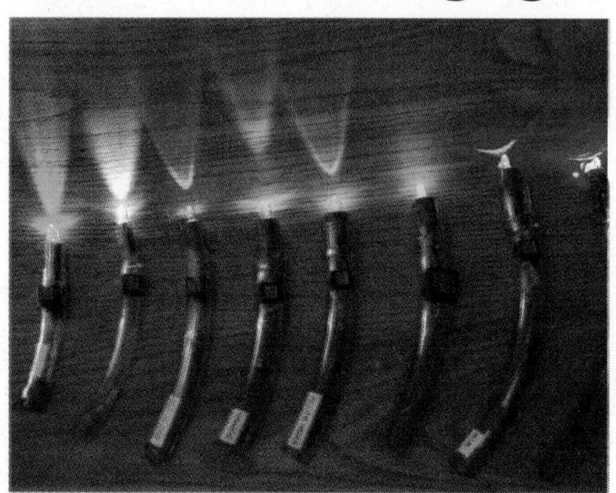

Figure A: Our collection of light doodle pens

Figure B: Lori in action

My wife Lori is an incessant doodler and I've played with long exposure photography for years. Inspired by the PikaPika light artistry group, and the ease of digital cameras, we took on the light drawing artform to see what we could do.

We have a large gallery of drawings on our website (LightDoodles.com). There you will also find a description of how we draw and a brief history of light drawing.

Any light source can serve as your creative implement and we shopped for every keychain flashlight, gimmick pen, and light wand we could find.

We finally sat down and asked what type of flashlight would accommodate Lori's most natural and comfortable hand position while drawing in mid-air. The answer was to hold the light like a pencil with instant on/off control directly under the index finger.

Since we wanted to complete each full drawing in one exposure, she needed to be able to switch between different colored pens quickly. We also found that when drawing a large picture we needed the light to be completely exposed on all sides to minimize fading around the edges.

With these parameters, I went hunting for parts at the local electronics and hardware stores and came up with what turned out to be a simple and versatile tool that resulted in some incredible art.

1. Parts list

I'll be creating a blue light pen. Attention to voltage requirements and current draw are important as different color LEDs have different ratings. Here is a list of the parts used.

- One LED
- One Normally Open (NO) switch
- A 20Ω resistor (size is determined using Ohm's Law)
- 1.5 volt button-cell batteries (3)
- Plastic tubing: 5/8" outside diameter, 1/2" inside diameter
- Plastic tubing: 1/2" outside diameter, 3/8" inside diameter

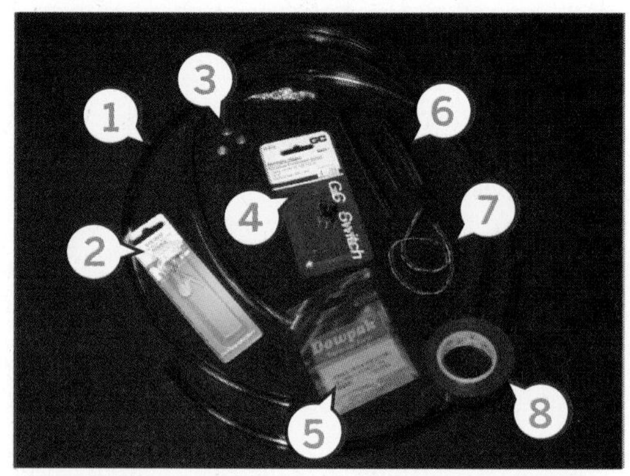

Figure C: The parts you need: 1. Plastic tubing 2. LED lamp 3. Button cell batteries 4. NO Switch 5. Resistor 6. Heat-shrink tubing 7. Wire 8. Electrical tape

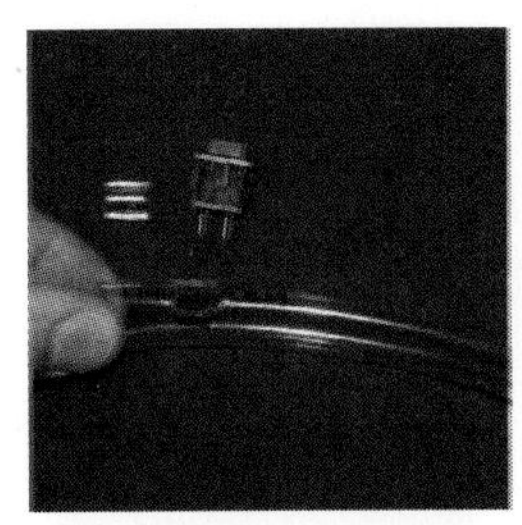

Figure D: The switch hole cut in the 5/8" tubing, cut near one end

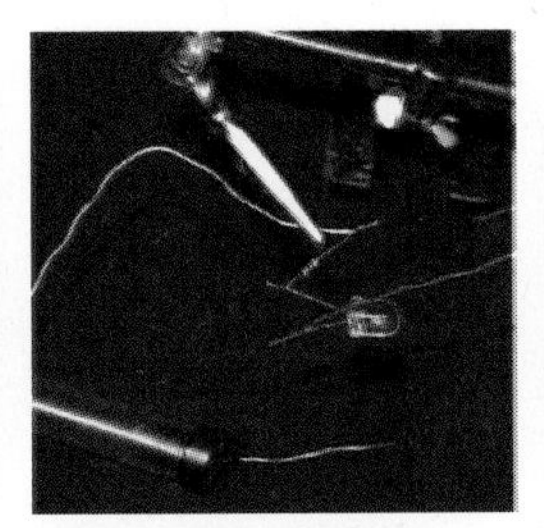

Figure E: The LED, the resistor, and the wires soldered in series

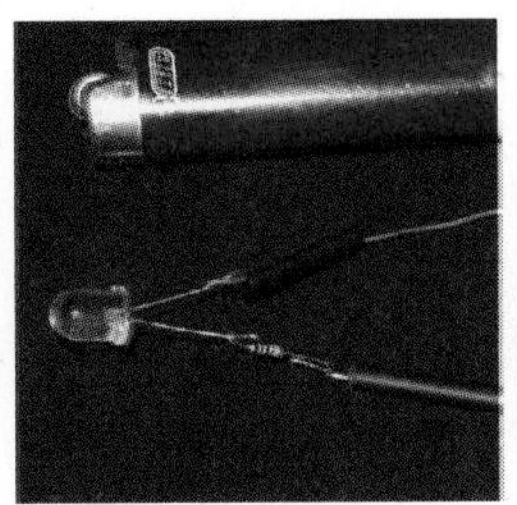

Figure F: Heat-shrink tubing being threaded onto the two wires

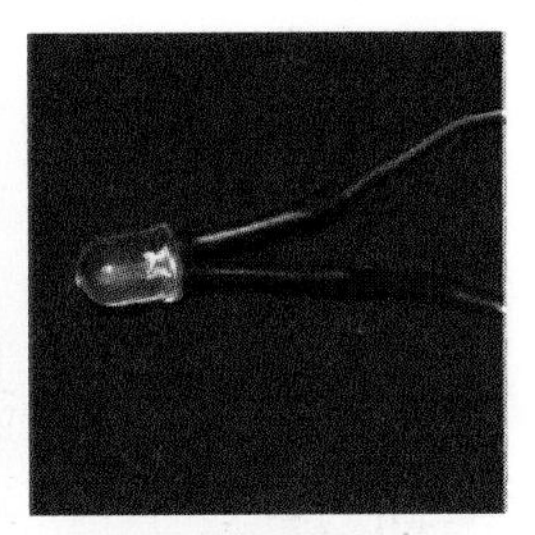

Figure G: The heat-shrink in-place, covering the solder joins and components

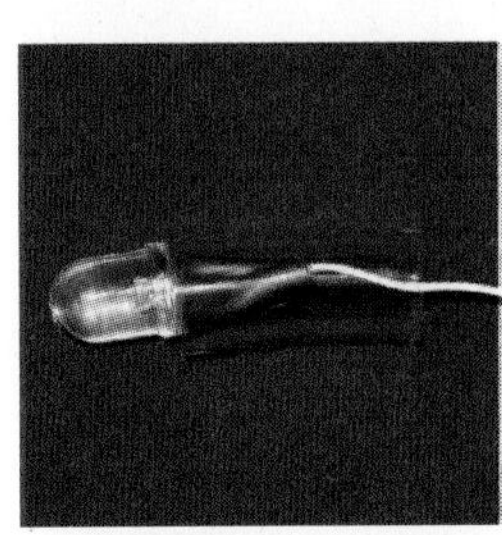

Figure H: A length of 1/2" tubing threaded over heat-shrink assembly

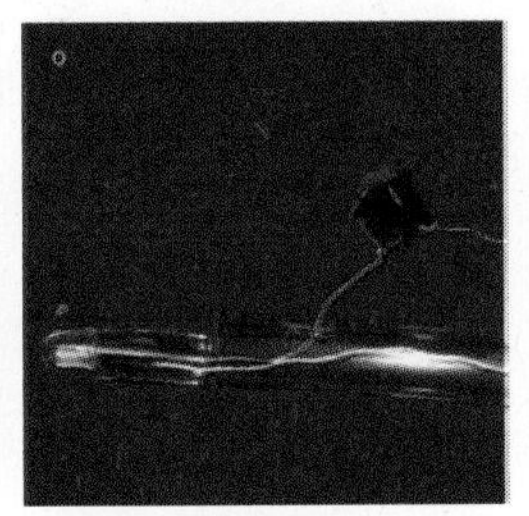

Figure I: The lamp assembled and switch wires threaded into the 5/8" tubing and the switch soldered to one of the wires

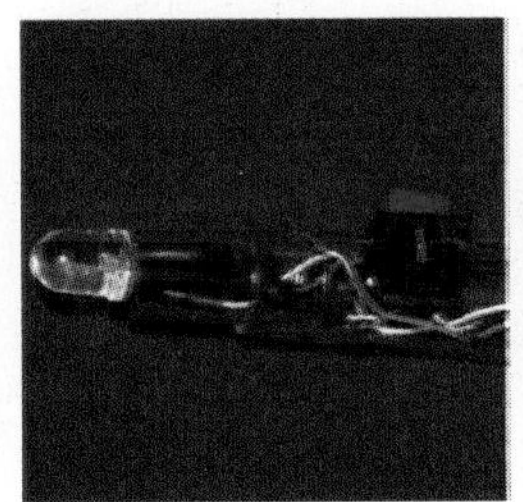

Figure J: The switch installed

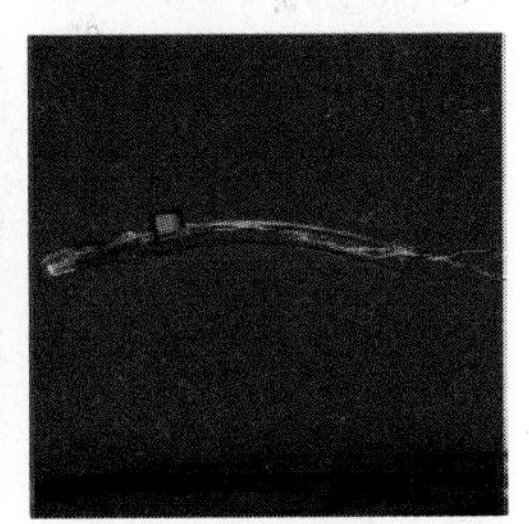

Figure K: The pen assembled, waiting for the power assembly

- Heat-shrink tubing
- 24-gauge wire
- Electrician's tape

LEDs, switch, resistors, heat-shrink tubing, and electrician's tape can be purchased at a local electronics store.

The plastic tubing was "discovered" in the hardware store. Many sizes are displayed on spools which are purchased by the foot. The 5/8" outside-diameter clear tubing best fit Lori's hand. The natural curve of the tubing turned out to be ergonomic and it helps keep the pens upright and stable when placed down.

The switch is a "Normally Open" type, which means that the circuit is complete and the light is on only when the button is pushed and held down. As soon as the button is released, the circuit is broken and the light goes off. Otherwise, I chose this switch for its size and shape, not for any of its other electrical properties.

Adding a resistor to the circuit is good practice in obeying Ohm's Law.

2. Warning: math content ahead

I picked up the basics of LED science from the "LEDs for Beginners" Instructable (www.instructables.com/id/LEDs-for-Beginners), reading not only the Instructable, but many of the follow-up comments. They supplied a wealth of theory and important links to everything you need to know about LEDs.

The back of the LED package provides the information you need to properly build the working circuit. Use this information to determine which type and quantity of battery and what size resistor to use.

The blue LED I used here requires a 4.0 Forward Voltage Drop (Vf) to light. It will pass 25 mA of Current (If). Three 1.5 volt batteries in series will supply 4.5 volts.

Any combination of batteries that add up to the required voltage will do. For instance, AAA batteries are 1.5 volts and three in series will give you 4.5 volts.

I found these tiny 1.5 volt button batteries inside of an A23 battery. Three of these work nicely.

Obeying Ohm's Law and using this Current Limit-

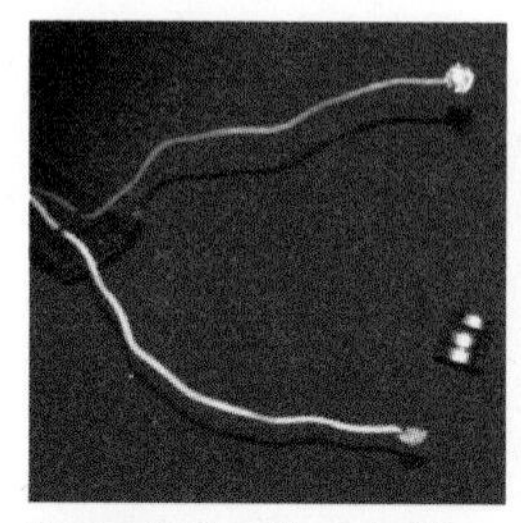

Figure L: The three button cell batteries in series

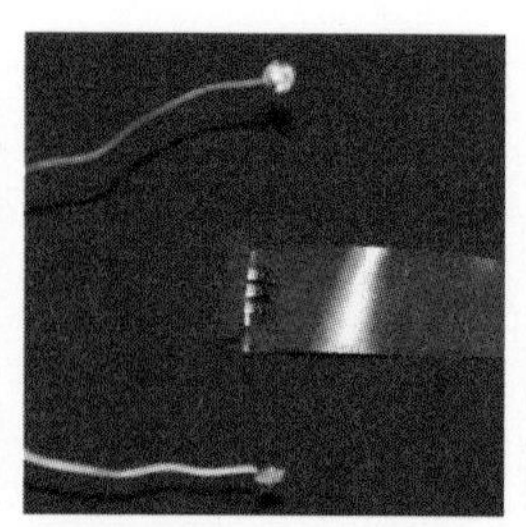

Figure M: The button stack being wrapped in electrical tape

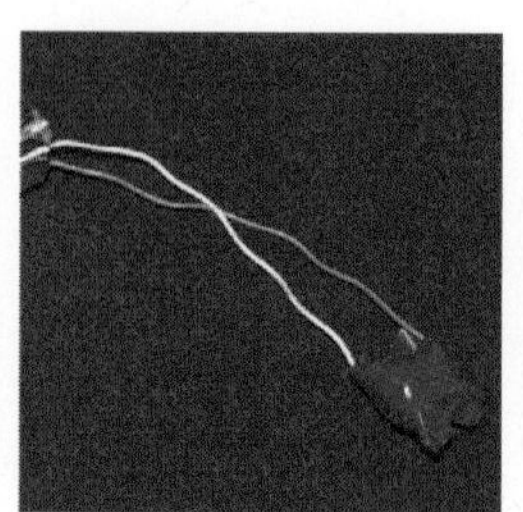

Figure N: The battery stack attached to the power wires

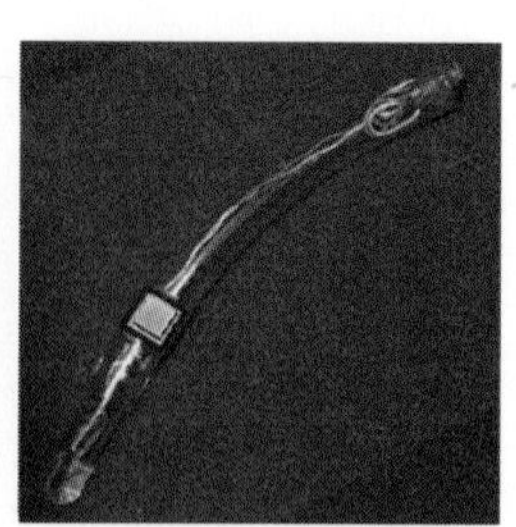

Figure O: Our finished LED pen, ready to paint with light

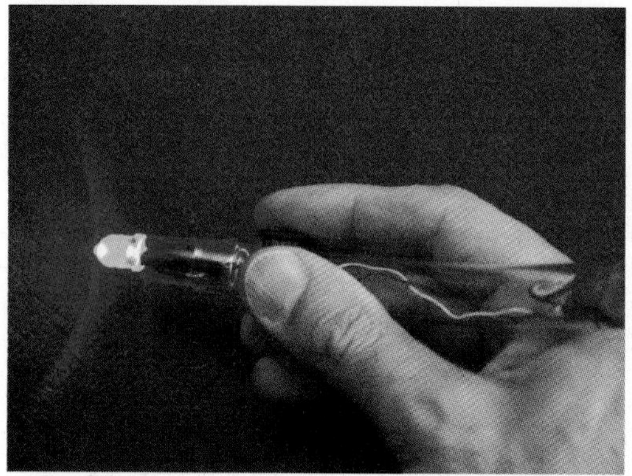

Figure P: The pen powered up

Figure Q: A finished piece of art created with LED light pens

ing Resister Calculator of LEDs (ledcalc.com), a 20Ω resistor should be placed in-line in the circuit.

3. Put it together

Everything in this simple circuit is placed in series and the parts can be arranged in any order with one exception: the LED will only light if the battery polarity is correct.

Cut an appropriate length of the 5/8" OD (outside diameter) tubing and cut a hole close to one end to accommodate the switch. Keep in mind how the pen will fit in your hand and where your finger will lie to operate the button. Since we hold the light like a pen, I placed the switch where we can easily push it with the index finger.

Solder the LED, the resistor, and the wires in series. Remember: the resistor can be placed anywhere in the circuit.

I placed heat-shrink tubing over the exposed wires and heated the tubing with a lighter, protecting against short circuits.

Next, fold the wiring up and slide inside of a 1" piece of the smaller 1/2" OD plastic tubing.

4. Add the switch

Feed the wires through the 5/8" tube, one through the hole, and solder the switch inline.

Continue feeding all wires through the tube. Squeeze the switch into the cut-out hole (I bent the leads to fit). Squeeze the 1/2" tube into the 5/8" tube.

5. Add the power source

Finally, add the batteries. This is very low-tech, but I have yet to find or build a battery holder that suits my purposes.

Strip the ends of the wires and wrap the bare wire and a small bit of aluminum foil together into a ball.

Use the electrical tape to hold the batteries together in series (positive terminal to negative terminal) and the wire ends in place on each end of the battery stack.

Polarity is important here. Test the light at this point and try reversing the connections if it doesn't work. (Remember that you have to push the switch while testing.)

Wrap a second piece of tape around the terminal ends. Stretch and wrap the tape tightly, insuring a positive connection. Fit the battery pack into the end of the finished pen.

6. Finished product and drawing example

It works for me! Now drawing is another story.

To see what we've done, visit us at LightDoodles.Com. You can also see more of our drawings on Flickr (flickr.com/photos/unklstuart/sets/72157601507669278).

To find out how to do the long-exposure photography, check out the "Write or Draw with Light" Instructable (www.instructables.com/id/Write-or-Draw-with-Light!).

Stuart Nafey is a mild-mannered IT guy by day, a bohemian long exposure photographer by night. He seeks ways to exploit his wife Lori's quick drawing skills and together they create fun art at Lightdoodles.com.

User Notes

We took our Light Doodles to the 2008 Bay Area Maker Faire (makerfaire.com). Here are some of our favorite images from the Faire sessions.

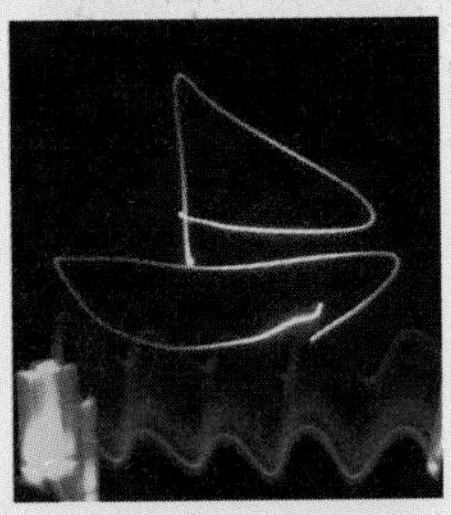

Home & Garden | Food | Photography | Science | Computers | Electronics

The Best of Instructables Science

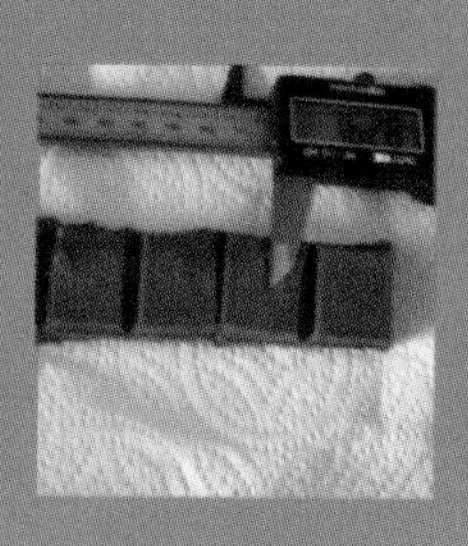

There are all sorts of geeks (software geeks, electronics geeks, internet geeks) but there is one class of geek that makes all other geekness possible—the science geek.

Hands up, I'm a science geek, but I'm a grass-roots science geek. I don't play in an underground lab, I play in my shed. I don't play with electron microscopes and synchrotrons, I use PVC and a Dremel.

Instructables has been an important part of my life since the beginning of 2006. It's been a sort of "permission to make," giving me a space to show off plans that have rattled around my head, unfulfilled, for decades. I've been able to contact people all over the world, help them with their projects, get help with mine. It's the most welcoming web community I have ever found, with the broadest spectrum of members, and I honestly believe we are more creative than any art forum.

It's surprising how much good science you can do with a few simple tools and an open mind—measure forces, bottle electricity, map the sensitivity of your skin. On Instructables we have measured the speed of sound with lumps of wood, and the speed of light with a bar of chocolate. We have made a spectroscope from an old DVD and modeled the structure of plant cells in cake. There are even schemes floating around for recreating Sputnik!

Browse through the Science section of Instructables and you'll find projects for recreating projects that are part real science, part Hollywood science—you'll be able to build batteries from scraps in the kitchen, breed crickets, or build Jacob's Ladders to send ten thousand volt sparks crawling around your hidden underground laboratory. Or your basement.

In this section of the book are projects encompassing all branches of science, from the gloopy biology of growing phosphorescent algae, to the mysteries of magnetism. It's only a taste of what anybody can achieve when they try.

See you in orbit...

—Mark Langford (Kiteman)

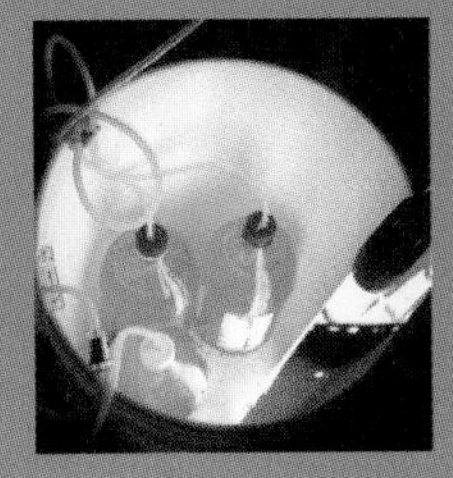

contest WINNER!

Light Bulb Greenhouse

Turn an old incandescent light bulb into a little green grow house

By Cyrus Ahmadi-Moghadam

The incandescent light bulb is the ultimate symbol of wasted energy. Since I don't use them anymore, I wanted to transform them into something green rather than throwing them away. This project was inspired by an ad I saw in *Popular Science* (June 2008).

Warning: This Instructable requires working with glass and broken pieces of glass. Proper hand and eye protection should be worn.

1. Materials

You will need:

- An old light bulb. Best if burned out, but if it's going to just sit and collect dust, a new one is fine, too.
- A 7/16" wrench socket head. Holds the bulb on.
- A shower flange (that thing that goes between the shower head and tiles)
- A piece of rubber ring material big enough for the bulb to fit into. (I got my rubber part from my bike.)

Figure A: Our finished light bulb greenhouse, planted with mint

- A small plastic cup (for holding dirt and the plant roots)
- Epoxy. Best glue ever.
- Soil

Tools:

- Pliers
- File
- Drill, drill bit

2. Prepare the light bulb

Start by using pliers to pull the bottom contact off of the light bulb. Next, use the file to break

Figure B: Ready to prepare the bulb

Figure C: Removing the contact

Figure D: The light bulb, filed, cleaned, and ready for planting

Figure E: The base with shower flange, socket head, and rubber ring, all epoxied together

For additional information, discussion, and more, please visit the Instructables project page:

Figure F: The roots and stem of the plant threaded through the socket head

Figure G: Placing the flange over the grow cup once the plant has been rooted

out the glass material between the two contacts. Break out the glass on the inside and use your file to smooth the edges. Clean the white powder from the inside of the bulb. You can do this with a rag and water. Careful.

3. Prepare the base

Using a drill press, widen the hole in the bottom of the socket. Use pliers to widen the hole in the bottom of the light bulb until the socket fits snugly. Epoxy the socket to the rubber piece. Let it dry for at least 20 minutes. Epoxy that to the shower flange. Let this dry for another 20 minutes.

4. Add a plant or plant seeds

This step is self-explanatory. Either add a plant or seeds.

5. Use it

Now all you have to do is fill the cup with soil and seeds (or small plants), put the base on it, then put the light bulb over it. Place it in a well lit window sill and water it occasionally.

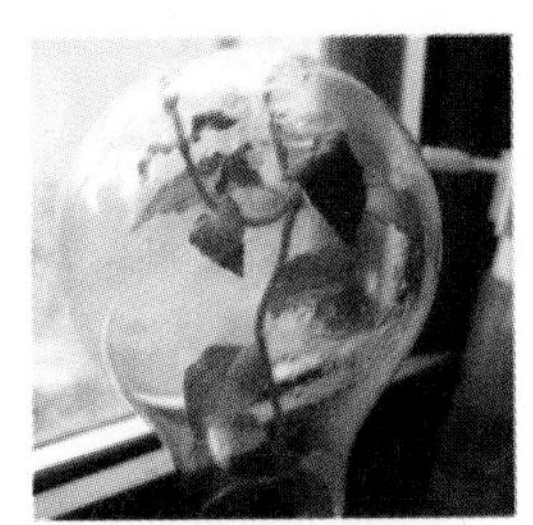
Here's my mint plant after one month of being completely abandoned (I was 1,200 miles away)

Cyrus Ahmadi-Moghadam says "I'm a long-term maker, been helping in the garage as young as I possibly could. Got into Ibles about a year ago. Lived in Shoreview, MN, for 15.5 years, now in Santa Fe, NM. Current project is an electric car (1980 Fiat 124 Spider)."

User Notes

In response to questions about watering the Light bulb Greenhouse, **Cyrus** posted these pictures in the Comments.

Ryan Chorbagian got a kick out of the project and made a steampunk-ified version of it for his girlfriend.

contest WINNER!

Grow Your Own Bioluminescent Algae

When you care enough to say it in glow-in-the-dark bacteria

By Chris Quintero

Figure A: Stating the obvious

You may have memories of running after fireflies with outstretched hands on a warm summer evening. You may have even watched some Discovery Channel documentary on the mysteries of the deep sea and marveled at those "glowing" organisms found there. Chances are however, you probably haven't heard too much about the plethora of bioluminescent creatures inhabiting this planet.

Bioluminescence (literally meaning "living light") occurs within many living organisms, although most are relegated to the deep sea. This chemical reaction involves the oxidation of Luciferin (just a name for a class of biological light-emitting pigments). While related, the name doesn't have devilish origins, but rather the same Latin origin "lucifer," meaning "light bringer."

Depending on the organism, the light can be used for camouflage, attraction, or even communication among bacteria, to name a few. Some of the more notable organisms that bioluminesce include fireflies, glow worms, bacterium, a plethora of marine life, and even mushrooms.

Today, we'll focus on a particular light-emitting algae known as *Pyrocystis fusiformis*. These dinoflagellates typically do not occur in high enough concentrations among marine algae to produce a vary noticeable glow. However, when the conditions are right (excess nutrients, enough sunlight, etc.) an algal bloom can occur and populations will explode. Chances are, you've heard of this phenomenon before, known as a Red Tide.

1. Gather the materials

A number of marine enthusiasts already grow phytoplankton at home for use in feeding various species of marine life. The method we'll use is similar.

To start, you'll need:

- A clear growing container (e.g., old soda bottle)
- Sea salt
- A grow light and timer
- Micro Algae Grow
- Optional: Small, motorized air pump
- A starter culture

Sea salt: No, not from your pantry, you gourmet fiend, you can get this at most pet or aquarium stores.

Grow light: You can pick up a plant fluorescent and rack from Wal-Mart for around $10. Make sure you have a light timer.

Micro Algae Grow: Our most crucial ingredient (besides the actual algae). There are a number of nutrient formulas people have experimented with, and truthfully, I've only had mixed results with this one. Experiment with what works best. You can get it here: http://tinyurl.com/3g2wdv.

Air pump: This is used to both circulate the nutrients and to keep vital CO_2 in the water. The algae need this to grow.

A starter culture: These can be obtained from a few places online. I've used Sunnyside Sea Farms, http://tinyurl.com/451gel.

2. Preparation and mixing

Sanitation is necessary so your batch doesn't crash. After you *really* wash out the soda bottle, make sure there is absolutely no residue left. Some people say swirl some diluted bleach around in the bottle. Others say to stick it in the microwave after it's completely dry (won't melt or deform if it's dry... wet is another story). Choose your preference.

Additionally, sanitize the tubing if you're using an air pump, and anything else you're using to prepare this batch.

For additional information, discussion, and more, please visit the Instructables project page:

Figure B: A "Lights from the Sea" *Pyrocystis fusiformis* starter culture

Figure C: Micro Algae Grow, nutrient formula

Figure D: Sea salt, of the aquarium store variety

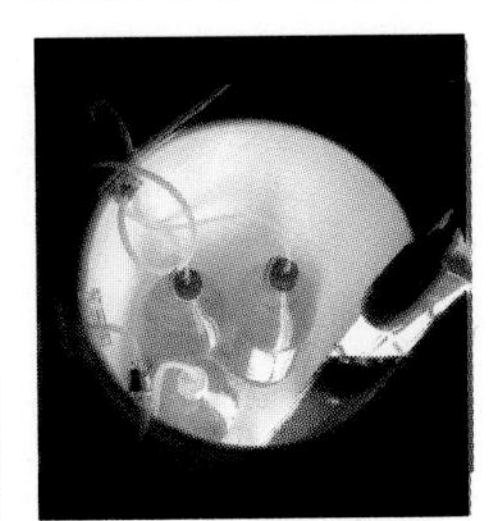

Figure E: A typical grower's set-up

Mix up a batch of salt water. I've used purified water, however some people swear by tap water as it adds in some minerals needed for the algae to grow. I'm sure the composition of minerals in tap water varies significantly from region to region, so I'll let you make that decision. For help, you can check out some of the discussions on the topic at http://tinyurl.com/53srjf.

Regardless, mix the salt to a 1.019 specific gravity (sg) concentration. Directions on how to do this are on the back of the package. You'll need a hydrometer if you've never done this before.

Add in ~1 ml of the Micro Algae Grow. In this case, less is more. The solution you received the culture in should already have enough nutrients to support sizable growth. If you don't want to mess with making your own solution (not necessarily a bad idea), many places that sell starter cultures will also sell culture solution.

Let both the solution and culture bag sit in the same area out of the sun for an hour or two. This is simply to let them reach room temperature. A sudden change in temp during transfer could shock the culture enough to significantly harm it. If your room temp is in the 70s, you should be okay. If my memory is correct, they grow ideally if the water temp is in the mid to upper 60s.

Finally, transfer the algae into your bottles. If you are using the air pump, be sure to cut a hole in the top of the bottle cap. When I say air pump, keep in mind I really only mean a few bubbles every 10-20 seconds or so. We don't want to disturb these dinoflagellites too much—just resupply their CO_2. I've even heard of people using another tank of yeast (which emit CO_2) as their air supply. Figure E shows a typical grower's setup. (Your bottles won't be green, though.)

3. Growth

These dinoflagellates need a constant cycle of light and darkness for optimal growth. Put your grow light and bottles in a dark place (e.g., a closet) where you can strictly control how much light they get. Set the timer so the grow light is on a cycle of 12 hours on, 12 hours off. Don't be worried if your starter culture doesn't emit light right after you receive it. They will only bioluminesce in their night cycle, so plan the light cycles for when you want to see it.

Monitor your cultures for any sudden changes in color, and give them a gentle shake every day or so or all the sediment will collect to the bottom. If you have a successful culture, you will eventually need to "split" the batch. Mix up another batch of saltwater/nutrients, and halve your culture between the new bottles.

Remember, these cool creatures will only brightly flash when disturbed, but too much disturbance can both harm them and wear them out. They have a "recharge" time so to speak between disturbances for optimal performance.

If you're looking for something that will constantly glow, you might want to check out bioluminescent bacterium instead. You can get some from Carolina Biological Supply. Culturing this is a rather different process, but you can find some guides on the Net. One bioluminescent strain is *Vibrio fischeri*. See here: http://tinyurl.com/3t2tx2.

Good luck and have fun!

Often seen with tools in hand, Chris Quintero is pursuing a B.S. in mechanical engineering at the Georgia Institute of Technology. In his spare time, he enjoys windsurfing, cycling, and the occasional unexpected explosion.

3D Magnetic Field Viewer

A super-simple device that shows magnetic lines of force

By Bill Sherman

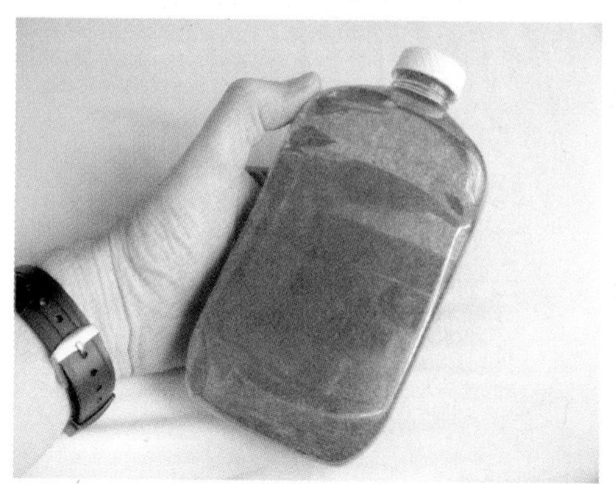

Figure A: A 3D Magnetic Viewer ready to reveal invisible magnetic lines of force

My job takes me to various research labs where I install powerful magnetic systems and train technicians in how to use them. This time, I needed a magnetic field viewer to show lines of force. This Instructable shows how to construct such a device using nothing more than steel wool fibers and mineral oil. I built this one while at my hotel.

1. Get your supplies

I went to Wal-Mart and bought the following items:

- Fine steel wool
- Bottle of mineral oil in plastic bottle
- Scissors

2. Cut the steel wool

Using the scissors, cut the steel wool into small, short pieces. About a tablespoon of cut wool should be enough. Spread out newspaper on your work surface and try not to get clippings all over the place.

3. Place in bottle of oil

Place the cut steel wool in the bottle of mineral oil. Clean the top of the bottle with a paper towel so that you'll get a good seal. Squeeze the sides of the bottle as you put on the cap to burp out the remaining air. Shake to disperse the fibers.

Important note: Clean up any mess, and dispose of excess fibers immediately. Place it in a plastic bag so it doesn't spread everywhere. The longer your mess sits around, the more likely it will spread.

4. Test your viewer

Place a magnet against the bottle. The viewer

Figure B: What you'll need to make your viewer

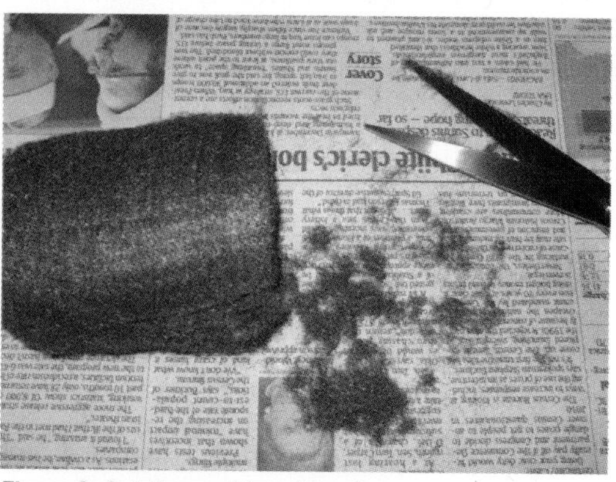

Figure C: Cut the wool into tiny pieces

For additional information, discussion, and more, please visit the Instructables project page:

Figure D: Tiny pieces of steel wool are dispersed in the oil and line up in three-dimensions to show the field

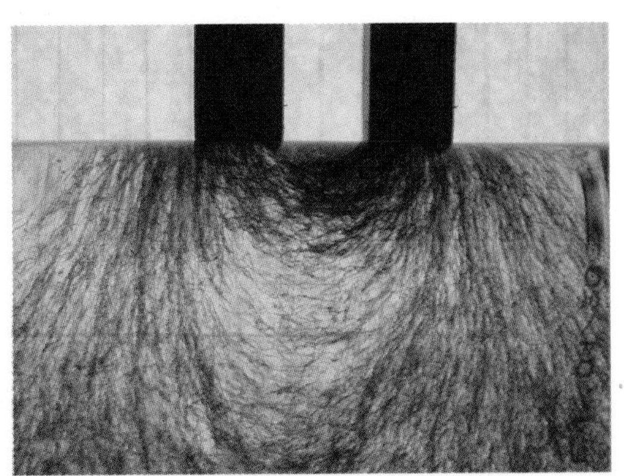

Figure E: Testing the effectiveness of the viewer

Figure F: Lines of force from the cell phone case magnet are shown

will display the magnetic lines of force as seen in Figure E.

5. Using the viewer

Placing the 3D Magnetic Field Viewer near a mobile phone shows the pattern of magnetic lines emanating from the phone's case.

Bill Sherman says: "All my life I've had a passion for science and how things work. Working as an electronics engineer has given me the tools and opportunities to learn even more. I spend my spare time building robots, artbots, and sharing these activities with my family."

User Notes

Chris Cashion wrote in to say that he thought this device was neat and that he planned on building one. He said that instead of using steel wool, he was going to use ferrous dirt particles.

Which led Bill to reply: Some of those particles may be micrometeorites. We get rained upon several tons a day around the world. I saw a project somewhere on the Internet that shows how you can collect them by using a funnel with a baggie-contained magnet. The collected rain will have micrometeorites that will stick to the outside of the baggie. You can use a large, clean plastic tarp to increase the area. Micrometeorites look like little rounded blobs and dumbells. This would make a good Instructable, but first I better find some references to cite the source.

After doing a quick search "collecting micrometeorites" I found this: tinyurl.com/6ycwom. And here is a lesson on collecting micrometeorites from California State University Fresno: tinyurl.com/5hn8cw (PDF).

And the ubiquitous Kiteman followed with: You could have checked a little closer to home: www.instructables.com/id/Give-your-loved-one-a-real-fallen-star-this-Chri

Measure the Speed of Light with Chocolate!

A delicious way to enjoy scientific constants

By Bradley Powers

Figure A: What you'll need: chocolate and a way to measure it

In this Instructable, inspired by the book *How to Fossilize Your Hamster And Other Amazing Experiments for the Armchair Scientist*, we use a bar of chocolate to measure the speed of light.

What you'll need:

- A bar of chocolate
- A microwave
- A metric ruler
- Safety glasses (not that this is dangerous, it just adds awesome-factor to any experiment)

1. Eat some chocolate

You know you want to.

Figure B: Let the munching begin!

2. To the microwave!!!

Remove the rotating tray thingy from your oven—we don't want the chocolate to cook evenly.

3. Zap the chocolate

Turn on the microwave until pools of chocolate start to form (about 40 seconds). Don't overcook it-burnt chocolate doesn't smell so good.

4. Measure

From "hot spot" to "hot spot" (where the chocolate has started to melt). Write down the measurement.

5. Now for the mathy stuff

Ok, now that we know the distance between hot spots, we'll use some math to figure out the speed of light. First, the distance we measured represents the half-wavelength of the waves being emitted by the microwave. To find the wavelength of the microwaves, we multiply by two. In my example, that gives us a wavelength of:

7.628 cm * 2 = 15.256 cm

Since the speed of light is equal to the wavelength times the frequency, we can figure out the speed of light. But we don't know the frequency of the microwaves. Most microwaves operate at 2.45 GHz, or 2,450,000,000 Hz, so, we take the product of the wavelength and the frequency:

15.256 cm * 2,450,000,000 Hz = 37,377,200,000 cm/s

which, given that we are doing this in a kitchen (and a small error in our measurements are multiplied by

Figure C: Remove the rotating platter from the oven

Figure D: Measure "hot spot" to "hot spot"

4,900,000,000), is shockingly close to the actual speed of light, which is 29,979,245,800 cm/s, or as it is typically defined, 299,792,458 meters per second.

6. Iterate

All good scientists know that repeating an experiment is good for making sure your results are statistically relevant, so do it again. Eat more chocolate. Have Fun!

Bradley Powers is an engineering student attending the Franklin W. Olin College of Engineering (Class of 2010) studying robotics. He is a former Instructables intern, and in his free time, can be found building robot helicopters and wielding the latest safety eyewear styles.

User Notes

Patrik says: Interesting idea! Of course, you're not going to get a nice linear standing wave inside a microwave oven, so depending on where you place the chocolate you might get a different distance. But it should be in the right order of magnitude at least.

Bradley responds: Sure. This isn't exactly a super scientific experiment, but if you perform it enough times...

Mike Dillemuth weighs in with: You're actually measuring the *wavelength*, or of the cycle of the microwave's tuned pattern of dispersion, not "the speed of light"...186,000MPS (miles per second). To do that, the photons, which is what light is essentially, would have to make a pattern. *Visible* light however will NOT. It merely stops. Infrared will make a heat pattern bloom, but then the pattern simply melts the chocolate, as the infrared is pretty hard to focus on one spot running lengthwise to make it "skip" until the waves show themselves as a cyclical/oscillating wave pattern.

Bradley responds: This is true. This was not intended to be an accurate measurement of anything, just a simple experiment that someone in (I dunno...) 6th grade could do easily, to start gaining some understanding of waves, light, etc. I know enough theory to know that we're not actually measuring the speed of light. But it's an interesting educational exercise nonetheless, and that was my intent in publishing this Instructable.

In response to some confusion over the term "speed of light," **Harry Collard offers:** All electromagnetic waves, (in order: Radio, Microwaves, Infrared (heat), Visible Light, Ultra Violet, X-Ray, and Gamma) travel at the "speed of light." The name is misinterpreted. "The speed of light" is actually the speed that electromagnetic waves travel, and visible light just happens to be part of this spectrum.

The Best of Instructables
Computers

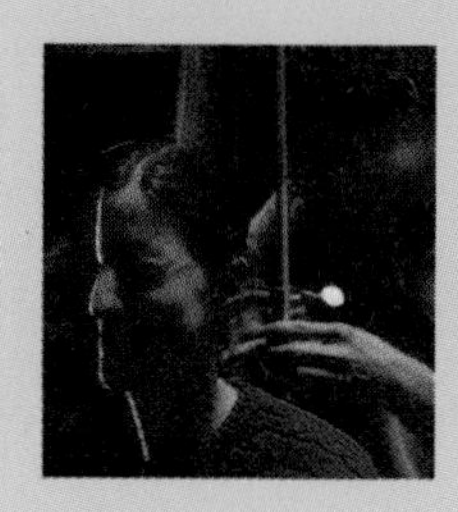
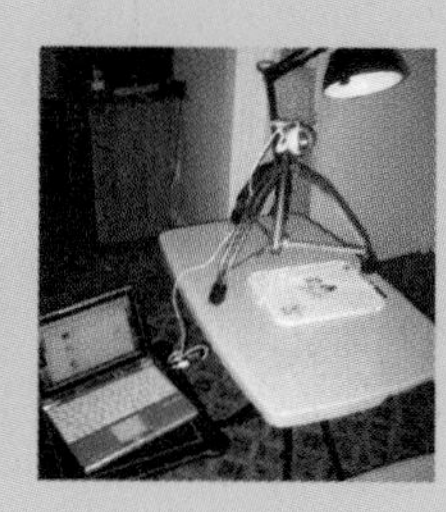

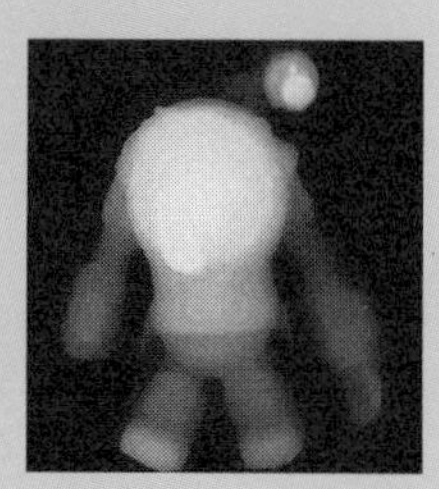

The secret behind the computer's success is its generality. With the right suite of sensors, software, and actuators, a computer can do virtually anything! The computer projects on Instructables are a prime example. Here is a smorgasbord of the practical, the playful, the creative, and the just plain weird.

Say you want to install a laboratory in your garage, but you don't have the money for the equipment. Fear not! Instructables teaches you how to turn your computer's sound card into an oscilloscope and your webcam into a microscope. Perhaps you want to impress your friends with your artistic talents? Follow the Instructable for the computer-controlled Etch-a-Sketch and you'll be churning out your own Mona Lisa in no time. Or maybe you just want to mod your computer to make it more personal? The site has a host of projects for pimpin' out your computer case, Lego-izing your USB stick, and even creating a Taxidermied Mouse mouse (seriously).

Over the past few years, the Instructables community has offered me valuable advice on some of my own computer-related projects. They've pointed me toward better programming libraries and helped debug my code. The best lesson I've learned is that, despite the power of computers, computer skills alone will not make a project fly, crawl, dazzle, or impress. You must combine these skills with solid electrical and mechanical engineering and couple that with your own creativity. Don't stop at hacking just the ones and zeros, keep going until you're hacking the physical world itself!

—Kenny Jensen (argon)

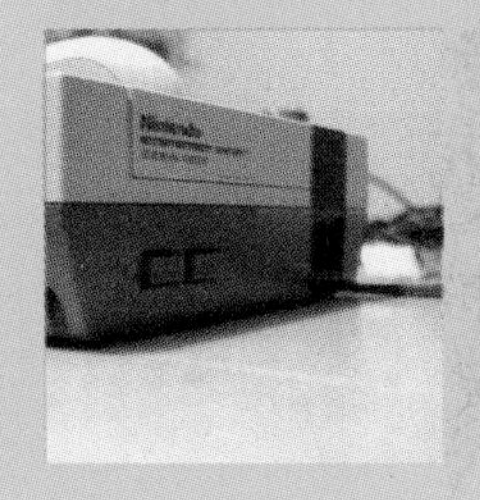

Nintendo NES PC

Turn an old NES into a functional PC

By hatsuli

Figure A: The NES PC

Ah, the Nintendo Entertainment System. Brings me back a lot of good memories: Super Mario Bros., Double Dragon, Megaman. It also brings back not-so-great memories. The agony of changing cartridges, blowing until you're dizzy and still getting nothing but a flashing screen when you start the console. When you finally got the cartridge to run, it could freak out at any time from the smallest dust particle in the connectors.

Luckily, those days are gone. NES emulators can be found for the PC. These nifty little programs are designed to run NES games as accurately as possible. All you need is the emulator itself, and a ROM for a NES game. You can obtain public domain ROMS for the NES from http://pdroms.de/files/nes.

I wanted to play NES and other oldish consoles on the NES PC, and also play Divx/DVD videos and more. Playing NES games on your computer is fine, but I wanted more of an original feel to it. I thought I'd be able to put a PC full with hard drive and DVD drive inside of a NES case, attach some controllers to it, and hook it up to my TV.

Here's the full list of console emulators my NES PC currently has installed:

- NES
- Super NES
- Sega Mega Drive/Genesis
- Sega Master System
- MAME (Arcade)
- Game Boy (Color)
- Game Boy Advance
- Sega Game Gear
- Turbo-Grafx 16 / PC-Engine
- Sony Playstation (games run from CD drive)
- Nintendo 64

The NES PC is used without mouse or keyboard! Everything is done using the gamepads, which makes it feel more like a console (like it should!).

1. Parts you'll need

An NES (duh)

You're free to use a non-working one, as the only part you'll be using is the case.

Motherboard and processor

You will need a motherboard and processor. Because of the tiny size of the NES case, you're not going to be able to fit a normal ATX motherboard in it. I used a Mini-ITX motherboard. They are 17cm by 17cm, so it's a great fit for the NES case. You can purchase Mini-ITX motherboards from www.mini-itx.com and also www.mini-box.com. I bought a Jetway 1.5GHz C7D motherboard. It was relatively cheap but powerful enough for my needs. Mini-ITX boards come with an integrated (built-in) processor, sound card, and video adapter. This is great when space is a luxury you can't waste. You should make sure the processor won't generate too much heat. There's little space for air to move around in the case, so it might get a bit too hot.

Note: It's important that your motherboard have a TV-out connection: either S-Video (preferred) or Composite. If you have an LCD screen you might want DVI or HDMI.

Memory

The motherboard needed DDR2 memory, so I got a 1 GB stick of that.

Storage

I already had an old 40 GB 2.5" laptop hard drive. It

Figure B: The Jetway Mini-ITX motherboard

Figure C: PicoPSU power supply

won't work with a standard IDE connector, so I got a 44-pin to 40-pin IDE adapter. I also had a slimline DVD/CD drive from the same old laptop. It also needed a slimline to IDE adapter to work. You can obtain these parts from some Mini-ITX vendors, but you should also search Google for them (try searching on "44-pin to 40-pin IDE" and "slimline to IDE adapter," without the quotes).

Power

You will need a power supply unit (PSU). There's a problem, though. ATX power sources are too big to fit inside the case. I ended up using an 80 Watt picoPSU. It's a tiny DC-DC power source. It works like a laptop's power source: you attach an external power brick that handles the AC/DC conversion and provides the picoPSU with 12V DC power.

Connectors

You will need leads to attach the power LED, power switch, and reset switch to your motherboard. I got them from an old computer, but you can purchase such cables from computer parts vendors. I also used some old case fans I had. If you've chosen a cooler motherboard/processor combination than I did, you might not need extra fans. There are some very cool fanless VIA EPIA boards, but they're not very powerful performance-wise.

Tools

You won't be needing any special tools other than a Dremel or something similar. It's used for cleaning out the case bottom and cutting out the hole for the backplate. You'll also need to solder some wires for the power/reset switches.

Note: Take care when handling the motherboard, memory, etc. They are extremely sensitive to static discharge, so make sure you're properly grounded!

2. Preparing the case

Following the example of other NES PC builders, I got rid of all the original NES hardware except for the power LED and power/reset switches. The NES power switch stays in when you press it. This can be fixed by removing a small metal part on the top part of the switch (compare the power and reset switches: the power switch has the metal part, the reset switch doesn't).

Next, I marked which plastic parts I'm going to need with a marker. Basically, you need only the four corner standoffs and the plastic parts that keep the reset/power switches in place. I also marked part of the case bottom to be cut off (marked in Figure D with a yellow line) to make space for the hard drive that will sit under the motherboard.

3. Preparing switches and power LED

Next, I unscrewed the switches and power LED from the case and soldered the motherboard leads for them. Make sure there are no shorts that could cause problems. The PCB is nice and big, '80s style, so you shouldn't have trouble.

4. Placing the hard drive

The hard drive will sit under the motherboard to maximize space efficiency. First, I covered the hole I cut (see step 2) with some plastic so the hard drive bottom wouldn't be seen from the outside.

Figure D: Prepping the enclosure—cut away everything except the corner standoffs and switches. You'll need to cut away the indicated portion of the case bottom to make room for the hard drive, too.

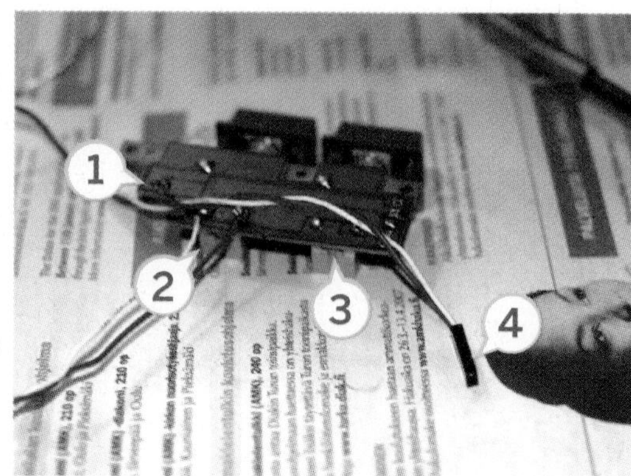

Figure E: 1. Power LED 2. Power switch 3. Reset switch 4. Attach this to the motherboard's connectors.

Figure F: The red area shows where the hard drive goes (trim away the standoffs and raised portion to make it fit)

Figure G: The hard drive in place

Figure H: Fitting the backplate into the case

Figure I: Connecting the optical drive

Next, I placed the hard drive (the location is marked in red in Figure F) and covered the top with duct tape so as not to short-circuit the motherboard, which will sit directly on top.

Note: I later found out the 2.5" laptop HD I had was broken, so I ended up using a regular 3.5" 160GB one. It fit just as well, but was a bit higher so the motherboard had less space vertically.

5. Cutting a hole for the backplate

Next I placed the motherboard on top of the hard drive (the other end of the board sits on top of the power/reset switches). I measured where the I/O backplate (this should be included with your motherboard) would need to be placed and carefully Dremeled a hole in the top and bottom halves of the case to fit the plate.

Figure H shows the hole. A tad ugly, but the picture was taken before I did any sanding. It's much nicer now. The fit was alright, so I used hot glue on the bottom half to make sure the plate stayed in place.

6. Placing the DVD/CD drive

I decided to use heavy-duty duct tape to fix the optical drive to the top of the case. Slimline optical drives are very light, so the tape worked fine. I had to cut off a part of the case (check Figure J) to fit the drive.

7. Putting it all together

I connected the IDE cables, the power for the hard drive and DVD/CD, drilled a hole for the PSU connector and squeezed the case-halves together. After some considerable violence, I managed to screw the case closed.

Note: I later noticed the processor was running too hot (over 70C!) so I added two extra fans: one on the top (see Figure L) and one where the original controllers were attached. Because of

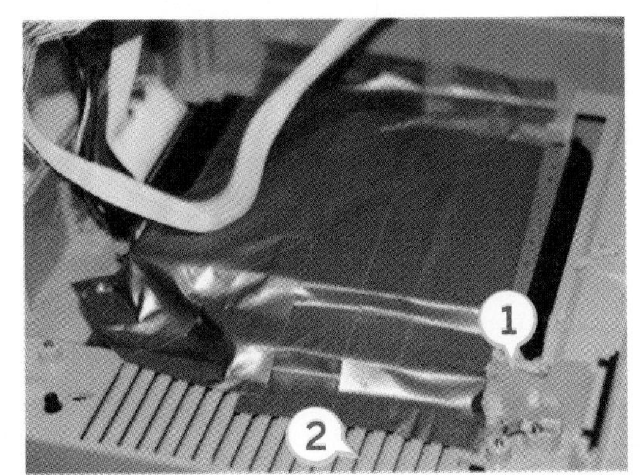

Figure J: Installing the optical drive—1. You'll need to cut away plastic here 2. This is a good place to install an extra fan.

Figure K: The drive, nicely hidden under the cover

this I can't put USB connectors into the controller ports...they have to be attached to the backplate. Oh well!

8. Testing it all and installing software

With trembling hands I attached the power, keyboard, and mouse. I then connected the video-out to my television and pressed "Power." Success! The red power LED happily turned on and I was greeted with the BIOS loading screen. I put my Windows XP installation CD in the drive and started installing.

After installing Windows, drivers, an Internet browser, etc., I moved all my games to the NES PC's hard drive. Next, I set up the front end that will work as my operating system "shell" (for a list of emulator front ends, see www.zophar.net/frontends/universal.html). As soon as Windows opens, the front end automatically starts fullscreen, hiding the Windows interface. I also went through some extra steps to make the NES PC seem less like a computer:

Boot screen

Using Stardocks Bootskin (www.stardock.com/products/bootskin), I switched the default loading screen to a more Nintendo-ish picture.

Login

My Windows booted straight to a Welcome screen, where you're supposed to select which user to log in as. I got rid of the log-in screen by following these steps:

1) Click Start Menu → Control Panel, then select User Accounts.
2) Select "Change the way users log on or off"
3) Un-tick the "Use the Welcome screen" and apply this setting. Close the User Accounts window.
4) Click Start Menu → Run, and type "control userpasswords2" and press Return
5) Un-tick "User must enter a username and password to use this computer"

Enter the password for the person you want to login as.

Customize startup

Next, I removed the "Loading settings" message that appears when Windows is starting up:

1) Click Start Menu → Run, type regedit, and press Return.
2) Navigate to the entry: HKEY_LOCAL_MACHINE → Software → Microsoft → Windows → CurrentVersion → Policies → System.
3) If there is an entry for "DisableStatusMessages" set it to 1. If there is no entry, right-mouse click the System word, and select New→DWORD value, and enter DisableStatusMessages, right-mouse to edit the value of it, and enter 1.

Disable pop-up balloons

To turn off the obnoxious pop-up info balloons in the right bottom corner of the screen:

1) Click Start Menu → Run, type regedit, and press Return.
2) Navigate to the entry: HKEY_CURRENT_USER →

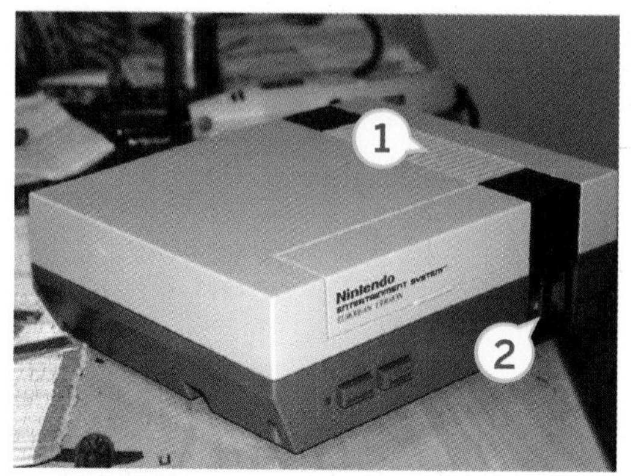

Figure L: The NES PC, all put together (1 & 2 are good locations for cooling fans)

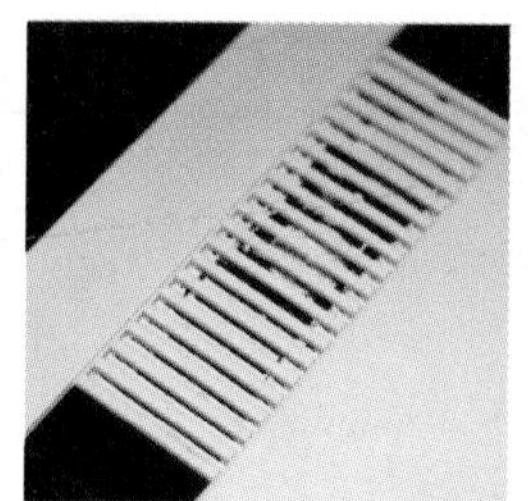
Figure M: Detail showing the fan described in Figure J

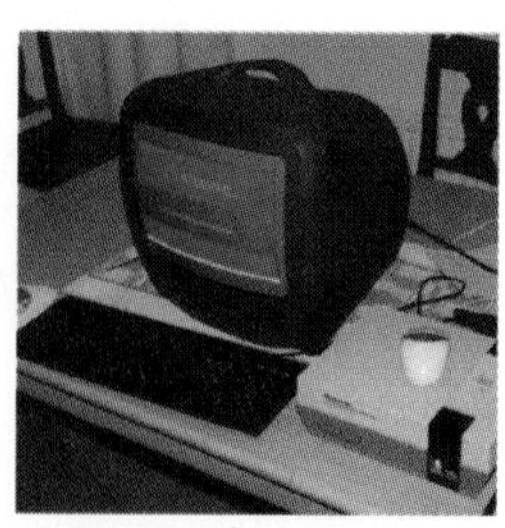
Figure N: Installing Windows XP

Software → Microsoft → Windows → CurrentVersion → Explorer → Advanced.

3) If there is an entry for "EnableBalloonTips" set it to the decimal 0 (the digit zero). If there is no entry, right-mouse click the "Advanced" word, and select New→DWORD value, and enter "EnableBalloonTips," right-mouse to edit the value of it, and enter the decimal 0 (the digit zero).

Lastly, and most importantly, I added the front end to the Startup folder in the Start Menu. That way, when Windows is started, the front end is automatically launched!

In Memoriam: Hashim "hatsuli" Al-Ani

Gareth Branwyn writes: When we were putting this book together, after choosing this project for inclusion, we sadly discovered that its author, a 19-year-old Finnish teen named Hashim Al-Ani, aka hatsuli, had lost a battle with cancer. Like any community, virtual or otherwise, people on Instructables were saddened by the news and offered stories on the forums of how he had helped them with their build of his project and offered other tech help. His mother, not an Instructables user, signed up to Instructables and posted this on the forum topic about her son's death:

Thank you for your kind words. To be honest, I cannot take any credit for what he was like. He just turned out to be a wonderful young man—all I did was love him.

I am so deeply impressed by your comments here. This world was something we were not part of. I mean, I knew all his friends, but the Web was his alone. Now I get to know some of you and to have a peek into this side of him.

He was very talented in so many ways. Computers were only part of his life. He was a talented guitar player and did wonderfully in his studies. For some reason, his life was meant to be this short but he seemed to have left a big footprint!
—Mamu

Part of his "footprint" is this Instructable. If you build it, take a moment to consider the life of the talented young man who dreamed it up. And when you have your NES PC up and running, be sure to savor the heck out of the experience. Hatsuli would have wanted it that way.

contest WINNER!

Lego USB Stick

Create a unique case out of Lego to house your flash drive

By Ian Hampton

Figure A: A Lego brick that never forgets

Figure B: Our 6x3 storage brick, made by hollowing out and combining smaller bricks

Figure C: Installing the flash drive inside our Franken-brick

Figure D: Filling the case with clear silicon to secure it in place

Figure E: Use sandpaper and metal polish to clean up the job

FIGURE F: Lego logic, hard at work

Here's a way to create a cool flash drive out of Lego bricks. I've seen such drives before, but I haven't seen one done quite this way.

1. Creating the case

Because my flash drive was quite large (in dimensions, not storage capacity), I had to create a 6x3 Lego brick. I chopped a 4x2 and a 2x2 brick in half using a pen knife and another brick as a guide. Pliers were used to remove the inner bits of the bricks and then the knife was used to remove the sides. Another 4x2 and a 2x2 brick were used to create the case.

The four parts were stuck together using super glue and a steady hand.

2. Installing the drive

A groove was cut into the case to locate the USB connector and after a small amount of modification to the PCB the drive was installed.

3. Securing the drive

I stuck an off-cut piece of one of the bricks at the bottom of the case as I found this made the drive sit at the right height. I then packed the whole thing full of clear silicon to made it stronger and to reduce any movement from the drive. Clear silicon was important as it allows the status light to shine through.

4. Stick it together and polish it up

I stuck a flat 6x2 and 6x1 brick to the top of the case to enclosed it all. Because of the amount of joins and glue lines some of the bricks weren't totally level so I used some fine wet and dry sandpaper to level the edges up. Two different grades of metal polish were then used to make the edges smooth and shiny again.

5. Finished

The finished product. Figure F shows the LED status light in action. I was tempted to make an end cap but it'd annoy me and I'd only lose it, anyway.

Ian Hampton lives in Oxford, UK. He's recently been building a *Back to the Future* Delorean replica for a banger rally. He also likes carrot cake.

Instructables project page: www.instructables.com/id/Lego-USB-Stick

Milkscanner

Using liquid to create a displacement map of an object for 3D computer imaging

By Friedrich Kirschner

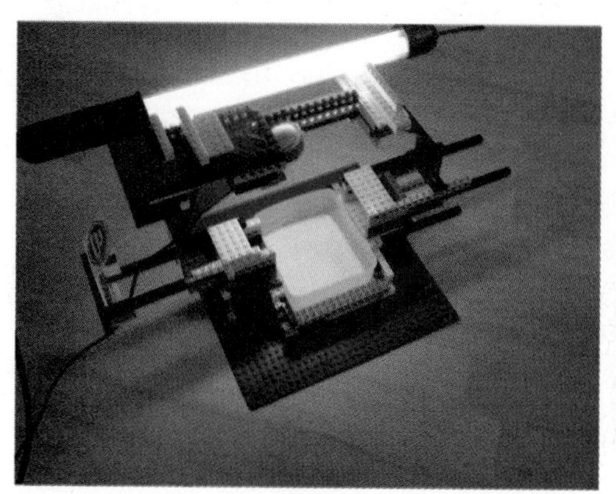
Figure A: Lego rig, with milk bath, webcam, lighting

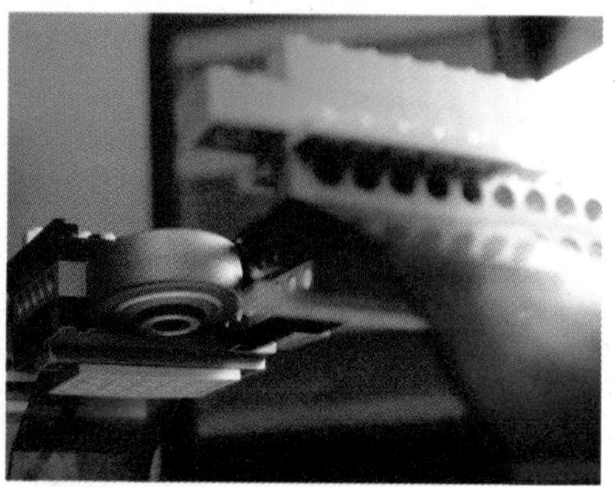
Figure B: Close-up of camera head on Lego rig

The Milkscanner is a simple tool that allows the scanning of physical objects, via liquid displacement, and creates a displacement map for use with Moviesandbox (moviesandbox.net) or any other 3D application that allows for the use of displacement mapping.

Ingredients:

- A webcam
- A plastic bowl
- 3 Cups of milk
- A custom Lego rig (or other means of holding camera)

You can download the Milkscanner PC application and source code (creates the displacement map automatically from a webcam image) from the Milkscanner webpage: milkscanner.moviesandbox.net. If you have questions, please feel free to leave a comment on the Moviesandbox forums at: forums.moviesandbox.net.

1. Scanning

Fix the webcam above the bowl (you don't necessarily have to use a Lego rig, but it is a fast way to build and adjust) and put the object to be scanned in it (Figure C).

Figure C: The object to be scanned in the empty bowl

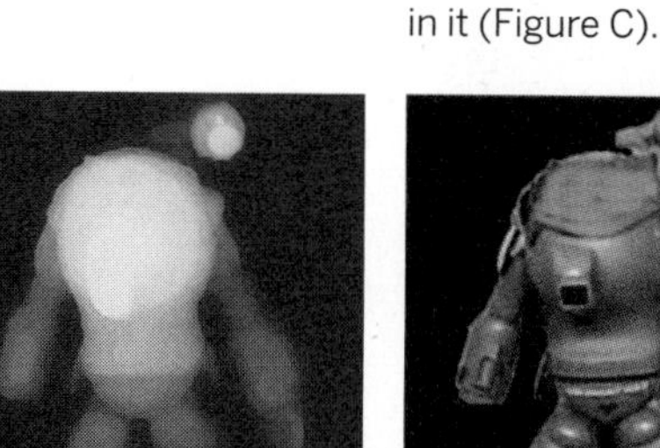
Figure D: The height map

Figure E: The color map

Figure F: The finished unreal 3D model

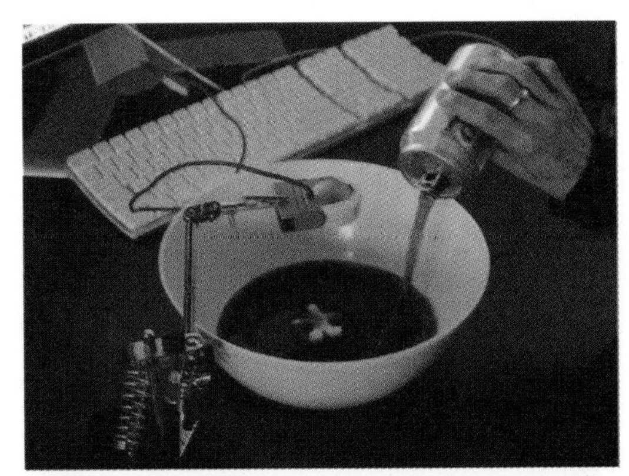

Figure G: Using soda as the liquid and a simple camera mount made from a soldering third hand

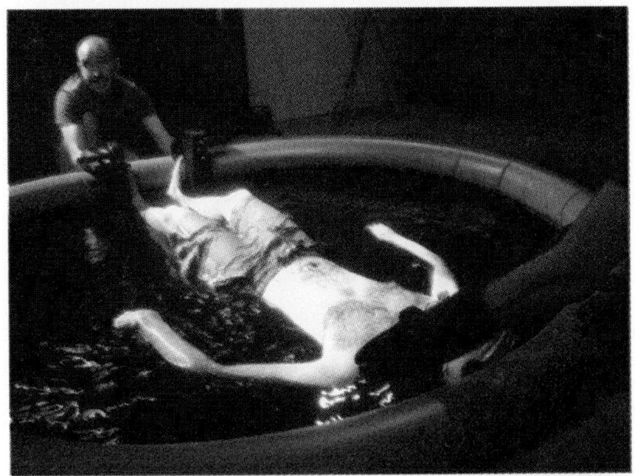

Figure H: Full body scanning using a kiddie pool and ink

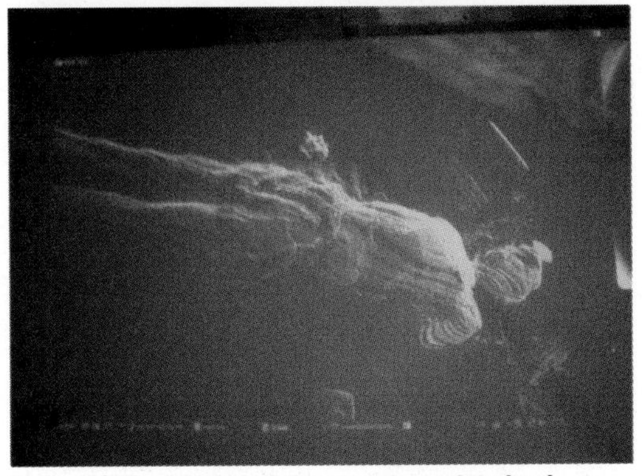

Figure I: 3D rendering of the scanning point cloud

Note: The scanner can only scan half an object at a time.

Now, begin the process of covering the object in milk. This displacement process basically create a series of visual "slices" through the object in your image-capturing software. You then subtract the white part from the pictures that the webcam takes, and the remaining images serve as the slices. From these, you build your 3D rendering. The process involves adding milk, three tablespoons at a time, and taking another picture. Through this sequence, a series of consistent visual components are created, three spoonfuls of liquid at a time.

2. Use your displacement map

After creating a displacement map from the scanning process, you can use it with the Trace Tool and bring it into Moviesandbox. How this all works will be much clearer after you watch the YouTube video at youtube.com/watch?v=XSrW-wAWZe4.

You can also import it as a displacement map in Blender or Maya or any other application that supports displacement maps. Remember: the displacement map can only cover half the object.

That's all there is to it. Enjoy!

Update: Check out a life-size version of the milkscanner, where people can scan their bodies using a cheap swimming pool, a webcam, and ink: vimeo.com/1190405.

Friedrich Kirschner is a filmmaker, visual artist, and software developer. He re-purposes computer games and real-time animation technology to create animated narratives and interactive performances.

30-Minute USB Microscope

A medium-resolution USB microscope you can make for under $100 in parts

By Mike Davis

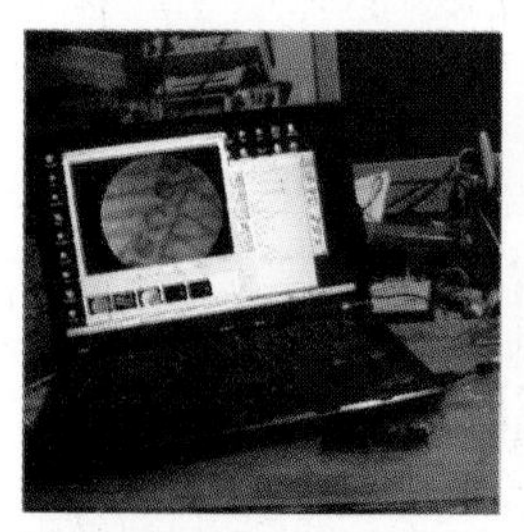

Figure A: The microscope in action

Figure B: Do you know where this is on a $20 bill?

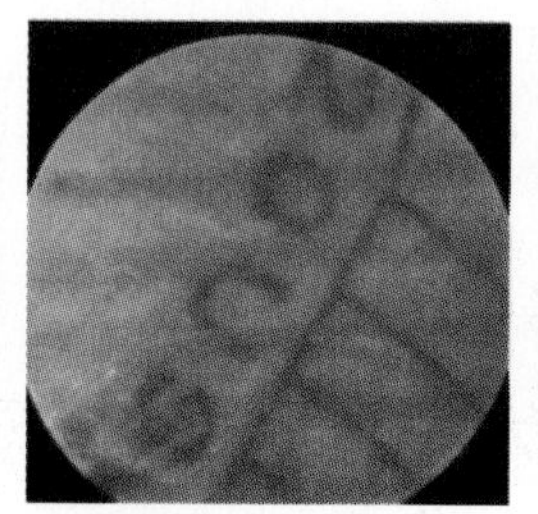

Figure C: More $20 bill microprint

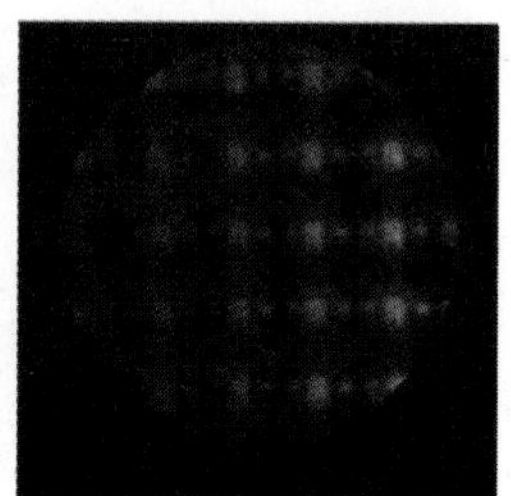

Figure D: Close-up of a liquid crystal display (switched off)

Here's a medium-resolution USB microscope I made for just under $100 in parts (you should be able to do the same, even if you buy the parts new).

Parts:

- A RadioShack pocket scope. Model MM-100, part number 63-1313
- A white LED
- A Logitech QuickCam Pro for Notebooks (Carl Zeiss optics), model 960-000045
- 30 AWG wire
- Heat-shrink tubing or black electrical tape
- Hot glue gun (or whatever glue your prefer)

Figure A shows the microscope in action. Figures B, C, and D show some views through the scope.

1. Modify the microscope

This is pretty easy: the microscope comes with an incandescent bulb that is normally powered by two 1.5v AAA batteries. Just rip these all out (see Figure E) and replace the light with a single white LED and extend its leads up through the case using the 30 AWG wire. Use heat-shrink tubing or electrical tape here to insulate the leads.

Test your light with a battery and make note of which lead is the anode, and which is the cathode.

On the camera board there is a small (freaking bright) orange LED. **Carefully** remove it and wire the leads from the white LED in its place. With this camera, the LED is under software control, and USB provides all the power. Make sure these leads have plenty of slack.

Apply a generous amount of hot glue to provide strain relief for the wires. Be careful positioning the white LED so that it points generally where the camera lens points. Figure F shows the location of both lights.

2. Remove plastic packaging from the camera

You might be able to do this step (Figure G) without taking it apart, but mine was already apart and things went well.

But from what I remember there is a metal shield with the Logitech logo on it; if you pry that up and away from the glue, there's a single screw holding the entire case together.

3. Assembly

OK, if you have the LED wired in well, put the microscope back together (you didn't lose those screws did you?).

Next, remove the little rubber bit from the microscope eyepiece, notice that the inside edge of the eyepiece has a graduated cone shape; this will help

For additional information, discussion, and more, please visit the Instructables project page:

Figure E: 1. Metal from the battery connectors; you need to rip all this out 2. Fine adjustment

Figure F: 1. White LED, hot-glued in place 2. Remove orange surface mount LED; in this orientation the cathode (negative lead) is on the left

Figure G: Remove the packaging from the camera

Figure H: 1. Webcam with plastic package removed 2. Screw

Figure I: Assembling the microscope

Figure J: 1. Neodymium magnets

Figure K: Scrap wood and steel base

Figure L: Attaching the microscope

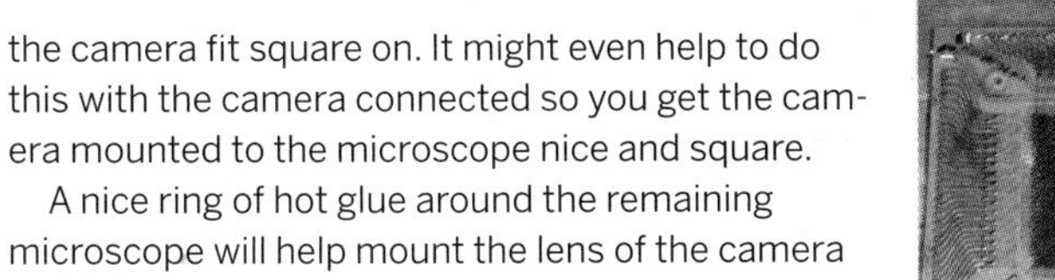

the camera fit square on. It might even help to do this with the camera connected so you get the camera mounted to the microscope nice and square.

A nice ring of hot glue around the remaining microscope will help mount the lens of the camera to the microscope eyepiece without getting any glue anywhere near the lenses. Figures H and I show the assembly.

Figure M: Core memory board from an old CDC-6600

Figure N: Magnified core memory

4. Make a base

This thing is really, really light so I glued a couple neodymium magnets (Figure J) to the bottom and created a wood base with a chunk of scrap metal on it (Figure K). You can find these magnets from many dealers, such as K&J Magnetics (www.kjmagnetics.com) and Applied Magnets (www.magnet4less.com).

The idea here is that the magnet will easily slide but otherwise won't move; this would be a frustrating problem with such tiny, tiny things.

Figure L shows the assembly in place.

5. Take some pics

Now you can take some pictures. At the beginning of this Instructable, I showed a few pics of things you might have around so you can get a sense of how much things get magnified.

One really neat thing I had around was a piece of core memory from an old CDC-6600 machine (classic machine nuts can begin going crazy now).

In Figure M you can see a broad picture of the board, and Figure N is a close-up of the toroid and wire mesh that make up the memory cells.

Since the camera is a 2 megapixel camera, it's got pretty good resolution, and the software from Logitech seems like it was made well for the job. The Zeiss lens has an electromechanical focus that adjusts well to the bizarre focal length we have here.

Mike Davis is a geek with the Make:NYC group.

Easy, Cheap, Animated Cartoon in 10 Minutes

Make your directorial debut with this easy process

By Lee von Kraus

Animate an average household bear!

I always liked to draw and was thinking of becoming a cartoonist as a kid, but the tedium of the technique at the time (1980s) drove me away. Now we have computers and everyone is making flash animations. Flash attracted me because you don't have to redraw the thing every frame, you just move its pieces. However, this severely restricts the expressiveness of one's creations. Also, you have to draw on a computer using a mouse or drawing tablet, neither of which ever gave me the results I wanted. I tried claymation because this allows expressiveness without having to redraw the character each frame. The problem with this is the clay is hard to work with and a pain. Finally, I came up with a very easy and fun way to do animations that can be expressive, does not require redrawing of the whole character each frame, and allows very easy modification without messy erasing and without having to use a computer mouse or computer drawing tablet. For this method, all you need is a dry erase board (whiteboard), a USB camera, a computer, and some free software.

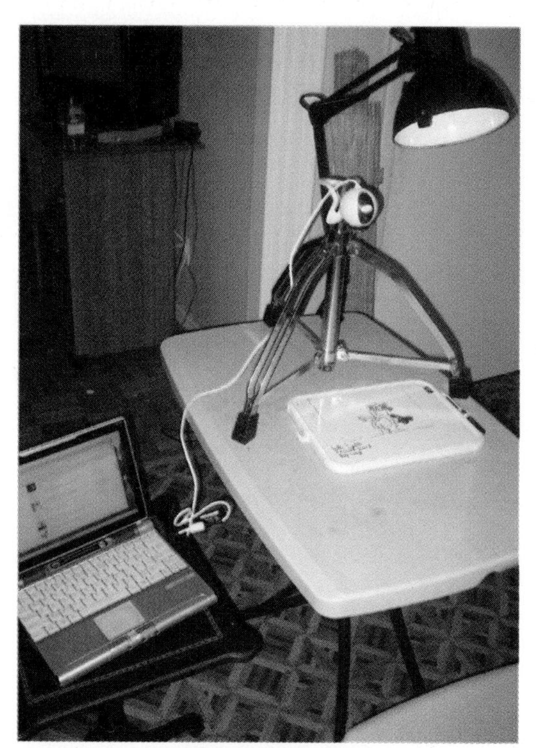

Figure A: Set-up for capturing drawings

1. Get your stuff together

You will need:

- **One or more dry erase boards**—More than one is nice because you can do multiple scenes at the same time without erasing the same board over and over.
- **USB camera**—I use a Logitech QuickCam Chat, it's cheap and does the job. It also comes with a weird bendy mount that allows me to mount it on a tripod that I have.
- **Computer**—Any old computer will work as long as it can run the software (see below).
- **Free software**—I use MonkeyJam to make the initial animations; it's free and it's great, does everything you need to make nice animations. (http://tinyurl.com/9khjt). I use Video Edit Magic to do the post animation editing (adding music, sound effects, narration, etc). It's a very cheap, simple, and effective program.

2. Set everything up

Set up the whiteboard on a table. Tape it down with double-sided tape so it won't shift while you draw and erase on it.

Mount the USB camera above it in a stable position. It's best to make some sort of mount that

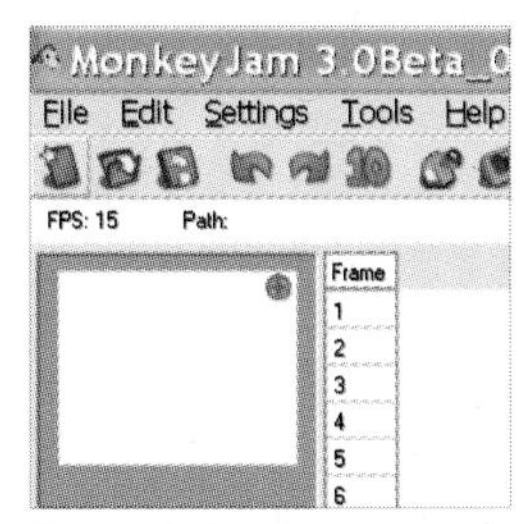

Figure B: Create a new layer in MonkeyJam

Figure C: Open Video Capture window to start capturing frames

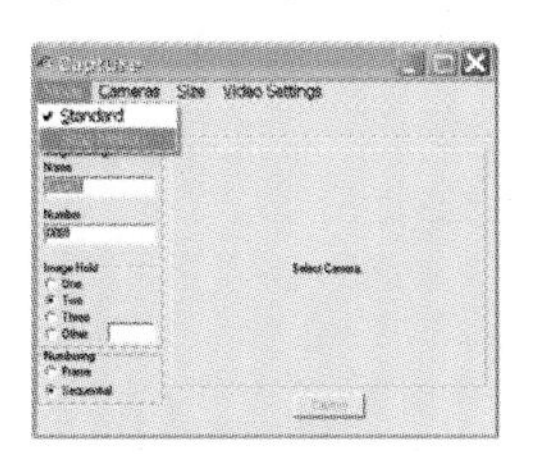

Figure D: Selecting stop motion mode in MonkeyJam

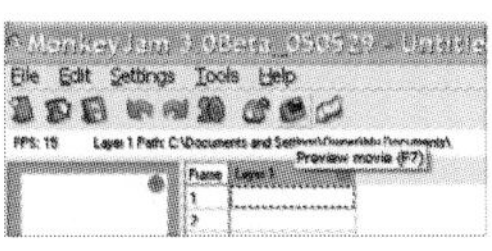

Figure E: Previewing your movie

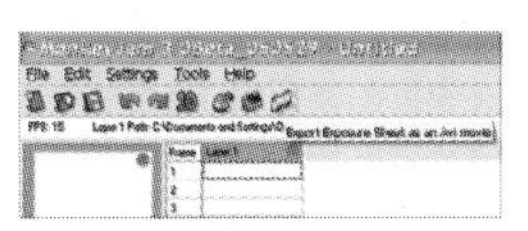

Figure F: Exporting the movie as an AVI

User Notes

j_l_larson made one www.youtube.com/watch?v=CPtV1Ca2T4E

David Tweeto made one with his little brother www.youtube.com/watch?v=iGj8Sb1PWng

you can keep in place permanently, or at least for the time that it takes you to do your animation. If it gets shifted in reference to the whiteboard, it's annoying to reposition it to the exact same position, and then your animation will have a "jump" in it. I mounted my webcam on an old tripod that I found on the street, I think the tripod was used for a high hat on a drum set. You will probably not want to mount the webcam more than about a foot above the whiteboard, 'cause webcam resolution isn't that great.

You can avoid light reflecting off the whiteboard if you use strong room lighting or reflect your light off another surface.

Plug the camera into your computer and start MonkeyJam.

3. Working with the software

Start MonkeyJam:

- Click the button on the upper left (piece of paper with a star on it). This will create a new layer in your animation. Just click "ok" in the box that comes up. An orange labeled layer will appear.
- Click the "Open Video Capture window" button (the thing to the right of the "10" button).
- In the window that pops up, select "Stop Motion" under the "Mode" menu.
- Select your camera from the "Cameras" menu.
- Then start drawing and click the "Capture" button whenever you want to make a new frame in your animation.

4. Working with the software, cont'd

You can check how your movie is looking so far by clicking the "Preview Movie" button.

When you're satisfied with your movie, click the "Export Exposure Sheet as an AVI movie" button.

5. Put your animations on YouTube

I was messing with Windows Movie Maker (comes with Windows), and I noticed a "Narrate" button. Use that to add voices/sound to your animations once they have been saved as videos.

You can also use Windows Movie Maker for basic editing (adding effects, etc.). You can also download VirtualDub from download.com; it's free and allows basic video editing.

Put your animations onto YouTube so people can watch them. There aren't many animations by nonprofessionals on there at the moment; I hope this will change after this Instructable.

Here are some excellent books on animation:
How to Draw Cartoon Animals by Christopher Hart
The Art of Animal Drawing by Ken Hultgren
Film Directing: Shot by Shot by Steven Katz
How to Write for Animation by Jeffrey Scott

Here are four very short animations I did so far:
www.youtube.com/watch?v=Fp_z-fCl7Qg
www.youtube.com/watch?v=MhJ8UQTRwEO
www.youtube.com/watch?v=GivRAVP6oPg
www.youtube.com/watch?v=9BJ4o4OTgEO

Lee von Kraus is currently doing research on augmenting animal intelligence via electronic brain implants. Afterwards he plans on becoming an astronaut and hopefully going to Mars.

3D Laser Scanner

Create your own hi-tech 3D laser scanner with stuff you have around the house

By Kenny Jensen

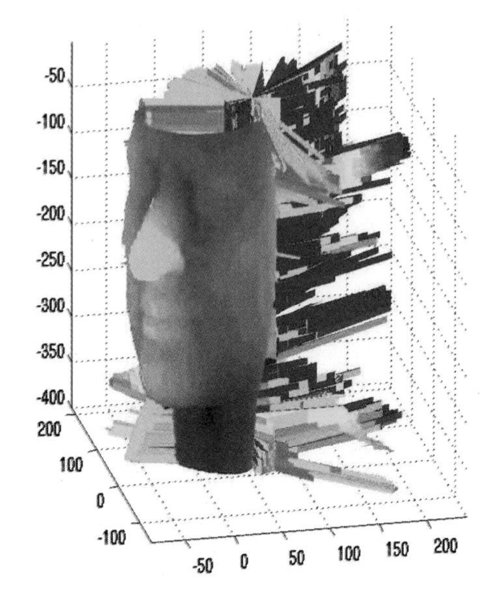

Figure A: The scanner data in the MATLAB program

Using just a laser pointer, a wine glass, a rotating platform, and a digital video camera, you can make accurate 3D models of an object or person.

Note: This Instructable was among the first posted to the site, in 2005. As such, it is dated, although the technique it describes is still relevant. To see a more current treatment of the subject, check out the episode of the *Know How Show* that is embedded in this project's webpage.

1. Position the camera, laser, and cylindrical lens

Align the laser so that its beam passes through the cylindrical lens, creating a vertical line rather than a point, and projects onto your target object. Initially, the lens I used was the stem of a wine glass, but in this picture (Figure B), I'm using a piece of acrylic rod. Position the video camera at a slight angle (~15 degrees) from the laser.

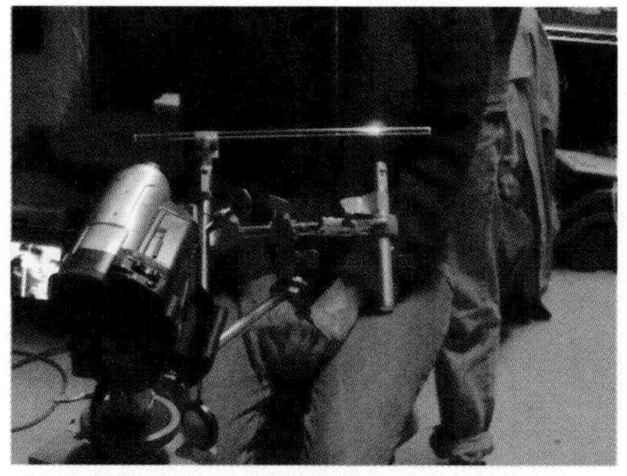

Figure B: The basic scanning set-up: the camera and a rig holding the acrylic rod "lens" and a laser pointer

Figure C: The subject being centered on stool and given a head rest to help remain centered

For additional information, discussion, and more, please visit the Instructables project page:

2. Position target on a rotating platform

Create a platform that rotates at a constant angular velocity. Record players, for example, are perfect for rotating small objects. For a person or heavier object, a motorized stool can be used. (Note: In the video that accompanies this Instructable, a chair motorized by an antenna rotor, used to rotate the antenna on a house, was used.) Position your target at the center of rotation. It's very important to "center" the person's head and to keep their head from moving during the scan. Seen in Figure C, I've created a sort of head rest with more lab clamps, something for the subject to keep her head against, centered and stationary, during the scan.

Warning: Make sure the subjects keep their eyes closed as shown in Figure C. Never look into the laser light.

3. Lights, camera, action!

Turn off the lights. To make the image processing easier, it's very helpful to get the room as dark as possible so that only the reflected laser light is visible. Rotate your target at a constant angular velocity. Record the video. Notice how protruding features displace the laser line (Figure D).

4. Process the video

Convert the video to an AVI file. Now, use an edge detection algorithm to find the location of the laser line. Reconstruct your 3D model. This Instructable's web page includes an early, uncommented MATLAB (a popular math computing program) script that was used to generate this image. Figure A shows what the video file data looks like after being rendered in MATLAB.

Note: A free Windows-compatible program for processing laser-scanned data is available from www.david-laserscanner.com. MATLAB is a commercial program, but there are some free, open source alternatives such as SciPy (www.scipy.org) and Octave (www.gnu.org/software/octave).

Kenny Jensen, Ph.D., is a physicist and inventor with degrees from MIT and UC Berkeley, whose research interests include a broad range of fields from carbon nanotubes to renewable energy.

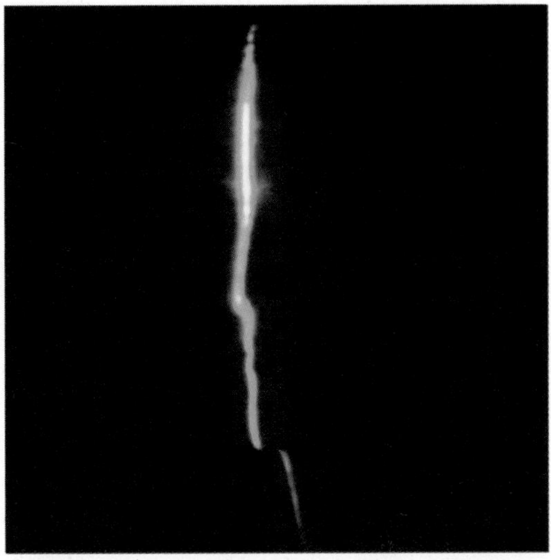

Figure D: The laser scan in progress. Make sure the room is as dark as possible.

User Notes

In response to the question in the Comments of how you get the MATLAB script to work with the recorded scanner data, **blueforce wrote:** You need to create the AVI file, and then put it in the same directory as the MATLAB script. When that's done, you need to change the "filename" to whatever the name of your AVI file is (make sure the name is within single quotes). e.g., 'test.avi'.

Inspired by this project, member **Steple did some scanning** and wrote a tutorial and a new, better MATLAB script. You can find both at n.ethz.ch/student/pleiness/en/index.php

Chromecow also wrote up a tutorial that includes some laser scanning jigs made out of Legos: www.chromecow.com/MadScience/3DScanner/3DScan.htm.

The Best of Instructables Electronics

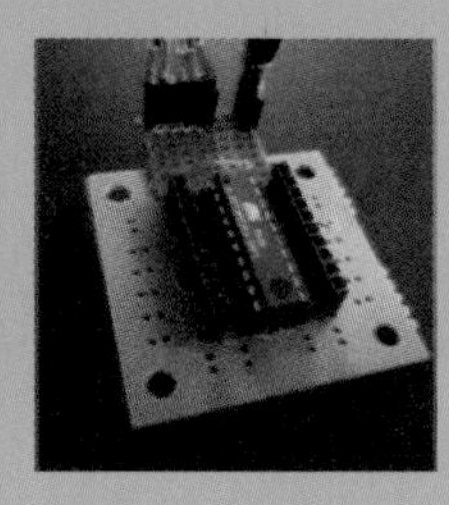

About 20 years ago, I was introduced to electronics by an inspiring high school teacher. I'm not really sure what first captivated me—in retrospect, the projects I did then seem a bit boring! I was drawn by the mysterious, seemingly magical, power that coursed through these mysterious circuits—and through the air! This was before the Internet, and that limited how much I could share my ideas.

There wasn't anyone at my school interested in hacking electronics, so that just left me alone in my garage with *Nuts & Volts* magazine. It's hard to imagine the enormous impact Instructables would have had on me if it had existed at that time. With the worldwide community of creative people that frequent the site, and thousands of amazing projects, you can always find something to inspire and people to engage on just about any topic.

This section has a range of modern projects for the budding electronics hobbyist, as well as those poking their heads into electronics from other hobbies. There's some "pure" electronics projects, and several mixed-media projects that feature clever ways of interfacing electronics with the world around you. Some of the projects show you techniques that are quite cutting-edge.

Many of the projects feature LEDs (light-emitting diodes) and microcontrollers (computers on a chip). I consider the LED the nearly universal spark that inspires many people to learn their first bit of electronics. LEDs are approachable, and the results are quickly realized. If the LED projects here are too complex, simpler ones are on the website. Microcontrollers take some commitment, but they're the key building block for modern electronics hobbyists. You will find a microcontroller at the heart of many of the projects here, as well as in nearly every commercial device you care to crack open. Thinking back to a digital counter that I built as a kid, it's a perfect example of how much more easily and more powerfully you can make things today using a single, tiny microcontroller chip. That counter required a dozen individual logic chips, plugged together with hundreds of wires in a precarious jumble. I could barely fit it into a shoebox!

So leave your shoeboxes in the closet, grab some microcontrollers and LEDs, and get ready to unleash some magical power of your own!

—Dan Goldwater (dan)

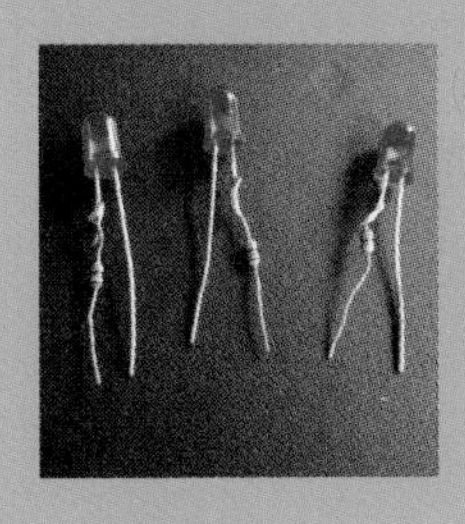

contest WINNER!

LED Throwies

LED Throwies are an inexpensive way to add color to any ferromagnetic surface! By Q-Branch

Figure A: That's a lot of Throwies!

Developed by the Graffiti Research Lab (graffitiresearchlab.com) a division of the Eyebeam R&D OpenLab (research.eyebeam.org), LED Throwies are an inexpensive way to add color to any ferromagnetic surface in your neighborhood. A Throwie consists of a lithium battery, a 10mm diffused LED and a rare-earth magnet taped together. Throw it up high and in quantity to impress your friends and city officials.

Check out this link to see the LED Throwies in action in NYC thanks to resistor and fi5e! graffitiresearchlab.com/?page_id=6

1. Materials list

LED Throwies consist of only a few inexpensive parts (see Figure B) and can be made for ~$1.00 per Throwie:

10mm diffused LED

- Vendor: HB Electronic Components (www.hebeiltd.com.cn/?p=leds.9.10mm)
- Average cost: $0.20 avg per LED
- Notes: Cost reductions for larger quantities. Comes in red, blue, amber, white in both diffused and clear. Diffused works better than water clear for the Throwie application.

Note: HB has even created a Throwie Pack (www.hebeiltd.com.cn/?p=throwies) page with deals on 10mm LEDs and lithium batteries!

CR2032 3V lithium batteries

- Vendor: CheapBatteries.com (www.cheapbatteries.com/coin.htm)
- Cost: $0.25 per battery
- Notes: Cost reductions for larger quantities. With the 2032 Lithium battery, depending on the weather and the LED color, your Throwie should last around 1 -2 weeks.

1-inch wide strapping tape

- Vendor: Your local hardware store
- Cost: $2.00 for one roll
- Notes: One roll will make many Throwies

1/2" diameter x 1/8" thick NdFeB disc magnet, Ni-Cu-Ni plated

- Vendor: Amazing Magnets (www.amazingmagnets.com/index.asp?PageAction=VIEWPROD&ProdID=63)
- Cost: $13.00 per 25 magnets
- Notes: Cost reductions for larger quantities

Conductive epoxy (Figure C)

- Vendor: Newark In One (www.newark.com/product-details/text/catalog/47973.html)
- Cost: $50.00
- Notes: The epoxy is optional. To use the epoxy, mix in equal parts A and B. Mix well. Then apply to leads and preload them into the battery contact surfaces. Allow a few hours to dry and 24 hours to cure.

2. Test the LED

Test your LED to determine color, brightness, and functionality. Pinch the LED legs, or leads, to the battery terminals (see Figure D). The longer LED lead, called the *anode*, should be touching the positive terminal (+) of the battery and the shorter LED lead, called the *cathode*, should be touching the negative terminal (-) of the battery. Figure E shows the cathode and anode.

Note: The positive terminal on the battery has a larger contact surface than the negative terminal, as shown in Figure F. The positive terminal extends around the sides of the battery. Don't let the cathode lead of the LED accidentally touch the positive terminal of the battery. This will create a short and cause the LED to function improperly.

For more information on LEDs, see www.kpsec.freeuk.com/components/led.htm.

For more information on batteries, see en.wikipedia.org/wiki/Batteries.

3. Tape the LED to the battery

Cut off a piece of 1" wide strapping tape approximately 7" long. Tape the LED leads to the battery by wrapping the tape two to three times around both sides of the battery. Keep the tape very tight as you wrap as shown in Figure G.

If taped correctly, the LED will not flicker.

4. Tape the magnet to the battery

Now, place the magnet on the positive terminal of the battery, as shown in Figure H, and continue to tightly wrap the tape. The magnet should be held firmly to the battery (see Figure I).

Note: If the magnet gets stuck to a ferromagnetic surface, don't pull on the LED Throwie. Apply a lateral force to the magnet and slide it off the surface while lifting it with a fingernail or tool.

Warning: Keep the magnet away from conventional hard drives, credit cards, and other data storage devices.

5. Toss your Throwie

The LED Throwie is ready to be tossed onto a ferromagnetic surface. Practice tossing your Throwies. Work on your accuracy and your own personal technique. Every Throwie won't stick every time, but if you toss them gently, they will stick eventually (Figure K shows the best way to hold them for the toss). Get them up high and in large quantities for greatest enjoyment.

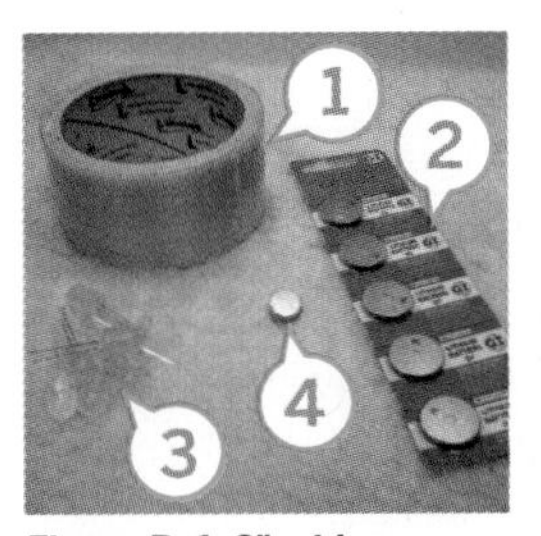

Figure B: 1. 2"-wide strapping. We tore it into two 1" strips. 2. CR2032 lithium batteries 3. 10mm LEDs, diffused, multiple colors 4. Strong little rare-earth magnet

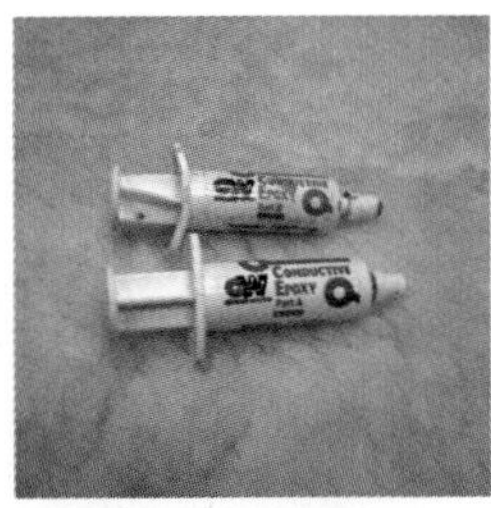

Figure C: Conductive epoxy is optional. It will improve brightness and robustness but it is time and labor intensive.

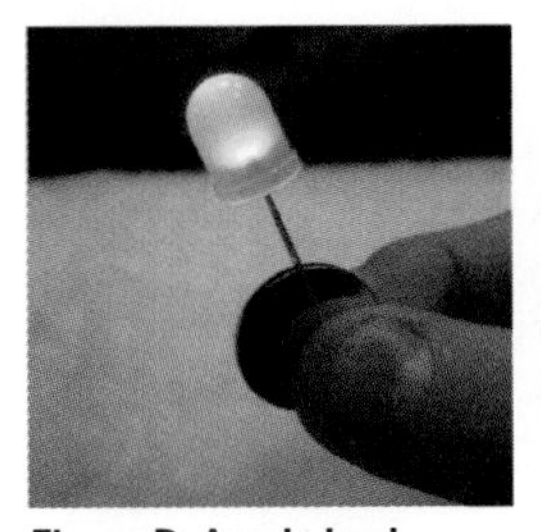

Figure D: Anode lead touching the positive terminal

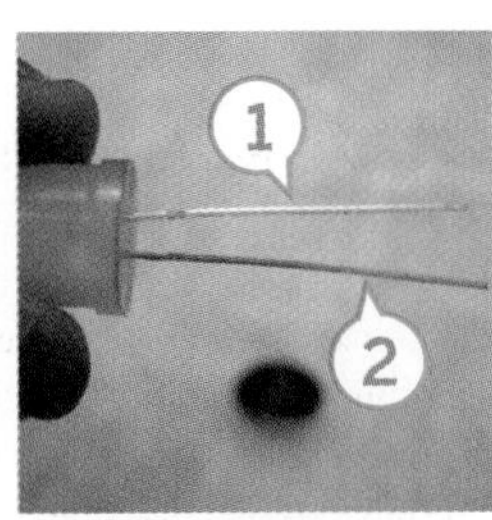

Figure E: 1. Cathode or K lead 2. Anode lead

Figure F: Notice how the positive terminal extends around the edge of the battery

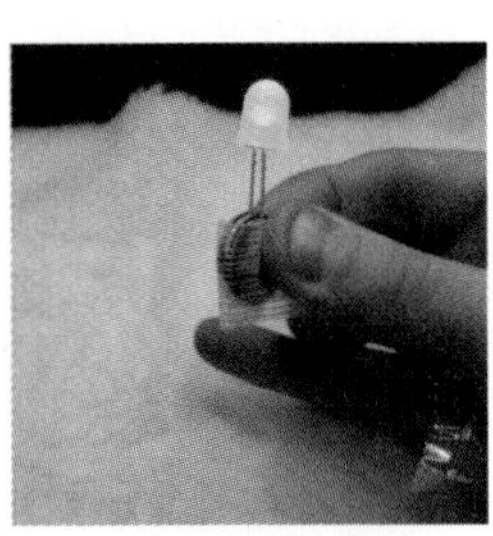

Figure G: Keep up the tension!

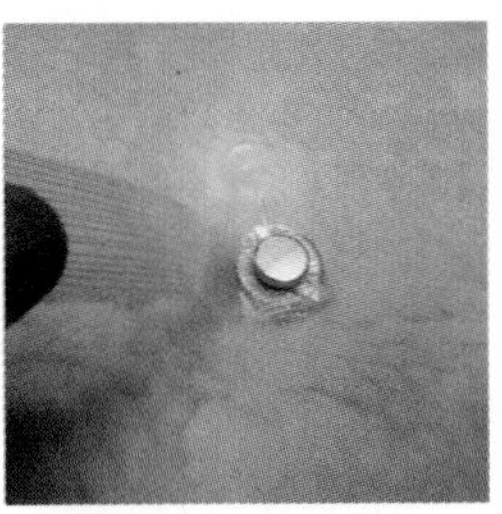

Figure H: The magnet is taped to the positive terminal of the battery

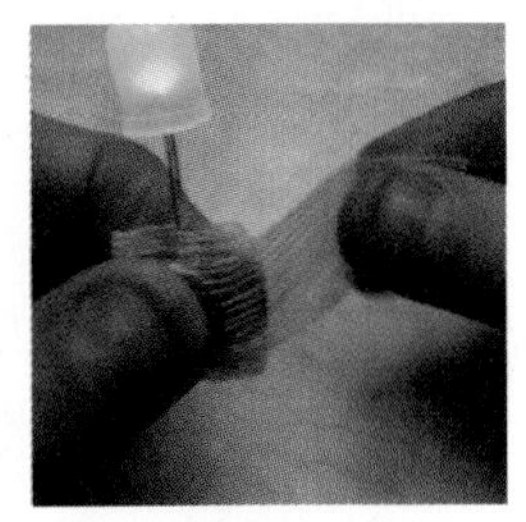

Figure I: Tightly!

Figure J: Look at J-pizzle's follow-through on this green Throwie!

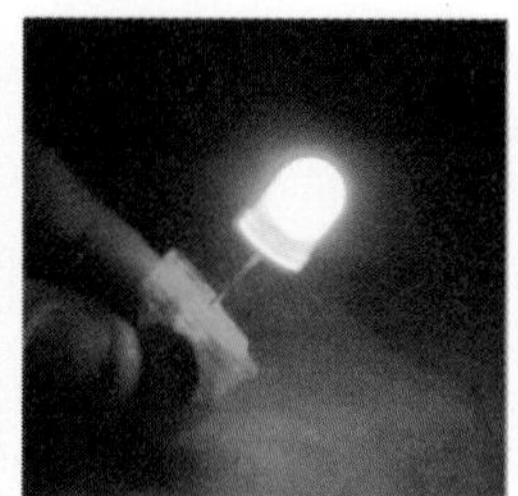

Figure K: Toss the Throwie by the battery/magnet bundle

Figure L: Wall of Throwies

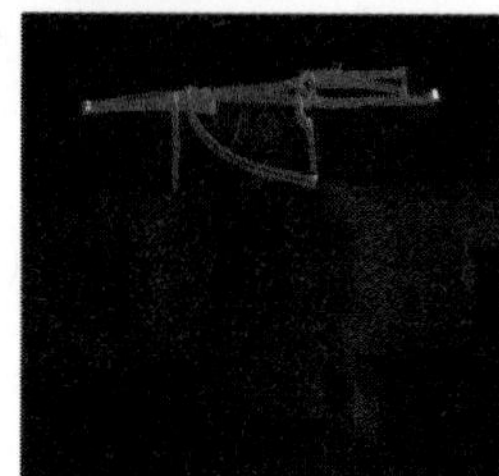

Figure M: Fi5e

6. Plan a campaign

Now, find a building or structure that will attract the magnets, form a crew, wait until night, and get some Throwies up. If you do it around a crowd of people, they will probably try to get into the act. It can quickly descend into chaotic fun. Give a handful of Throwies to a stranger and let them get up too. Remember, Throwies are only a temporary alteration of your local environment. Depending on the color, Throwies can last up to two weeks.

7. Other applications and upgrades

Other applications: Other than tossing it, you can also use your LED Throwie to write in the air with light while taking a long exposure picture, as shown in Figure M. You can put them on your bike as an additional reflector. You can put them on surveillance cameras to make them more visible at night. You can use them to play a version of bocci ball on a magnetic surface in the dark.

Upgrades: You can make a better LED Throwie by using heatshrink tubing on each lead to make sure they don't short to each other or the battery. This upgrade will allow you to bend the LED so it faces in the direction you choose. You can also dip the Throwie in epoxy, silicon, or potting compound to make an all-weather LED Throwie. A resistor in series would allow you to increase the Throwie shelf-life. Bigger batteries = longer life. Stronger magnets = increased stick probability. You could add a solar panel, photocell, etc....Have fun.

Note: See the Flickr instructional set at www.flickr.com/photos/everythingdigital/sets/72057594069888500 for a Throwie on/off switch mod by A. Joyce, aka EverythingDigital.

User Notes

In the Comments, **Instructables member Vndr** asks about the life of the Throwies, guessing that, since 2032 batteries are 240mAh and LEDs need 20mA, they'd only last for about 12 hours.

Dan, Instructables co-founder, replies: It depends on the LEDs voltage/current curve. For a typical green/blue/white with 3.5V nominal, driving at 3.0V is probably only 1-5mA so you get a few days or week of life. For a red/orange/yellow LED with 2.0V nominal, you are basically shorting out the battery and your current is limited by the 2032 cell—haven't tested them but I would not be surprised if they maxed out at under 100mA, so you still get a couple hours at least with the lower voltage LEDs.

Programmable LED

A microcontroller-controlled LED inspired by various LED Throwies, blinking LEDs, and similar Instructables By Alexander Weber

Inspired by many LED Throwies and other LED projects on Instructables, I wanted to do my version of an LED controlled by a microcontroller.

The idea is to make the LED blinking sequence reprogrammable with light and shadow, e.g., you could use your flashlight. You can see a video that demonstrates how to reprogram it here: http://9600baud.blip.tv/file/198088.

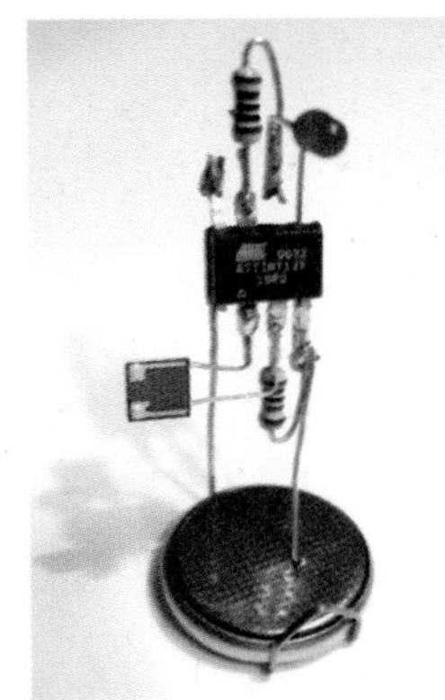

Figure A: The completed programmable LED

Figure B: 1. Soldering iron 2. Solder 3. Solderless breadboard 4. Multimeter 5. AVR programmer

1. How it works

An LED is used as a visible output. For input, I used an LDR (Light Dependent Resistor). The resistance of the LDR changes as it receives more or less light. The resistor is then used as analog input to the microprocessor's ADC (Analog to Digital Converter).

When programmed with the software that I introduce later, the controller has two modes of operation: one for recording a sequence, the other for playing back the recorded sequence.

Once the controller notices two changes of brightness within half of a second, (dark-bright-dark or bright-dark-bright), it switches to recording mode. In recording mode, the input of the LDR is measured multiple times a second and stored on the chip. If the memory is exhausted, the controller switches back to playback mode and starts to play the recorded sequence.

As the memory of this tiny controller is very limited, 64 bytes (yes, bytes!), the controller is able to record 400 bits. That is enough space for 10 seconds with 40 samples per second.

2. Tools and materials

Tools (Figure B)

- Soldering iron
- Solder
- Breadboard
- AVR programmer
- 5V power supply
- Multimeter

Materials (Figure C)

- 1K resistor (2)
- LDR (Light Dependent Resistor) such as the M9960 (1)
- Low-current LED, 1.7V, 2mA (1)
- Atmel ATtiny13V, 1KB flash RAM, 64 Bytes RAM, 64 Bytes EEPROM, 0-4MHz@1.8-5.5V (1)
- CR2032 battery, 3V, 220mAh (1)
- Some hookup wire

Software

- Eclipse (optional)
- CDT plugin (optional)
- GCC-AVR

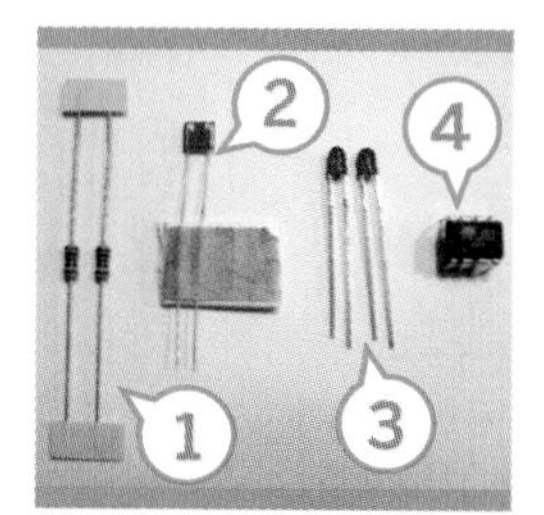

Figure C: 1. Two 1K resistors 2. Light Dependent Resistor 3. Two LEDs 4. Atmel ATtiny13V

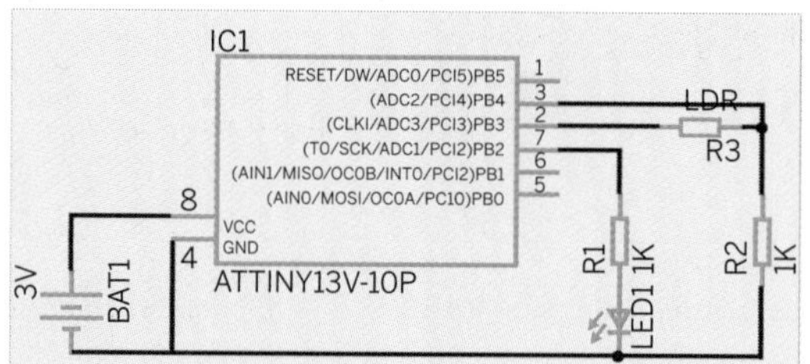

Figure D: Schematic

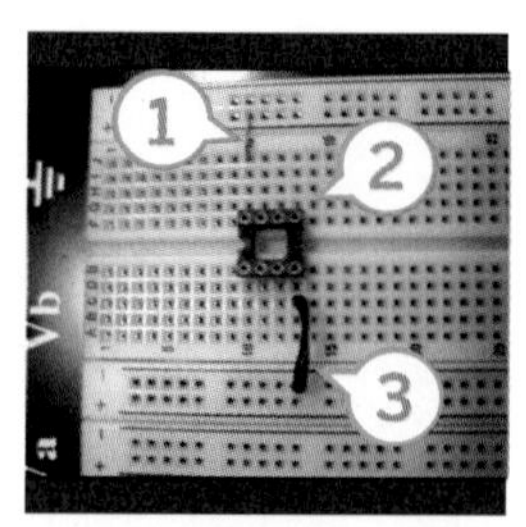

Figure E: 1. Wire to positive voltage 2. IC socket 3. Wire to ground

Costs overall should be below $5 without tools.

I used the ATtiny13V because this version of this controller family is able to run at 1.8V. That makes it possible to run the circuit with a very small battery. To have it run for a very long time, I decided to use a low current LED that reaches full brightness at 2mA.

3. Schematic

The schematic for this circuit is shown in Figure D.

Note: The AVR's reset input is not connected. This is not the best practice. It would have been better to use a 10K pull-up resistor. But it works fine for me without and it saves a resistor.

To keep the circuit as simple as possible, I used the AVR's internal oscillator. That means we avoid having to include a crystal and two small capacitors. The internal oscillator lets the controller run at 1.2MHz which is more than enough speed for our purpose.

If you decide to use another power supply than 5V or to use some other kind of LED, you have to calculate the value for the resistor R1. The formula is:

R = (Power supply V - LED V) / 0.002A = 1650Ω
(Power supply = 5V, LED V = 1.7V).

Using two low current LEDs instead of one, the formula looks like this:

R = (Power supply V - 2 * LED V) / 0.002A = 800Ω

The value of the resistor R2 depends on the LDR you choose. 1KΩ works for me. You may want to use a potentiometer to find the best value. The circuit should be able to detect light changes in normal daylight. To save power, PB3 is only set to high when a measurement is made.

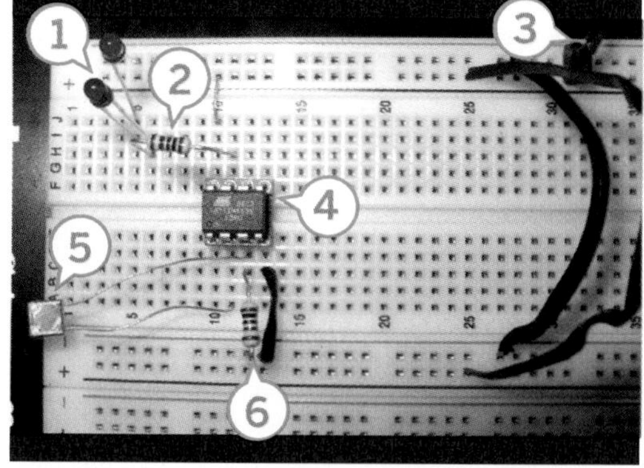

Figure F: 1. Low current LEDs 2. Resistor 1K 3. Power supply 5V 4. ATtiny13V 5. LDR 6. Resistor 1K

4. Assemble on a prototype board

You should first set up the circuit on a breadboard and test it out. That way, you can assemble all parts without having to solder anything. Figures E and F show how to put it together on a breadboard.

5. Program the circuit

The controller can be programmed in different languages. Popular programming languages include Assembler, BASIC, and C. I used C as it matches my needs the best. I was used to C ten years ago and was able to revive some of the knowledge (well, only some...).

To write your program, I recommend Eclipse with the CDT plugin, shown in Figure G. Get eclipse from www.eclipse.org and the plugin from www.eclipse.org/cdt. However, you can use whatever source code editor you are comfortable with.

For compiling C language to AVR microcontrollers you will need a cross compiler. Lucky as we are, there is a port of the famous GCC (GNU Compiler Collection). Windows users can download WinAVR

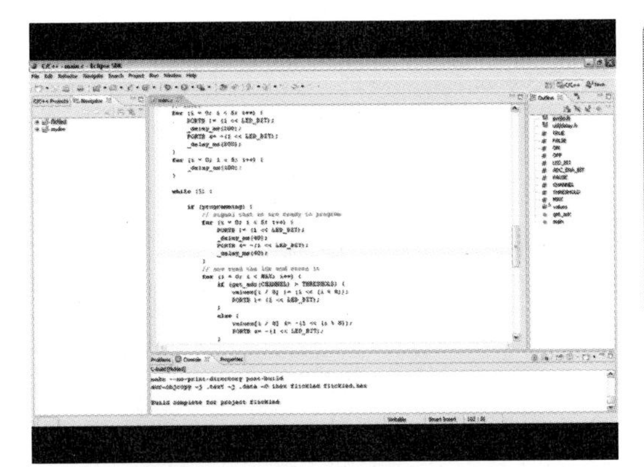

Figure G: Eclipse with the CDT plugin

Figure H: AVR ISP programmer

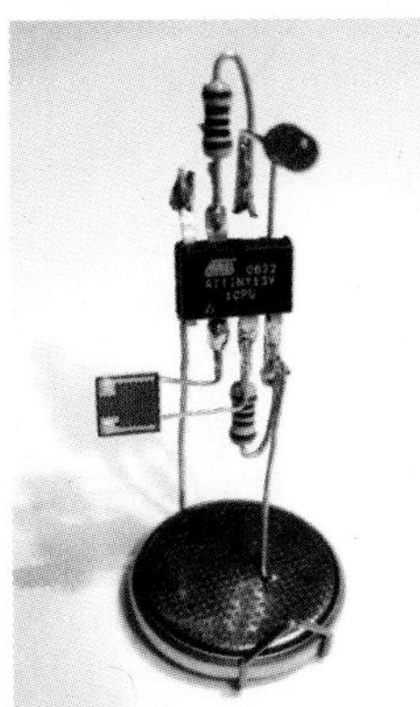

Figure I: The assembled project

from http://winavr.sourceforge.net. If you are on Linux, you should install the avr-gcc, avr-binutils, and avr-libc packages (these packages are available for many Linux distributions). On Mac OS X, first install MacPorts (www.macports.org), then use MacPorts to install avr-gcc, avr-binutils, and avr-libc.

For more information on the GNU AVR tool chain, see www.avrfreaks.net/wiki/index.php/Documentation:AVR_GCC.

After you compile the source code, you have to transfer the resulting hex file to the controller. That can be done by connecting your PC to the circuit using an ISP (In-System Programmer) or by using a dedicated programmer that you can drop the chip into. I used a dedicated programmer so I wouldn't have to wire up the header pins needed to connect the circuit to the programmer. The drawback is that you have to swap the controller between the circuit and the programmer every time you want to update your software. My programmer, shown in Figure H, comes from www.myavr.de and uses USB to connect to my notebook. There are many others around and you can even build one yourself. For example, Adafruit Industries (www.adafruit.com) sells the USBtinyISP AVR programmer kit.

User Notes

Guilherme Souza says: If you make a lot of those, would they interact?

Alex replies: Yes, they would. Take a look at my other instructable for an example: www.instructables.com/id/Synchronizing-Fireflies

For the transfer itself I used AVRDUDE, which is part of the WinAVR distribution (you can install the AVRDUDE under Linux or MacPorts). Here's an example of how you might invoke AVRDUDE:

```
avrdude -F -p t13 -c avr910 -P com4 \
-U flash:w:flickled.hex:i
```

Check the documentation for your AVR programmer and for AVRDUDE to determine the correct commands.

Note: You can download the source code, main.c, from this Instructable's web page (www.instructables.com/id/Programmable-LED). The compiled hex file, flickled.hex is also available there.

6. Soldering

After you get the circuit working on the breadboard, you can solder it.

Warning: Do not solder directly to the battery. You should construct a battery holder out of wire and use the weight of the circuit and some creative bending to ensure contact.

You can do this on a PCB (printed circuit board), on a prototype board, or even without a board. I decided to do it without as the circuit consists only of a few components. Figure I shows the completed project. If you are not familiar with soldering I recommend that you search online for soldering tutorials and videos first.

My soldering skills are a bit rusty but I think you get the idea.

I hope you enjoyed it.

Alex Weber lives in Hamburg, Germany, and develops software by day and tinkers with electronics by night. Visit him at http://tinkerlog.com.

Ghetto AVR Programmer

Getting started with AVR microcontrollers on the cheap By Elliot Williams

Microcontrollers are so cheap these days. If only there were a way to program them up just as cheaply....

In this Instructable, find out how to build up a complete AVR microcontroller toolchain: compiler, programmer software, programmer hardware, and some simple demos to get your feet wet. From there, it's just a hop, skip, and a jump to world domination with the help of servo-based 4-legged walkers: www.instructables.com/id/EYLB2BCA9BEUGCWIZL.

Note: I've gone USB! www.instructables.com/id/EDRQZ56F5LD8KDX is a better picture of what I'm using now, and is a real improvement because it will work with computers that don't have parallel ports. That said, the basics here are still applicable, especially all the software details.

The endpoint is not quite as swanky as the suite from Atmel (the maker of the AVR microcontroller), but it's gonna run you about $150 less and takes only a little more work to get it set up.

This Instructable is based on the Atmel ATtiny 2313 chip, mostly because it's one of the smaller chips (in size) while still being beefy enough to do most anything. And at $3 a pop (non-bulk), they don't break the bank.

That said, most of the steps are applicable across the AVR family, so you'll be able to re-use most everything when your programming needs outgrow the ATtiny and you reach for the $8-$12 ATmegas. Figure A shows the completed project.

Figure A: The Ghetto AVR Programmer

Total cost, around $10, even if you go deluxe. Even less if you can find an old parallel port cable.

1. Go order parts!

Required parts list (stuff to buy or scrounge):

- Atmel ATtiny2313 chip: $2.50-$3
- Socket for chip: $1
- Parallel port connector (DB25): $3 at RadioShack or $0.95 at www.sparkfun.com
- Header pins: $1-$2 (at least one strip each male and female, maybe 2 female)
- Some LEDs and resistors: $1

Add-ons:

- A pushbutton switch or two: $1
- A piezo speaker: $1-$5
- A few light sensitive photocells: $2-$6
- Breadboard for making complex circuits: $8-$10

Other stuff you oughta have:

- Computer: the older, the better because it needs a parallel port
- Hookup wire, solder, soldering iron
- Super-duper glue
- Source of ~5V DC: batteries will work, and old computer power supplies are perfect.

For additional information, discussion, and more, please visit the Instructables project page:

I get a lot of stuff from SparkFun Electronics because they're fast, reliable, and fairly priced. They carry AVR chips and have everything on the "required" list in one place and a nice website to boot.

If you're gonna be ordering a lot of chips, you can get a deal from Digi-Key or similar. For instance, you can find the ATtiny2313 (you need the DIP package form factor) for $2.26 each: http://search.digikey.com/scripts/DkSearch/dksus.dll?KeyWords=ATtiny2313+dip.

Go order the parts now, and let's set up the software side while you wait.

The waiting is the hardest part.

2. Get and install software: Linux version

If you run Windows, skip this step. Or install Linux, then come back. :-)

The Linux AVR software toolchain used is:

1) An editor of your choice (emacs, gedit, kate, etc.)
2) AVR-GCC compiler and libraries
3) AVRDUDE programmer
4) A nice Makefile to tie it all together

Ubuntu and other Linux flavors has packages for the toolchain. To install it, run the command:

```
sudo apt-get install avrdude \
   avr-libc binutils-avr gcc-avr
```

Bam! The versions that come with Ubuntu 8.04 are good all around. Woot Ubuntu! For other Linux versions, check out the documentation at the AVR-Freaks Wiki: www.avrfreaks.net/wiki/index.php/Documentation:AVR_GCC.

The following script sets up your parallel port so that you don't need to be the superuser to use it. If you're Linux-savvy, you can add the same commands to a boot script to enable the parallel port every time you boot up. Otherwise, just run it by hand:

```
sudo modprobe parport
sudo modprobe ppdev
sudo chmod 666 /dev/parport0
```

3. Get and install software: Windows version

The Microsoft Windows software toolchain is based on the same compiler as Linux, but with some optional extra GUI friendliness.

WinAVR (http://sourceforge.net/projects/winavr) is the compiler and programmer software. It also includes a nice programming editor, Programmer's Notepad.

Download and install it. It's that easy. Almost.

As the last step in setup, you'll need to make the parallel port user-accessible. WinAVR includes a program for you to do this. If you accepted the default installation options, run *C:\winAVR\bin\install_giveio.bat* by double-clicking on it. A window will flash open and close, and then you're done.

4. Make a programmer cable

The cable you're going to make is a "Direct AVR Parallel Access" or DAPA cable.

The web site www.captain.at/electronics/atmel-programmer has a nice schematic of the parallel port pins for your reference.

Here are the pins I used:

Table 1

Parallel Port	AVR Function	Color
2	MOSI	Orange/Grey
11	MISO	Orange
1	SCK	Green
16	RESET	Brown
18	GND	Brown/Grey

Note: That pin 1 (SCK) is on the upper-right hand side when you are looking at the solder pins from the back (Figure C). It's upper-left when you're looking at it in Figure B and in the schematic at the previously mentioned link.

Also, the schematic has ground connected to 20 and 21 while mine (and others) use 18. Many of the pins connect to ground, and it doesn't matter which of them you pick, as long as you get ground.

If you look around the Web, you'll find that most people put resistors in either the cable or the cradle (next step) to protect their computer's parallel port from excessive voltages on the AVR chip for use when programming it in-circuit. We will be using strictly 5V here, so there's no such worry, and I leave them out for simplicity.

Note: Sometimes you'll find yourself programming these chips in a working circuit. In that case, you'll want to make sure that there isn't any way that higher voltage in the device can make its way

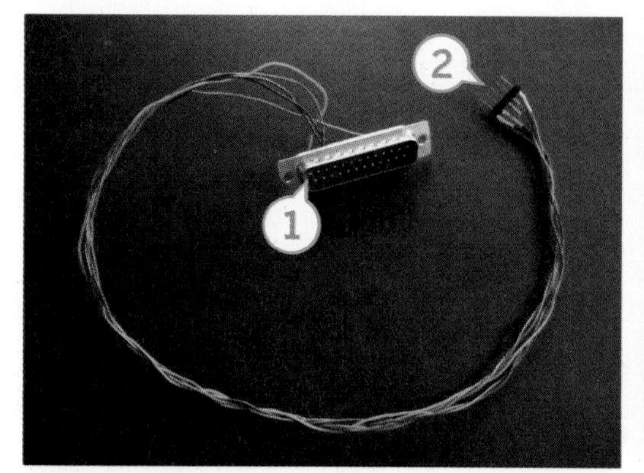

Figure B: 1. Pin one goes here 2. Top to bottom: MOSI, MISO, SCK, RESET, GND

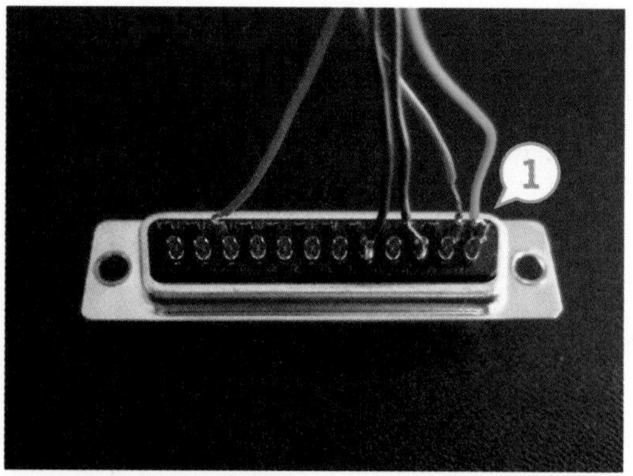

Figure C: 1. Pin one: Reset

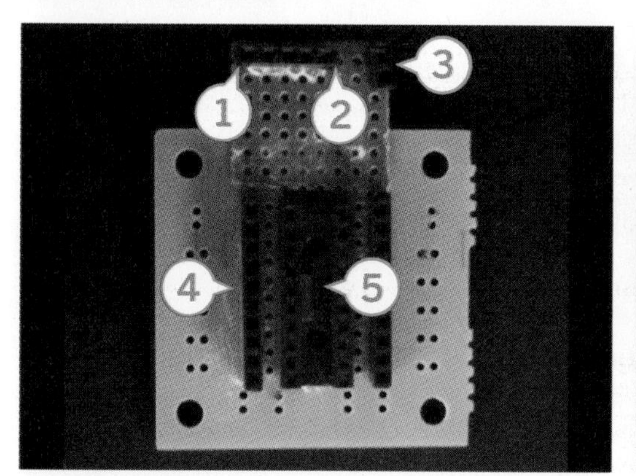

Figure D: 1. GND 2. MOSI 3. 5V (top) and ground (bottom) 4. These connect directly to the AVR pins to make them easily accessible later 5. Socket for AVR

Figure E: 1. Clock 2. +5V 3. Reset 4. MISO 5. MOSI 6. Ground

to your computer. So if you have a robot with a 12V motor drive, double-triple check that the motor power lines don't connect to any of the programming pins. But if you're programming in the cradle, you'll be fine.

5. Make the programming cradle

Now use the 20-pin socket and female headers to make a cradle that connects the pins from the cable to the corresponding pins on the chip.

The first thing to do is superglue all the headers and the socket to the circuit board. That way, it's easier to solder it all together. Figure D shows the sockets from above, and the solder joints are shown in Figure E. You can even make an extension for the header like I did if you need more room on your circuit board.

The wiring is as follows:

Table 2

Cable	ATtiny2313 Pin
MOSI	17
MISO	18
SCK	19
RESET	1
GND	10

Consult the ATtiny2313 datasheet at www.atmel.com/dyn/resources/prod_documents/2543S.pdf to double-check the pinouts.

You're soldering up the wires on the underneath, and it's mirror-image. It might help to mark where pin 1 is on the bottom side before soldering. (I did it wrong once. Once.)

That said, the wiring is very simple. It just goes

Figure F: Cradles for the ATtiny2313 (left) and ATtiny13 (right)

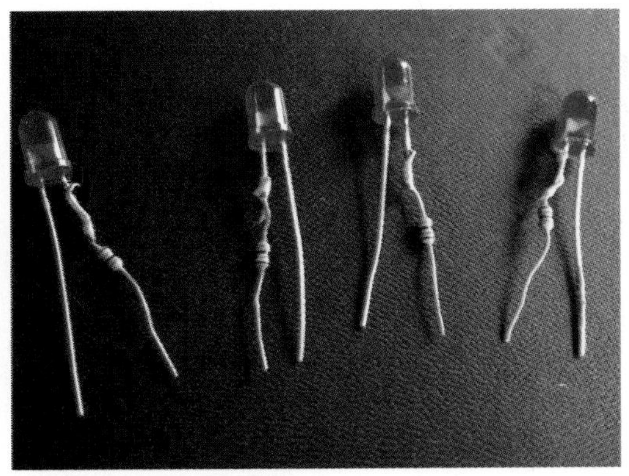

Figure G: Four blinkenlights

pin to pin, and then connect all the header pins to the closest pins on the socket. Fortunately for me, a pre-printed circuit board from Radio Shack did the trick. The cradle is versatile too. Figure F shows my ATtiny13 programmer cradle to the right of the one we're building in this Instructable.

6. Make some Blinkenlights

An integral part of the Ghetto Development Environment is the plug-in LED (see Figure G). The AVR chips can all source a fair amount of juice (50mA or more), which is enough to burn out your standard red LED, so it's a good idea to put protective resistors inline with your LEDs. Enter the Ghetto Blinkenlight.

To make them, simply solder a smallish (150Ω) resistor to the negative lead of an LED, then you can just plug it straight into 5V and it won't burn out. (5 - 1.4) v / 150Ω = 24mA, which is just about right. Make a few; I made eight. You only need one here.

7. Set up the programming project

In Linux and Windows, the procedure for starting a new project is basically the same:

1) Make a new directory called LED_Demo
2) Copy the Makefile (www.makezine.com/go/avrprogmakefile) into that directory
3) Download the code, LED_Demo.c (www.makezine.com/go/avrprogdemoc), into the directory, or start writing new code from scratch.
4) Edit the Makefile to reflect the chip you're using and the name of the project

Then you're ready to program.

Editing the Makefile

Windows: The Makefile's set up for you. All you have to do to make the Makefile work is remove the ".txt" from the end of the filename. Open it up in Wordpad or Programmer's Notepad anyway, just to have a look at the options you can change later.

Linux: Rename the Makefile.txt to Makefile. Then open the Makefile up in an editor. Un-comment the line that has "/dev/parport0" and comment out the line that has "lpt1" by placing a # in front of that line.

8. Set up your first demo circuit

Put the chip in the cradle. You'll notice that the pins all angle outwards a bit when the chips come from the factory. Bend them in (gently) by pressing flat against a table so that they're all even and parallel as shown in Figure H. Then it fits in the socket nicely as shown in Figure I.

Notice the alignment of the chip. The little dot marks pin 1. It must go to pin 1 on the cradle.

Plug in your 5V power and the programmer cable. Plug the programming cable up to your computer.

Take a Ghetto Blinkenlight and put it between pins 8 and 10 so that the resistor (negative) is in pin 10, the ground for the chip (see Figure I).

9. Program the chip

If you're using Linux:

1) cd to the LED_Demo directory
2) Cross your fingers
3) Type "make"

Windows:

1) Open up LED_Demo.c with Programmer's Notepad.

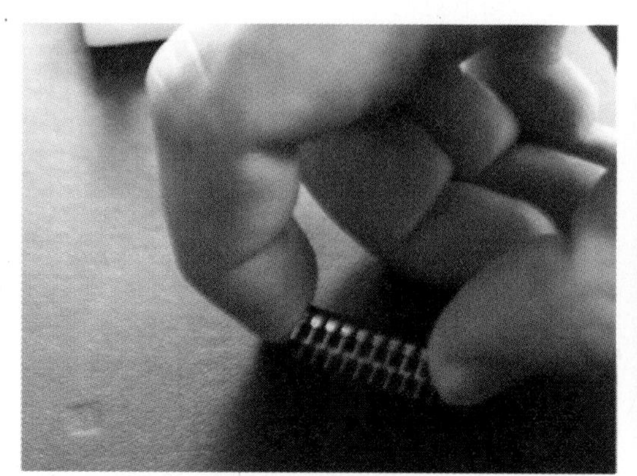
Figure H: Gently straightening the pins

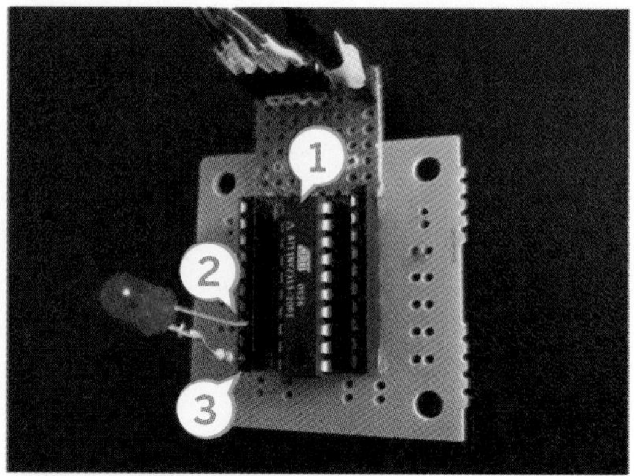

Figure I: 1. A small notch and round dot mark pin one 2. Pin eight, Port D4 3. Pin ten, Ground

2) Cross your fingers.
3) Select "Tools...Make Program"

On both Linux and Windows:
Watch the log as the text goes scrolling by. There are two things you're looking out for here:
1) Did the program compile successfully? If there are no errors, it did. If it didn't, why? Did you make changes to the original code?
2) Did the chip program successfully? In this step, the computer writes the code to the chip, then verifies the chip's memory. It should say "Contents Verified." If it says something about the parallel port or "giveio.sys", confirm that you enabled the parallel port back a couple steps ago.

Success? Yay! There's nothing like the sweet smell of blinking LEDs in the morning.

10. Understanding the software

The code's pretty simple, as far as AVR code goes, but still displays a few tricks of the trade. If you're still basking in the glory of your blinking light, you can come back later.

The `#include` lines load up some of the extra functions and definitions in the AVR-libC suite. In particular, the delay function `_delay_ms()` is in delay.h. Interrupt.h has all the pin definitions that make life easier. You're always going to want to include it.

An AVR program always starts from the function main(). In this simple case, it's the only function we have.

The first command in main(), `DDRD = _BV(PD4);`, seems pretty cryptic, but here's what it's doing. DDRD is the Data Direction Register for port D: all the input/output pins are broken up into different ports for easier access. The one we're using happens to be in D. We need to enable the pin PD4 for output for the LED.

The DDRs are set up so that they have 8 bits, one for each pin in the port. You set a pin up for input by writing 0 to its bit in the register. You can set it to output by writing a 1. We want the contents of DDRD to be 00001000, or output only on pin 4 (read right to left). So how do we do this?

`_BV(i)` takes a variable, i, and converts its value to an 8-bit binary number where the i'th bit is a 1 and the rest are zeros. Just exactly what we need. And PD4 is the number corresponding to pin 4 on this port. So we set DDRD to _BV(PD4), and then pin 4 (and only pin 4) is set up for output: blinking our LED.

The rest of the program repeats forever in a loop. It alternates between turning pin D4 on and off, with a delay in-between.

You can turn the individual pins on (and off) by writing a 1 (0) to the PORT register. The syntax is just like above with the DDR: `PORTD = _BV(PD4)` sets the fourth pin in port D to 1.

The _delay_ms() function then waits for roughly 1 millisecond (ms). It may not be quite 1ms, though. Depending on what clock speed your chip is set at, it may be a lot faster. The timing's not critical here, so let's overlook that for now.

Finally, `PORTD &= ~_BV(PD4);` turns pin PD4 off without affecting the rest of the values in PORTD. Let's look in detail at how it does it:
1) _BV(PD4) creates a binary number with the 4th bit (from the right) as a 1 (00001000).

Figure J: My 2400 baud modem

2) ~ is the logical complement operator. It turns the 00001000 into 11110111.
3) & is the bitwise "and" operator. It compares two bits, and returns a 1 if and only if both bits are 1. If either is 0, it returns a 0.

The &= in `PORTD &= ~_BV(PD4)` is a very common shortcut. It stands for `PORTD = PORTD & ~_BV(PD4)`. This last command compares the current value of PORTD (00001000) and the value (11110111) described above. The zero in the 4th place in `~_BV(PD4)`, when used with the &; operator, always makes the 4th bit of the result = 0, effectively turning off bit PD4.

The 1s in the rest of `~_BV(PD4)` make it so that the & operator doesn't clobber the rest of the contents of PORTD. Since the & returns 1 only if both inputs are 1, the remaining bits in PORTD are re-assigned whatever value they already have, leaving them unchanged.

Could we have set `PORTD = 0`? Sure. Since we're only using the one pin in D, it would turn it off just fine. But it would have the side-effect of turning off *all* the pins on port D, and it wouldn't have provided such a nice example. The bit-masking techniques (`&= ~_BV()` and its opposite, `|= _BV()`) are pretty useful to learn for chip-level programming.

The last part of the code, return(0), never gets reached. The while(1) command ensures that the chip is always going to be stuck in the while loop. I just included the return() command because the compiler complained when I didn't include it: the main() C function in a program is always supposed to have a return value, even if the chip will never get there.

That's a lot of programming for one day. Take some time to admire the simple beauties of the flashing LED.

11. The end and web resources

So there you go, your first AVR application. If you want to learn more about the AVR chips, here are some good web resources:

- AVR Freaks (www.avrfreaks.net) is the motherlode: a community of friendly users with a forum.
- Cornell's EE476 class webpage (http://instruct1.cit.cornell.edu/courses/ee476) is a tremendous source for info, and their final projects (http://instruct1.cit.cornell.edu/courses/ee476/FinalProjects) are a treasure-trove of crazy, cool project ideas, all well-documented.
- Psychogenic.com (http://electrons.psychogenic.com/modules/arms/sec/1/AVR) is good for AVR programming on Linux.
- The AVR-libC demo program (www.nongnu.org/avr-libc) is not a bad one to learn from either, but maybe a bit advanced if you're just beginning with microcontrollers or C.
- www.sensorwiki.org/index.php/Building_a_USB_sensor_interface has good instructions, and heads in the direction of USB connectivity.
- If you run Linux with KDE and want an integrated GUI environment, have a look at KontrollerLab (www.cadmaniac.org/projectMain.php?projectName=kontrollerlab). It's very similar-looking to Atmel's AVR Studio (www.atmel.com/avrstudio/) for Windows.
- www.avrtutor.com has a short getting-started-with-AVR tutorial written in C. www.avrbeginners.net has a complete tutorial, but it's based on assembly language instead of C, so I'd consider it an advanced tutorial. You might consider looking at them in order.

Figure J shows me using the ghetto dev kit to run a 2400 baud wireless transmitter. Takes only a couple minutes to run a few wires to the breadboard, and off you go!

By day, Elliot Williams is a mild-mannered government economist. By night he's a hacker, alchemist, professor, roboticist, bartender, skater, and charter member of HacDC, a Washington, DC, tech cooperative.

Arduino Charlieplexed Heart

The Arduino-controlled Charlieplexed LED heart uses a total of six wires to control 27 LEDs By Jimmie Rodgers

Here is an Arduino-controlled Charliplexed LED heart. It uses a total of 6 wires to control 27 LEDs. Either PIC or AVR could easily be used for this project, but Arduinos are easy to program, very common, and have to have plenty of power left over for input.

For those not familiar, Charlieplexing is a technique for using N inputs to control N x N-1 LEDs. Check out www.instructables.com/id/Charlieplexing-LEDs--The-theory for a great explanation of this technique. Basically it allows you to drive a bunch of LEDs with very few pins.

1. Required materials

- LEDs (27)—I would suggest red, but pink is cool, too
- Resistors (6) to match your LEDs. I used 100R, but check out http://led.linear1.org/1led.wiz to calculate it based on your LEDs. You can then split that number in half, as there will be one on each wire, but round up as needed. 100R is a good place to start.
- A 6-wire cable. I used some 8-wire Category 5 (4 twisted pairs) that I already had.
- Soldering iron, solder, wire stripper, wire cutters
- Some type of perforated circuit board. I used RadioShack part number 276-148 and it fits the heart perfectly.

Figure E shows the schematic as viewed from the top. You will want to review that before, during, and after building. Once it is built, you will need to test thoroughly. Each two-wire combination to a battery should light up one, and only one, LED. If two LEDs light up, then you have a short. Some ghosting can occur in a second LED when being driven by a strong current source (that is ok, this will not be noticeable when animated). There will be three combinations that won't light anything up, so don't worry and just make a note of those. If more than three combinations fail to light up, then you probably have a short, or a fried LED.

2. First layer

Now to begin the soldering. Follow Figures F through I to assemble the first layer. You are going to create 4 sets of two pairs. They will all be aligned in the same way. Then you will connect the middle sections of both the top and bottom. I've already started making a connection between the top two, but that will be covered in the next step.

3. Start the layering

Continue to solder and start layering the connections. This is where it starts to get complicated. You are going to mirror the connections you've made so far. Whatever you do on the right, do the opposite

Figure A: The heart, connected to the Arduino-compatible Bare Bones Board

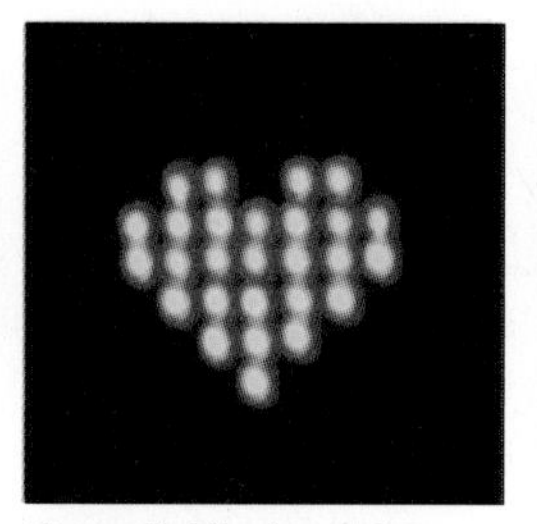

Figure B: The heart in action

Figure C: I first placed all the LEDs in the board, and then marked it for the LEDs on the outer corners of the heart

Figure D: Here the LEDs are set right over the marks I made

For additional information, discussion, and more, please visit the Instructables project page:

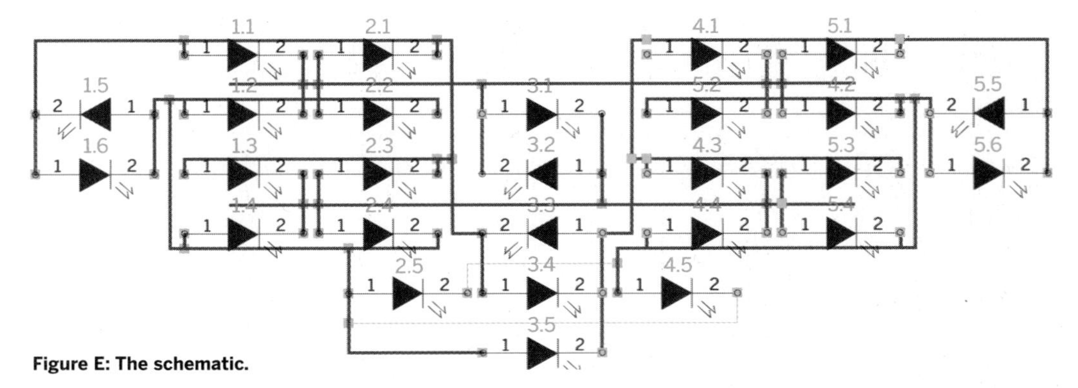

Figure E: The schematic.

on the left. Just be careful not to create any shorts, and give the bent wires enough room so they won't short if moved a little. And be sure to refer back to the schematic often.

- Top right positive lead bends over left to the top right negative lead.
- Top right set's bottom negative lead bends to the right to make contact with the positive lead.
- Do the same for the bottom set.
- Bend the top set's top left negative lead down to the bottom set's top left negative pin.
- Bend the lower right positive lead of the top set down to the lower right positive lead of the bottom set.
- Do the opposite on the other side.

It may look bad, but you are going to end up with something that looks like two C's facing each other on each side. The top middle sections will be joined, as well as the bottom middle sections.

You then add the two center LEDs in opposite alignment of each other. You want the top one to be in the same alignment of all the others so far, and the bottom one to be the opposite. Then you connect the negative of the bottom one to the positive of the top one, and then to the top center connection. You do the opposite for the other two, and connect them to the bottom center connection.

It will be a total of two separate connections on each side, with the two in the middle. The start of the 6 total connections. Now you just need to fill in the other possible combinations (a total of 30).

4. Final layers

The two side pairs need to be installed the same way the middle pair was, but now you connect them to the C-shaped connections closest to them. One to the outer, and the other to the inner of their side.

Then you install the next lower middle pair of two in the same fashion of opposite alignment, and connect them to the two middle C connections. However, don't connect the bottom-most LED into its set yet, as you are about to bend a lead under that connection (Figure O).

I aligned the next two in the same direction on opposite sides of the bottom middle LED. I bent the positive lead of the right one between the bottom center LED and its mate. Connecting to the negative

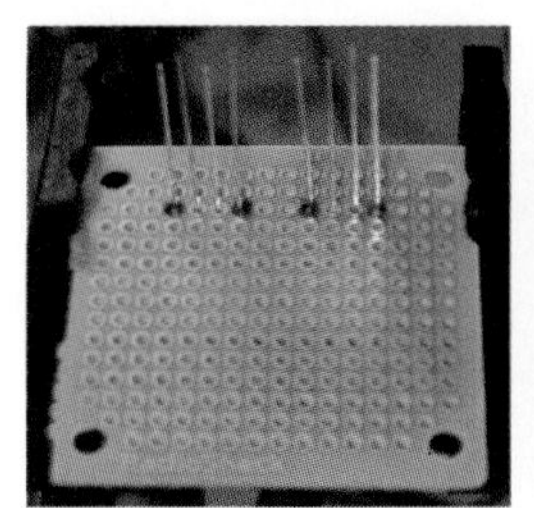

Figure F: Solder the top 4 LEDs into place. Align them all the same way. The row will go: -+ -+ -+ -+

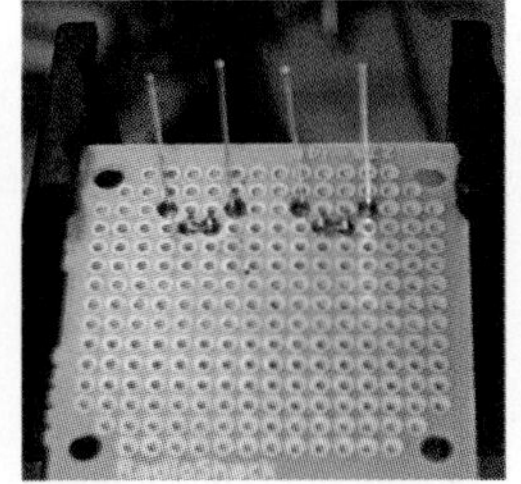

Figure G: Bend the middle pins down and solder them to the board

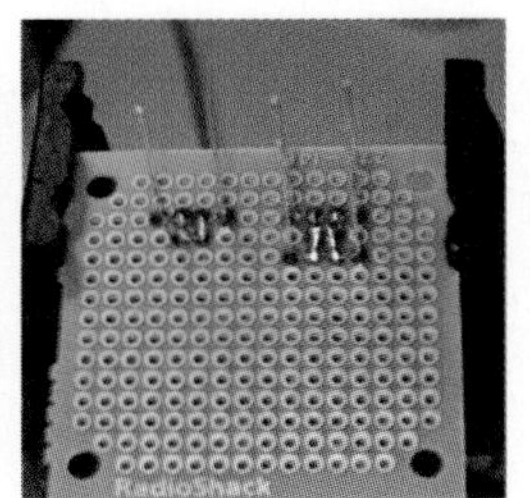

Figure H: You need to solder all 4 of the middle leads together

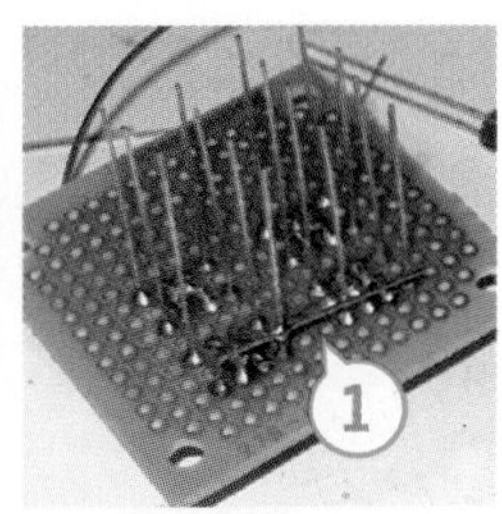

Figure I: Shows a solid connection across the middle sections of the top sets. Do the same across the bottom two sets.

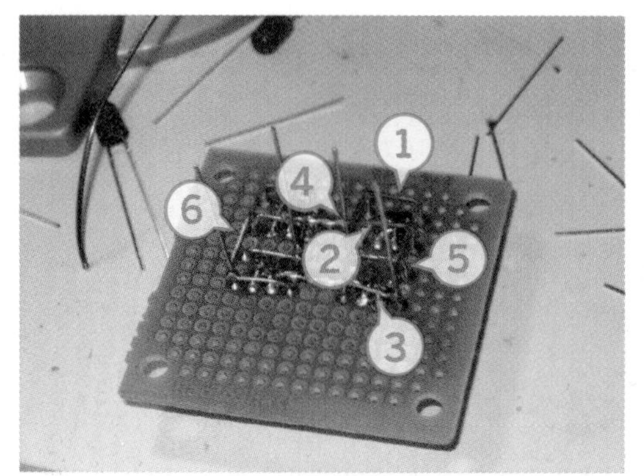

Figure J: 1. Top right positive lead bends over left to the top right negative lead 2. Top right set's bottom negative lead bends to the right to make contact with the positive lead 3. Same as top right set 4. Bend the top set's top left negative lead down to the bottom set's top left negative lead 5. Bend the lower right positive lead of the top set down to the lower right positive lead of the bottom set 6. The mirror opposite of the right set

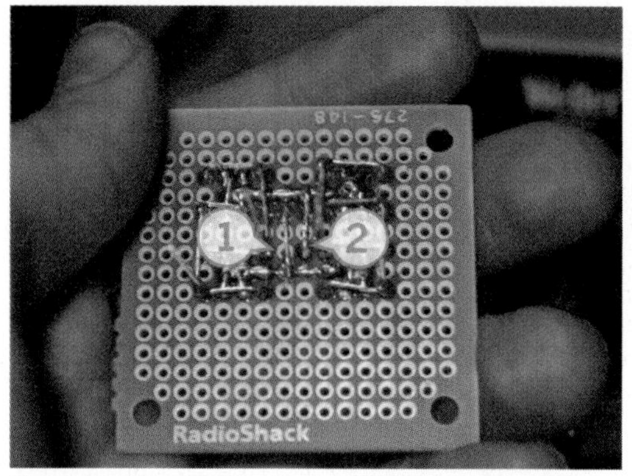

Figure L: 1. The negative leads of both the middle LEDs goes to the bottom center connection 2. The positive leads of the two middle LEDs go to the top center connection

lead of the bottom left LED, and the outer left C shape. You then take the bottom left LED's positive lead, and bend it to the bottom right LED's negative lead, and then to the opposite outer C shape.

For the final LED, I bent the positive lead to the bottom middle's positive lead, and the negative lead to the outer left C shape. Next, clip any stray leads, and make sure there aren't any shorts.

You will end up with the pairs covering all but three of the possible combinations that six solid connections will give you. So poke around with a 3V positive and negative wire and watch the different LEDs light up one by one.

Figure K: Another view of the connections

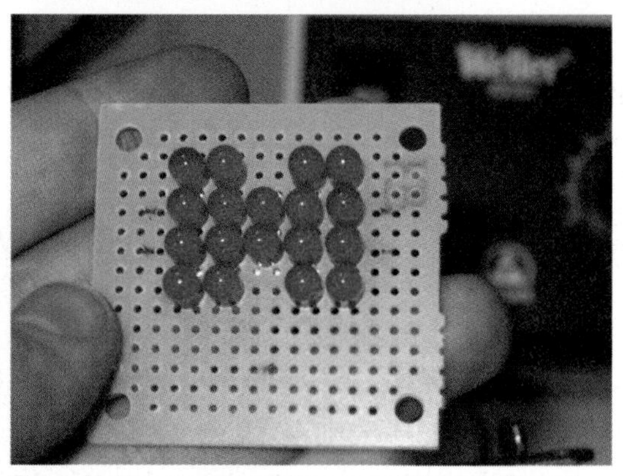

Figure M: You can now address any of these LEDs with different combinations of wires

5. Connecting and programming

All that's left to do it to wire it up and connect it to an Arduino. Since I'm using Category 5, I'm going to follow its color scheme (see Figure R):

Orange: bottom middle
Orange w: top middle
Blue: inner right C shape
Blue w: outer right C shape
Green: inner left C shape
Green w: outer right C shape

Next, test the connections from the other end of the cable. This makes it easy to touch random combinations to the 3V batteries.

I created a spreadsheet to track the connections (www.makezine.com/go/openheart_spreadsheet). If yours are different for some reason then keep track of the differences, since changes will have to be made when you load the software.

Figure N: 1. Connected to the outer C shape
2. Connected to the inner C shape.

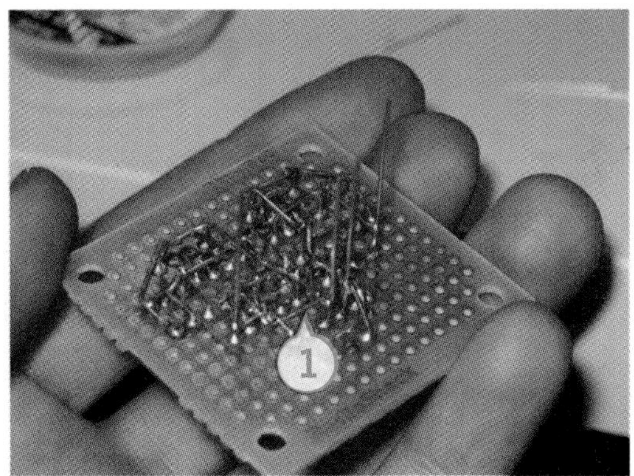

Figure O: The lead (1) goes under the connections made by the lower pair

Figure P: The finished connections from behind

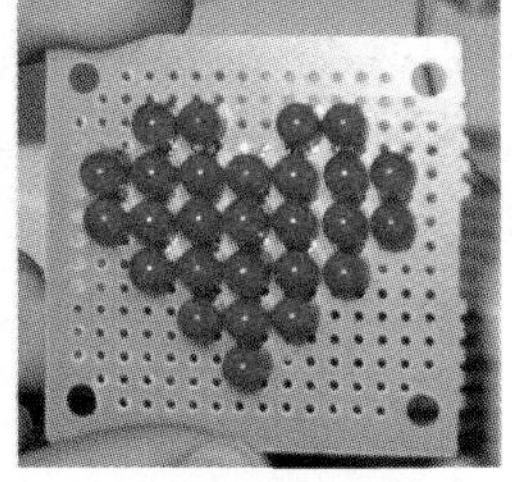

Figure Q: The finished heart

Figure R: Connections for the heart

Figure S: 13 LEDs lit

Figure T: All 27 LEDS lit

When I started out, I used a number scheme based on the columns. There were a total of five, with the outer ones having six each (65556). Ignore that for programming, go with the 1-27 one, as it is much easier to address in the program. However, it is still useful while testing, as trying to remember which one is 16 can be crazy, but finding the third one in column 4 (43) is easy. Both are there for you.

The source code that I used for the two animations in the videos is available at www.makezine.com/go/openheart_source. Remember to put a resistor between your Arduino pins and your wires!

Unfortunately there is a limitation on the Arduino's array size. I'm not sure what it is, which is why I made an array of bytes to save space. Basically you can have any two of the animations, but not all of them. If the Arduino doesn't respond, comment out some lines of the animation, and then send that over.

Using the latest version of the software from the programmer, you should be able to squeeze at least 500 frames of animation out of the Arduino. Now you've got an Arduino-powered animated heart! Check out my blog at http://blog.jimmieprodgers.com for updates on this and other projects.

Jimmie Rodgers is a part-time electronics hobbyist and active Boston Dorkbot member whose dream is to become a full-time Maker. "I love anything dealing with blinking lights and/or sounds, which is why I am starting www.InteractiveLightAndSound.com."

I've developed a kit for the Charlieplexed heart named "Open Heart" that is available in the Maker Shed at www.makershed.com.
A Flash programmer is available to generate the animations on my site at www.jimmieprodgers.com/OpenHeartProgrammer.html. The programmer generates the newest version of the code, making it useful even if you make a heart of your own design. You will just need to change out the heartpin array with one that maps to your own.

Ultra TV-B-Gone

Turn off a TV in style

By Jacob McKenzie

This TV-B-Gone uses a 9V battery to send its signal through a matrix of 20 infrared (IR) LEDS. This extends the working range of the device to about 90 feet (line of sight). Using this in a regular-sized room, you are pretty much guaranteed to zap the TV no matter where you point it.

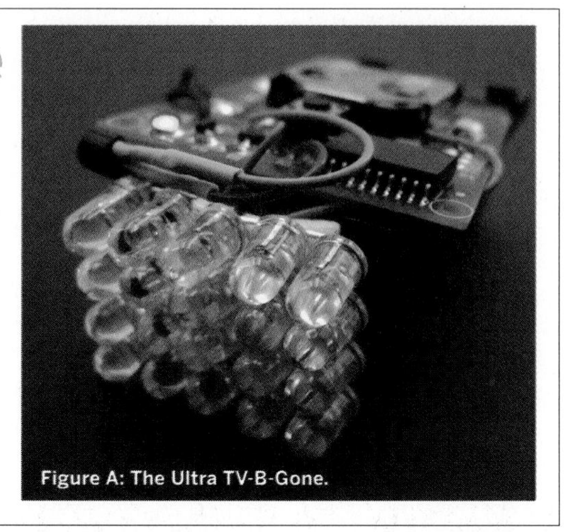

Figure A: The Ultra TV-B-Gone.

1. Get the stuff

You won't need much to build this; here is a list of the materials shown in Figure B.

- TV-B-Gone (keychain version)
- one 2N3904 Transistor (if you have any transistors around, experiment with what you have—it will probably work)
- one 9V battery
- one 9V battery holder
- 20 IR LEDs

You'll need these tools as well:

- Soldering iron
- Some electronics solder
- Desoldering pump
- Hobby knife
- Pliers
- Wire cutters and strippers

If you don't have the TV-B-Gone, you can get one at the Maker Shed: www.makershed.com.

2. Modify the TV-B-Gone

Take apart the TV-B-Gone and examine the board. Notice that it uses two sets of batteries. The two 3V batteries (Figure C) on top drive the LEDs and the bottom 3V battery (Figure D) powers everything else. To save a little space you need to move the 3V battery to the top holder and use the 9V battery in place of the 6V supply.

To get rid of the lower battery holder you have to use a sharp cutting tool to break the connection on the right side of the top battery holder, as shown in Figure E. Next, solder a wire on the left side from the big pad through the hole that is right next to it (Figure F). Now you can remove the lower battery holder and move the bigger 3V battery to the top holder.

3. Add wires

Remove the IR LED that is on the TV-B-Gone and replace it with a pair of wires. Then solder wires for ground and +9V in the two places shown in Figures H and I.

4. Make the LED array

Start with two LEDs and decide which direction you will stitch. Bend the inside lead towards the second LED and solder it (Figures J and K). Repeat until you have a string of four LEDs (Figures L and M). Then repeat the entire process five times.

Now bend the leads of one set to the side and attach another set between the two bent leads (Figure N). Repeat this until you have filled out the entire grid (Figures O and P).

NOTE: Always check the polarity of the LEDs you are soldering. This configuration creates five parallel blocks of four LEDs in series.

5. Complete the circuit

If you are looking at the flat side of a 2N3904 with the pins down, the pins are called Emitter, Base, and Collector from left to right. Attach the Collector and the LED connection from the TV-B-Gone PCB to the

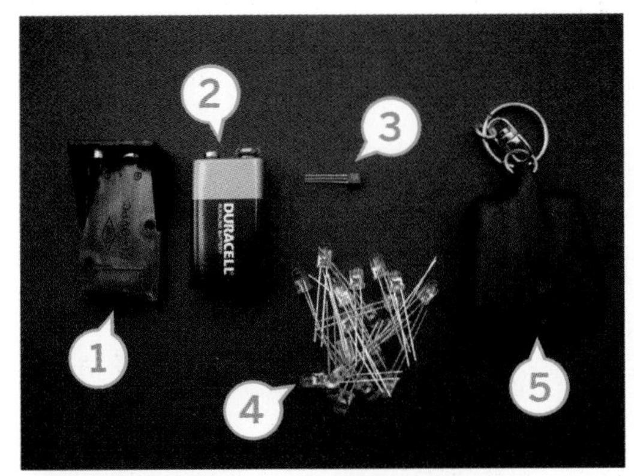

Figure B: 1. 9 volt battery box 2. 9 volt battery 3. 2N3904 Transistor 4. 20 IR LEDs 5. TV-B-Gone

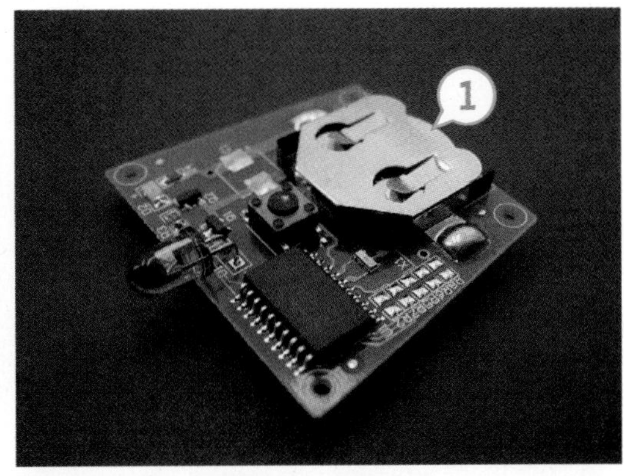

Figure C: 1. The 6 volt power source

Figure D: 1. The 3 volt power source

Figure E: 1. Cut the solder trace here. Be sure to completely break the connection.

Figure F: 1. Solder a wire from this side to the other side of the board

Figure G: After removing the bottom battery holder

Figure H: Replace the existing infrared LED with a pair of wires

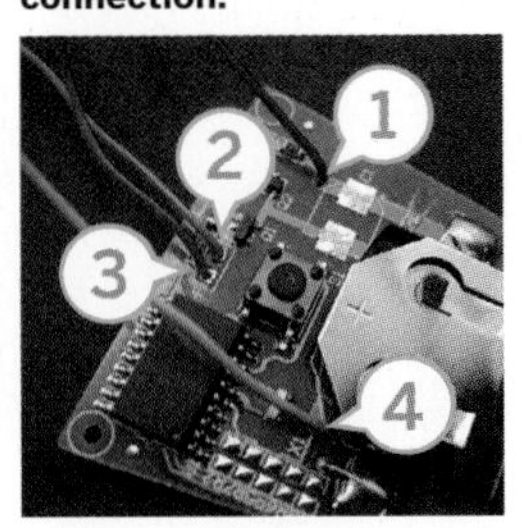

Figure I: 1. Positive 9 volts 2. LED negative 3. LED positive 4. Ground

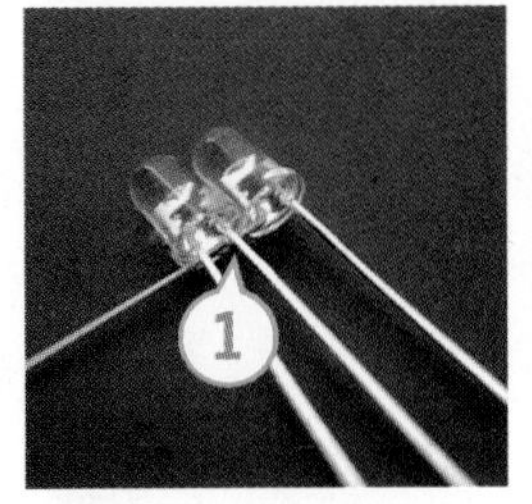

Figure J: 1. Solder here; note that all LEDs must be oriented in the same direction

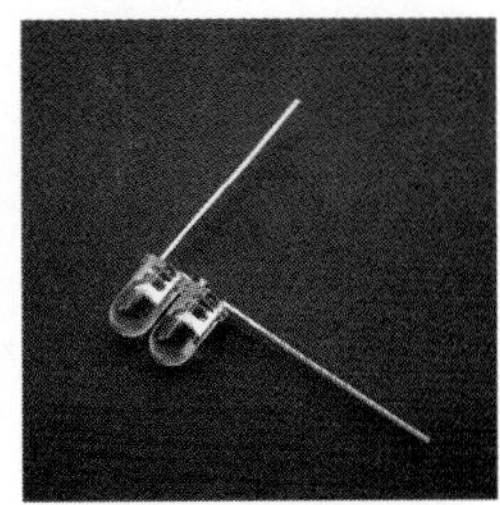

Figure K: A pair of connected LEDs

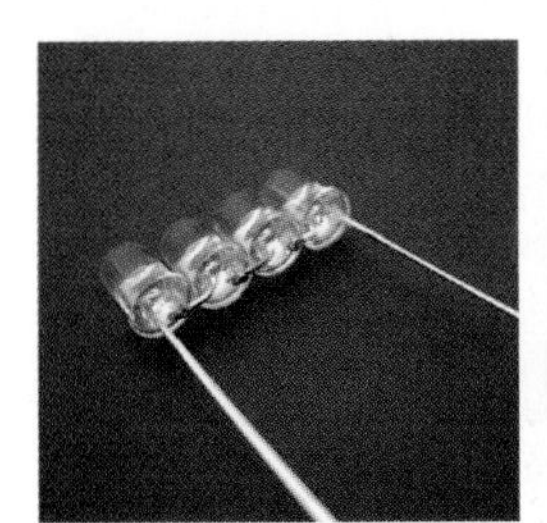

Figure L: Four in a row

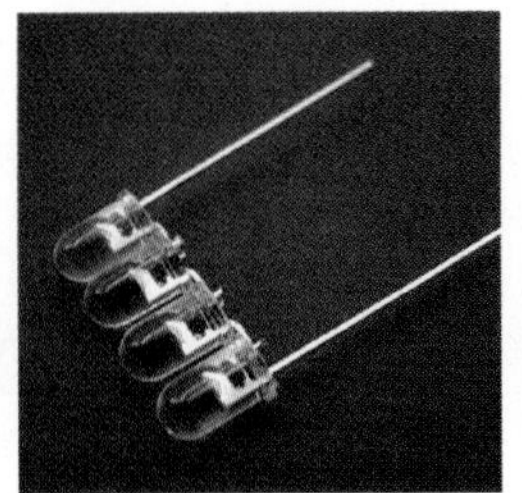

Figure M: Note that they are all oriented the same way

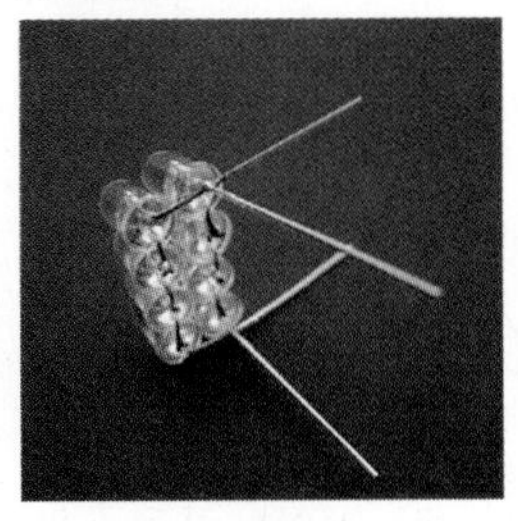

Figure N: Eight-up

Figure O: Viewing the array from the top

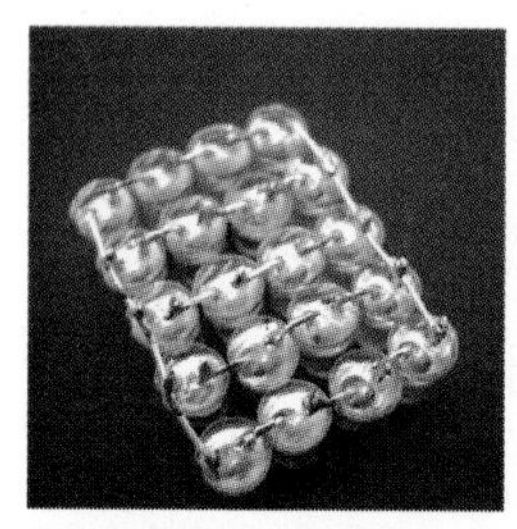

Figure P: Viewing the array from the bottom

Figure Q: 1. Collector, 2. Base, 3. Emitter

Figure R: Connecting the Base and Emitter

Figure S: 1. To battery negative 2. To transistor base 3. To LED array negative 4. To Transistor emitter 5. To LED array positive 6. To battery positive

Figure T: The completed TV-B-Gone

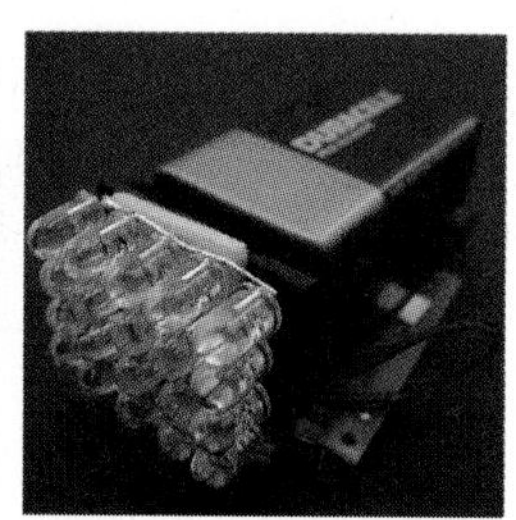

Figure U: Completed TV-B-Gone showing 9 volt battery pack

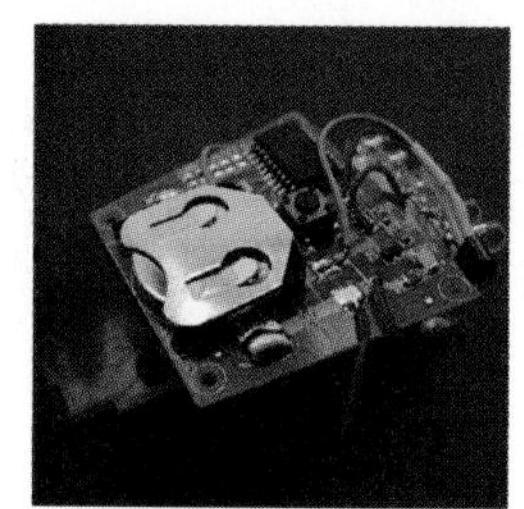

Figure V: Completed TV-B-Gone showing 3 volt battery pack

negative side of the LED array (see Figure Q).

Now, connect the Base to the LED+ wire. Then connect the emitter to ground on the circuit board as shown in Figure R.

Wire the positive side of the LED array to the 9V supply. Finally, connect the ground and 9V wires from the PCB to the 9V battery clip. Figure S shows the connections.

You're done! To keep it all together, attach the LED array and PCB to the battery box (use double-stick tape or whatever you have handy).

Jacob McKenzie is a mechanical engineering student at UC Berkeley and former MAKE magazine intern. He thinks bicycles and Lagrangian mechanics are both awesome.

User Notes

Mitch Altman, creator of the TV-B-Gone, says: R2, R7, and R8 are pull-up resistors that are used for some versions of TV-B-Gone, but not others. Most versions of TV-B-Gone use microcontrollers that have its firmware (i.e., the controlling software) manufactured into the chip (these are called Masked-ROM chips)—these also have built-in pull-up resistors in the chip. Some versions of TV-B-Gone use a One-Time-Programmable (OTP) microcontroller (the firmware is programmed into the chip after the chip is manufactured), and since these OTP chips do not have built-in pull-up resistors, they need to be added externally. One pull-up (R2) is for the switch. One pull-up (R7) is for jumper R5, and the other (R8) is for jumper R6. The R6 jumper isn't used for anything, but the R5 jumper tells the TV-B-Gone whether to use its North American or European database. If R5 is 15KΩ or less, the TV-B-Gone uses its European database, otherwise it uses its North American database.

The first versions of TV-B-Gone did not have a visible indicator light. As Jake said, the TV-B-Gone works fine without it. If you have one of these old TV-B-Gone remotes and would like a visible indicator, then you can add Q2 (2N3904), R3 (1K), R4 (any value from 270 through 1k), and D2 (any visible LED).

Dance Messenger

Attach this fun persistence of vision toy to your shoe and write messages or patterns while you walk, run, or dance! By Dan Goldwater

The Dance Messenger is a fun persistence of vision toy. Attach it to your shoe and write messages or patterns while you walk, run, or dance!

Figure A: Write a message in the air just by moving your foot!

Figure B: The Dance Messenger

1. The circuit board

I used a fairly generic circuit board I had lying around for this project. The circuit is really simple though; it's just an Atmel AVR microcontroller, 10 LEDs, a programming header, and a few resistors and capacitors. You might want a button so you can turn it off. The board I used is mostly surface-mount components, but all the parts are available in through-hole form so you could easily build this on a proto-board. Figure C shows the assembled board.

On the web page for this Instructable (www.instructables.com/id/Dance-Messenger), I've attached the EAGLE CAD (www.cadsoft.de) files for the circuit board, the C source code for the microcontroller, and the Gerber files of the circuit board in case you want to manufacture it. Figure D shows the board layout, and Figure E shows the schematic.

Note: EAGLE CAD is free for small-size boards like this one.

Parts used:

- Atmel ATmega8L microcontroller, Digi-Key (www.digikey.com) part number ATMEGA8L-8AU-ND
- 6-pin .1" male header, Digi-Key WM26806-ND
- 1206-size surface mount LEDs in color of your choice, Digi-Key 160-1406-1-ND, 160-1404-1-ND, 160-1402-1-ND
- 150Ω resistor array/networks, Digi-Key EXB-V8V150JV (4)
- 10μF 0805 size capacitor, Digi-Key 587-1299-1-ND
- Tactile switch, Digi-Key CKN1835CT-ND
- 3.7 volt lithium-ion battery from Batteries America www.batteriesamerica.com/newpage8.htm

Note: A 3v coin cell is a convenient choice for the battery. Since it is flat, you can just tape or glue the circuit board to it, and then tape or glue the whole thing to your shoe.

Want an easy way to get all the parts? Buy a kit.

Adafruit Industries has an open source build-it-yourself kit of a similar persistence-of-vision toy, the MiniPOV. Their version uses only through-hole components, so it is easier to build, and you can program it directly from your computer's serial port (Adafruit Industries sells a USB-to-Serial converter as well if you don't have an RS232 serial port). They also have very detailed instructions for novice electronics hackers, instructions on how to use the GNU C compiler as well, and links to easy-to-use applications for programming your own messages.

The MiniPOV is available directly from Adafruit Industries (www.adafruit.com) and also MAKE magazine's Maker Shed (www.makershed.com).

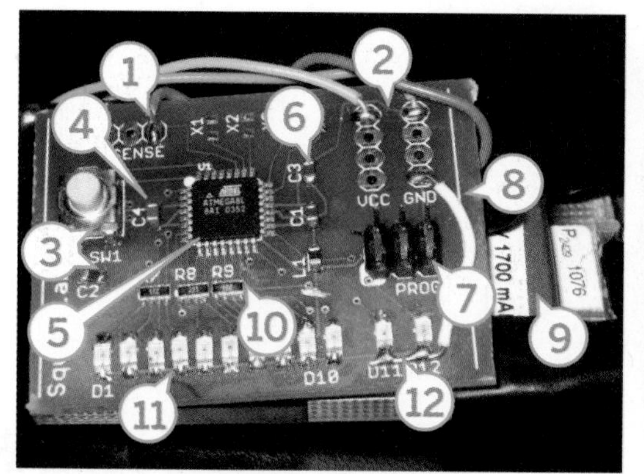

Figure C: 1. Not used 2. Power and ground connection to battery 3. Push-button switch and capacitor 4. 10μF capacitor 5. Atmel ATMega8 microcontroller 6. Don't need L1, C1, or C3 for this project 7. Programming header 8. Optional connection 9. Rechargeable 3.7V lithium polymer battery 10. Resistor networks—four resistors per package 11. Ten yellow surface-mount LEDs 12. Red and green status LEDs (optional)

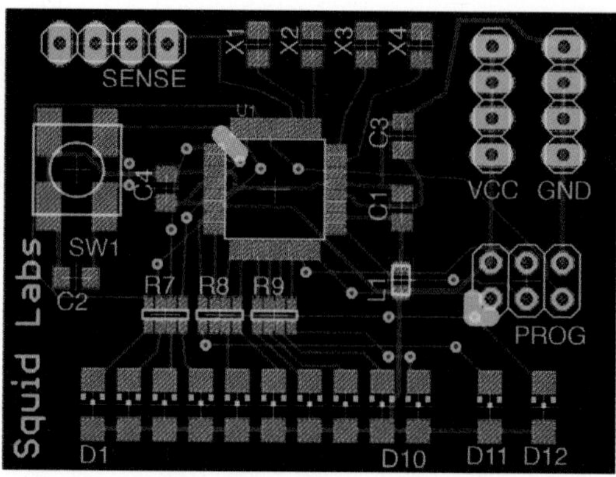

Figure D: EAGLE board layout

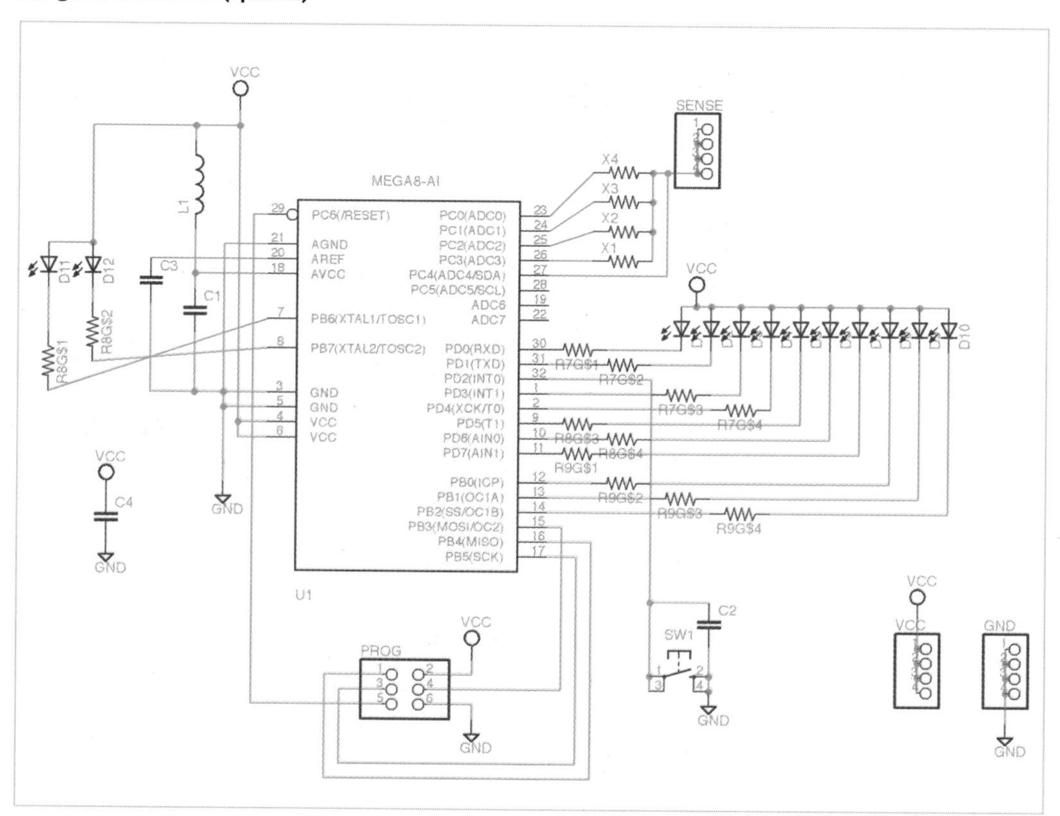

Figure E: EAGLE board schematic

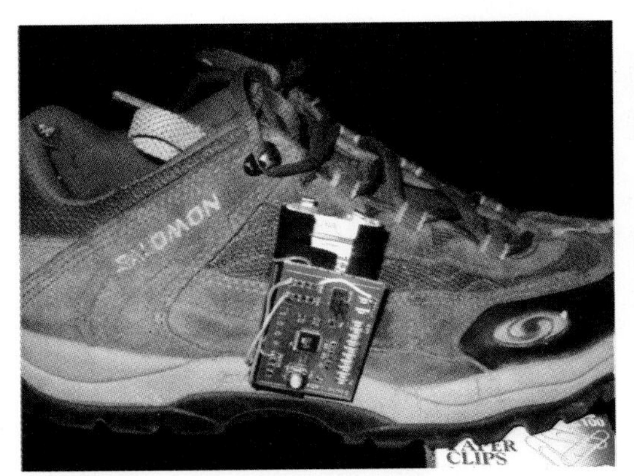

Figure F: Connect it securely because you will want to shake that leg when you're done!

Figure G: Dancing message

Figure H: A close-up

2. Programming the microcontroller

To program the microcontroller you will need a programmer such as the Atmel AVR ISP mkII programming kit (Digi-Key part ATAVRISP2-ND) or the USBtinyISP programmer kit (you assemble it) from Adafruit Industries (www.adafruit.com). The programmer connects your USB port to the programming header on the circuit board.

You will need the GNU AVR tool chain to compile and transfer the code to the microcontroller. You could also use Atmel's free AVR Studio (www.atmel.com/avrstudio). The supplied code was written for the Imagecraft C compiler (www.imagecraft.com), but compiling it with the GNU tool chain is simple (the source code does not include a Makefile, but you should be able to use a Makefile from another AVR project such as the MiniPOV from Adafruit Industries).

Here is where you can obtain the GNU AVR tool chain for Windows, Linux, or Mac OS X:

- Windows: sourceforge.net/projects/winavr
- Linux: Install the AVR-GCC, Avr-Binutils, and avr-libc packages (these packages or something similar are available for many Linux distributions)
- Mac OS X: Install MacPorts (www.macports.org), then use MacPorts to install the packages listed above for Linux

For more information on the GNU AVR tool chain, see www.avrfreaks.net/wiki/index.php/Documentation:AVR_GCC.

The C code supplied does not include on/off button functionality, but this should not be hard to add.

Note: You should add a battery-voltage tester to sense when the battery is dead (this is important if you are using lithium-ion batteries, since they are permanently damaged by discharging too much). To make a battery-voltage tester, you may be able to use a 3.0V zener diode and 220k resistor across the battery, and then use the analog-to-digital converter on the ATmega8 to compare when the battery voltage falls below the zener reference voltage.

3. Attach to shoe

Just tape or glue the board & battery to your shoe (see Figure F).

4. Do some dancing!

In Figures G and H, my friend Corwin shows us some moves!

How well does this device work? It is a bit hard to notice it at walking speed (that's because if you look straight at it the effect is reduced). If you avert your vision slightly, and look away at a fixed object, the effect is much clearer. At running speed or dancing it works nicely.

Dan Goldwater loves beams of light and the devices that make them. Dan is a co-founder of Instructables.com and SQUID Labs and now makes practical digital light art for bicycles at Monkeylectric.

Home & Garden | Food | Photography | Science | Computers | Electronics

The Best of Instructables Robotics

There's a popular saying (among robot geeks, anyway) that a human is a way that a robot builds a better robot. If that's true, there couldn't be a more revolutionary tool to aid our future robot overlords than the Internet, especially vibrant communities of human collaborators, like the one found at Instructables.

Before the Net became ubiquitous, competitions and conventions were prime ways that robot builders showed off their designs and where the designs of successful robo-competitors were incorporated into other builders' future designs, all in a spirit of open source development (before it was called that). This process worked, but it was slow. Today, a bot builder can post an innovative design, and within days, even hours, builders from around the world can be trying their hand at the build, asking questions, sharing what they've learned, improving upon the design, and sharing their results.

I was involved in one little robo-evolutionary branchlet when I used a simple light-sensing circuit, known as Herbie, in a project in MAKE Volume 2. Herbie was designed by Randy Sargent at MIT in 1996. He entered Herbie in a line-following competition. He came in dead last, but the circuit was clever and soon found its way online. Dave Hrynkiw, of Solarbotics.com, used it in a light-seeking robot in his book *Junkbots, Bugbots & Bots on Wheels* (McGraw-Hill 2002). I took his design, made some changes, and built my bot inside a computer mouse. Mousey the Junkbot was born. Soon after my piece appeared in MAKE, Jake McKenzie posted his "Mousebot Revisited" Instructable (found here) with even more improvements and tips for making the build easier. Solarbotics then came out with a Herbie the Mousebot Kit, a beautiful kit that adds even more features to the circuit (e.g., mousebots that can chase each other around a room). And I realized, as I was typing this, that I'm wearing a Solarbotics Herbie the Mousebot T-shirt. Careful. These robots might be sneakier than we think.

—Gareth Branwyn (garethbranwyn)

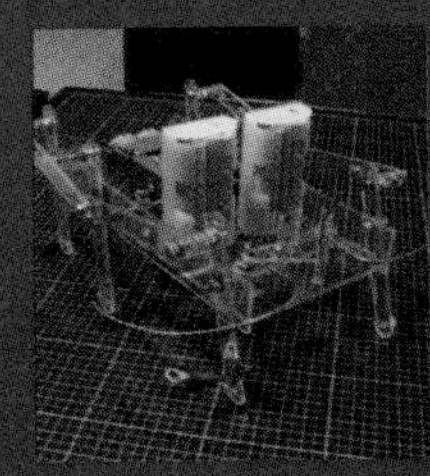

How to Build Your First Robot

Build a full programmable robot in a few hours for under $100

By Frits Lyneborg, aka "fritsl"

Figure A: Look, it's your first robot. Isn't he cute?

Figure B: The PICAXE-28X1 Starter Pack

This is a walkthrough of how to build a programmable, autonomous (i.e., not remote-controlled, not strictly pre-programmed, but reacting to its surroundings) robot in a few hours. It's surprisingly easy and doesn't involve advanced knowledge of electronics to get started. You *will* have to stay focused as we cover the basics here and a few things will be foreign to you at first. This is meant to be an eye-opening, learn by exploring, project. After building this, you'll be able to build and control all sorts of electronic devices! Sound crazy? It's true, you just need to try it to understand how much power is contained in some of the chips you can buy for a few bucks these days. Welcome to the wonderful world of microcontrollers!

The programming examples I show in the end are to make this robot "wall avoiding" (i.e., it will sniff around and explore based on which objects it meets, what is on the left, right, and ahead of it). But it can be programmed to do all sorts of things—easily. If interest is shown, I will provide more programs for it.

1. Buy the materials (project board, microcontroller, and starter pack)

Shopping list:
A PICAXE-28X1 Starter Pack

The 28-pin project board in this package is like a game of Mario Bros; fun and full of extras and hidden features, making you want to play with it over and over again. This includes the main brain, the PICAXE-28X1. At around $40, this is the most expensive part, but you're buying a computer, and it includes a lot of nice, useful stuff. You get a CD-ROM with lots of manuals on it, cables, a project board, and the microcontroller itself. Actually, it's cheap for all you get. Be sure to get the USB-version—images in online stores often show a serial-cable version, though they're selling the USB. When buying the USB-version, it is *not* necessary to also get the USB cable as an extra item. Once you have this Starter Kit, you just have to buy a new board and microcontroller for future projects—much cheaper.

Note: A version II with cool code can be found at letsmakerobots.com. If you have any problems or questions regarding this project, please feel free to contact me. There are also videos of this project available on its Instructables page.

For additional information, discussion, and more, please visit the Instructables project page:

Figure C: The venerable L293D Motor Driver

Figure D: A servo motor (left), header pins (top right), and a DIL resistor chip (bottom right)

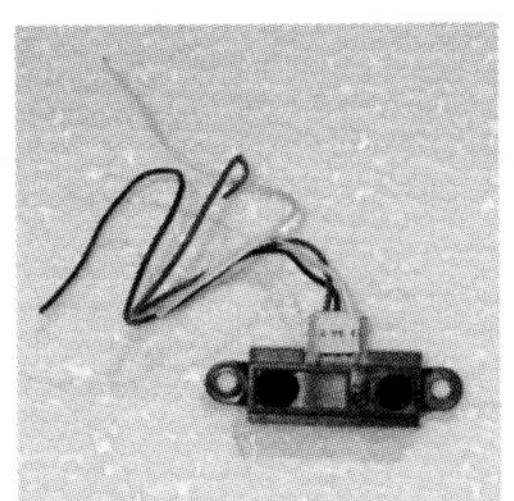

Figure E: The GP2D120 IR sensor

Figure F: Gear motors, wheels, and tires

An L293D Motor Driver

The name says it all, more about this chip later :-)

A PICAXE Servo Upgrade Pack

This is an easy way to get a servo topped with some small parts needed for this project. You can also get any standard servo, a row of header pins (shown in Figure D), and a single 330Ω resistor instead of the DIL Resistor chip that comes with the Servo pack, if you wish.

What is a servo?

A servo is a cornerstone in most robotic applications. It is basically a little motor and gearbox with wires to it, and an axle that can turn some 200 degrees. On this axle you can mount various-shaped "servo horns" onto which things you want to control can be attached. There are three wires on a servo: two for power, and one for signal. The signal-wire goes to the device that controls the servo, in this case, our microcontroller. The result is that the microcontroller can decide where exactly the axle should turn and position itself. This is pretty handy, as you can program something to physically move to a precise position.

A Sharp GP2D120 IR sensor—11.5"/Analog

An IR Sensor in the 11.5" range will do. Do not buy the "digital version" of the Sharp sensor for this kind of project. They do not measure distance as the analog ones does. Be sure to get the red/black/white connection wires for it. These are not always included, and it's a non-standard connector!

I usually use ultrasonic sensors, such as the SRF05 (find it anywhere via Google). The SRF05 is much more reliable and precise than the GP2D120 IR Sensor. It is also faster, but costs a little more, is a little more complicated to write code for, and a little more complex to install—so it is not used here for beginners. If you go for the SRF05, I've made a small walkthrough to connecting the SRF05 on letsmakerobots.com/node/66.

Two gear motors with wheels

The higher the gear ratio, the stronger the robot; the lower, the faster the robot. I recommend a ratio somewhere between 120:1 to 210:1 for this kind of a project.

2. You will also need:

- Double-sided adhesive tape (for mounting, the foamy kind is best)
- Some wire
- Ordinary adhesive tape (to isolate a cable perhaps)
- Simple soldering equipment (any cheap kit will do)
- An ordinary small nipper or scissors to cut things
- A screwdriver or multitool

You could also get (while you're at it):

- Some LEDs, if you want your robot to be able to signal to the world or make cool flashing-effects
- More servos to make your robot move more... er... arms?
- A tiny speaker if you'd like your robot to produce sound-effects and communicate with you
- Some sort of tank-track system. Robots with tank tracks are way cool, and the controller and the rest can be the same. Tamiya makes cool tank-track-systems.
- Any kind of line-sensor kit, to turn your robot into a Sumo, a line-follower, stop it from driving off tables, anything else that needs "a look down"

Figure G: Attaching the wheels

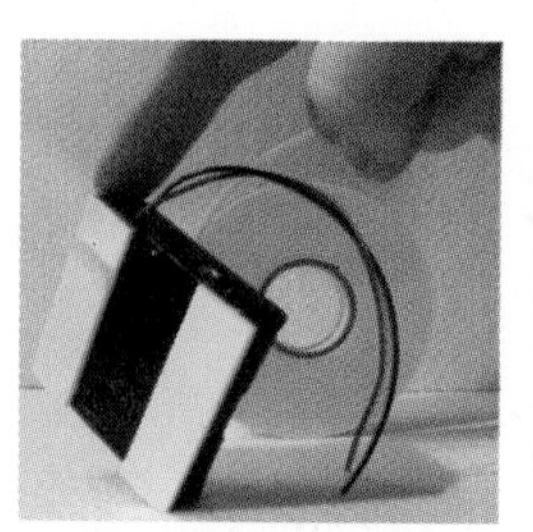

Figure H: Using double-sided tape to attach the battery holder

Figure I: Assembling the main components using the battery holder as the "body"

Figure J: Side view with wheels, servo, gearmotors, and battery holder assembled together

Figure K: Top view of servo, looking down on "servo horn" disk in center

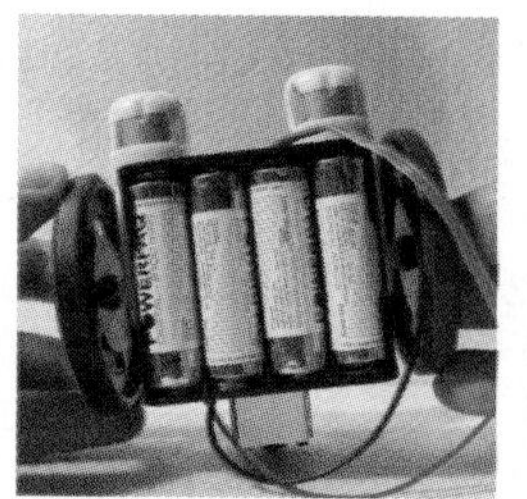

Figure L: Bottom view, showing battery pack

Figure M: The PICAXE project board

Figure N: Installing the 330Ω DIL (or Dual InLine) resistor (the yellow chip)

3. Okay, let's make a robot!

You've ordered the stuff, received your package(s), and now you want to build. Let's get started!

First, mount the wheels to your geared motors. Add tires (rubber bands in this case) (Figure G).

4. The double-sided adhesive tape trick

An easy way to mount stuff for fast (and amazingly solid and lasting) robots is double-sided adhesive tape (see Figure H).

5. Build the body out of...nothing. Really!

Insert the batteries, so you have a realistic idea of weight and balance. With the batteries below the axle of the wheels, you can make it balanced, but it's no problem if it's not exactly balanced. Just make sure that if anything drags the ground, it is behind the wheel axles. Add some double-stick tape to the bottom of the servo as well.

6. Design your robot

Choose your own design, you can also add extra materials if my "design" is too simple. The main thing is that we have it all glued/taped together: batteries, servo, motors, and wheels. Make sure the wheels and servo can turn freely, and it can stand on its wheels (Figures J–L).

7. Disconnect!

Take out the batteries, to avoid burning something unintended! They were just a test load.

8. Get started with the board

You should have a project board similar to the one in Figure M. Notice that it has a chip in it. Take it out. That chip is called a Darlington Driver. It's quite handy there on the board, but we'll not need it for this project, and we need its socket space, so away with the chip! It is easiest to get chips out of their sockets by inserting a small, flat screwdriver just below the chip, rocking it gently, and tipping it up carefully. Don't bend the pins!

9. Insert the chips

A fresh, brand new chip usually does not fit into a socket without some adjusting. Place the chip on one side, so that a row of its pins is flat, on hard, level surface, and gently bend all its legs (pins) slightly inward, applying the same pressure along

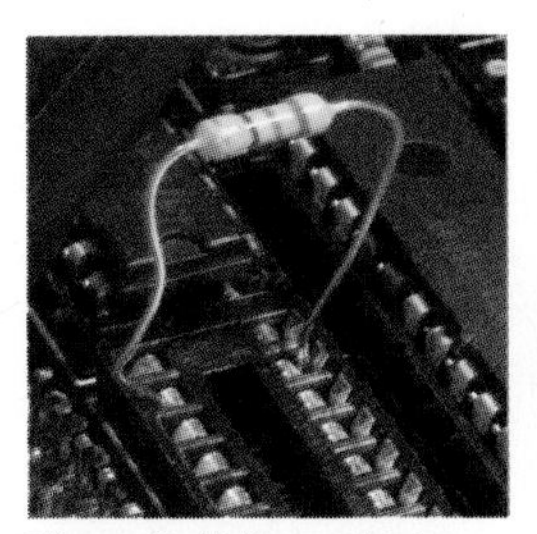
Figure O: Using a single 330Ω resistor

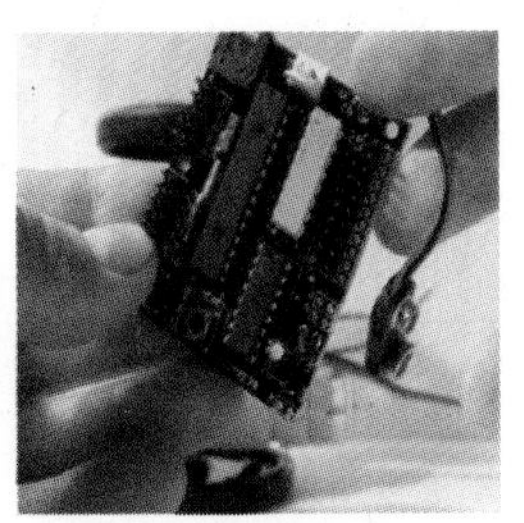
Figure P: The PICAXE-28X1 microcontroller chip installed. Brains, baby!

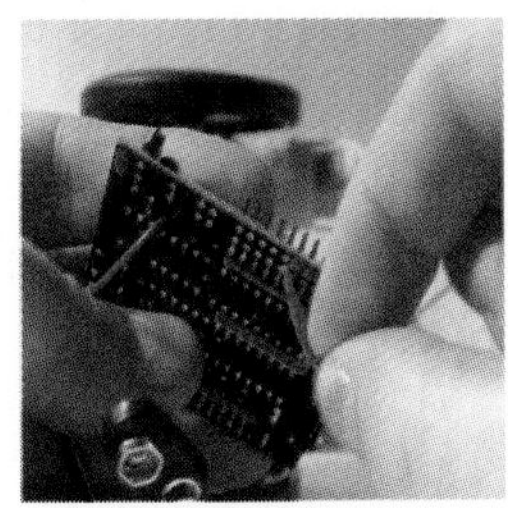
Figure Q: Plastic left over from the manufacturing process. Remove.

Figure R: Wiring up the two gearmotors

all the pins. Keep doing this on both rows of pins, checking against the socket on the board, until all pins line up with the holes on the chip socket and it can easily be pressed in. Take your time as the pins can easily snap off.

Make sure all the pins are firmly in the sockets.

If you bought the Servo Upgrade from PICAXE, you have a yellow chip. Put it in the socket in place of the Darlington (Figure N). This yellow chip is actually called a DIL (or Dual InLine) Resistor. It is really eight 330Ω resistors in a neat IC package. If you happen to have a single resistor, you can insert it instead, in the slot numbered "0" (see Figure O), as this is the only one we'll actually use for our single servo. Insert the large chip, the brains, the big daddy o' IC, the microcontroller—our PICAXE 28—into the project board (see the 28-pin chip in Figure P). It is important to put this in the correct way. There's a little dot, called a dimple, on one end of the chip, and also a corresponding one on the project board. They must go together. This chip will get power from the board via two of its pins. All the remaining 26 pins are connected around elsewhere on the board, and they will be programmable for you, so you can send current in and out to detect things and control things with the programs you upload into your microcontroller. (Cool!)

10. Insert the motor controller

Insert the L293D motor controller into the last socket (Figure P). Be sure to install this the right way just as with the microcontroller.

The L293D motor controller will take four of the outputs from the microcontroller and turn them into two. Sound silly? Well, any ordinary output from the microcontroller can only be "on" or "off." So just using these would only make your robot able to drive forward or stop. No reverse! And reverse comes in handy when facing a wall. The board is so smart that the two (now reversible) outputs get their own space, marked (A) and (B), next to the motor controller. More about this later.

11. The red plastic on the back of the board

On the backside of the board, you may find some strange plastic strips. These have no use, they're left over from the manufacturing process. They "dip" the boards in warm tin (yum!), and parts they do not want tinned are sealed with this plastic material. Just peal it off when you need the holes they cover.

12. Connect the motors' wires to the board

Take four pieces of wire and solder them to the four holes on the board marked "A & B". Alternately, you can use some other means of connecting the cables—you can buy all sorts of standard plugs and pins that'll fit. But if you're like me, just solder them onto the board. You can strengthen them with tape or use heat-shrinking tubing to support the wires (Figures R-S).

13. Connect the wires to the motors

The two "As" go to one motor, and the two "Bs" to the other. It doesn't matter which is which, as long as "A" is connected to one motor, and "B" to the two poles of the other. (And yes, I know, my soldering iron is really, really dirty. Hey, as long as it works, you know?) (Figure T)

14. Hooking up the servo

Now let's hook up the servo. If you read the PICAXE documentation, you will see that you're supposed to use two different power sources (one for the microcontroller and one for the servos). For our purpose, and this is a simple robot, I think a single source is fine.

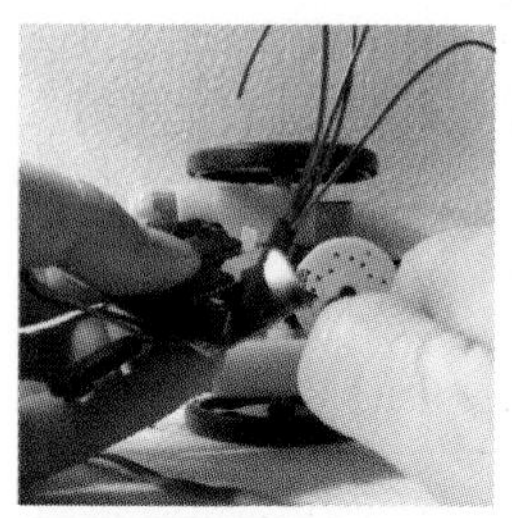

Figure S: Using heat-shrink tubing to neatly bundle the motor wires

Figure T: Soldering up the motor wires to the terminals on the gearmotors

Figure U: Location of servo motor pins, showing S (signal), V (voltage), G (ground)

Figure V: The three servo wires hooked up

You will need to solder an extra pin to output "0", if you want to use the standard servo connection. Such a pin comes with the PICAXE Upgrade pack—you only need one for one servo, and they can be bought in any electronics store. If your servo's cable is (black, red, white) or (black, red, yellow), the black should go to the edge of the board. Mine was (brown, red, orange), and so the brown goes to the edge.

Red is the power and usually referred to as V (or V+, +, V1, V2). The black (or brown in my case) is G (or Ground, GND, "-"). So we have our two poles, positive and negative, voltage and ground. Remember your physics lessons? The last color is the "signal," the electronic pulses that control the movement of the servo shaft and tell it where to go. This wire is usually white, yellow, or orange. A servo needs both (+) and (-) or (V) and (G), and a signal (Figures U and V).

Some other devices you'll attach to your board may only need "Ground" and "Signal" (G and V), and some may need V, G, as well as Input and Output. This can all be confusing in the beginning, and too often, things are named differently. But after a while, you'll start to see the logic behind it all. It's actually rather simple—even *I* get it!

15. Hooking up the head.

Now let's hook up our Sharp IR sensor (or the SRF05, if you went for that option). If you bought an SRF005, or similar, check out letsmakerobots.com/node/66 for details on how to hook up this sensor, as it's different than the Sharp.

There are many different ways of hooking up the Sharp IR sensor. Following are the basics of what you need to know:

- Red needs to be connected to V1, that's (in this setup) anything marked "V" or is connected to this
- Black goes to G, anywhere on the board
- White is to be connected to Analog Input 1

If you read the documentation that comes with the project-board, you can find out how to attach the accompanying ribbon cable and use it.

What I've done in Figure W is cut off a cable from an old burned-out servo, soldered in a pin, and connected the whole thing just as with a servo. You can use it to see which colors of the Sharp go to which row on the board, or at least one way to do this (Figures W-X).

Whether you use the ribbon cable or my method of connecting the Sharp IR, you should also connect the three remaining analog inputs to V (look at the little pins connected in Figure X, next to the plug). I had some jumper blocks lying around, and you can see that all three leftover analog pin connections are shorted (connected V to G). The last pair, not shorted, are just two "ground" pins, not Inputs. No need to bother with these. If you use the ribbon, you can connect the inputs to V (or ground, for that matter) by connecting the wires in pairs. The reason it's important to shortcut the unused analog inputs is that they are otherwise left "floating." This means that you will get all sorts of weird readings on your sensor(s) if these unused Input pins are left open and with no input signals on them.

16. Let there be life!

You now need to get the red wire from your battery's (+) hooked up to the red wire on the microcontroller board's (V). And the black wire (-) to the (G) on the board. How you do this depends on the equipment you bought. If there's a battery connector on both

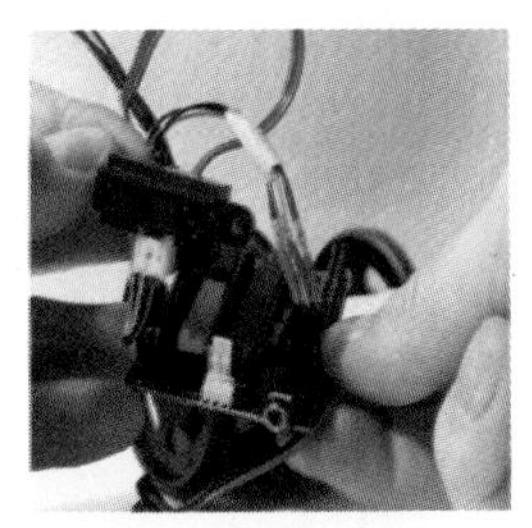

Figure W: The IR sensor hooked up to analog Input 1

Figure X: The other analog inputs are jumpered (shorted) so that they don't interfere with the IR signal

Figure Y: Plugging in the serial jack so the fun can truly begin

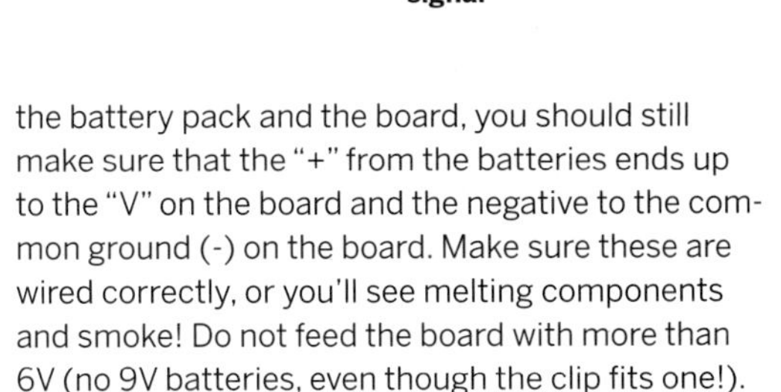

the battery pack and the board, you should still make sure that the "+" from the batteries ends up to the "V" on the board and the negative to the common ground (-) on the board. Make sure these are wired correctly, or you'll see melting components and smoke! Do not feed the board with more than 6V (no 9V batteries, even though the clip fits one!).

Note: Remember, we are only working with one power supply here. You will need to use the same ground (there always must be a *common* ground), but you'll use both V1 and V2 power pins. That way, your logic chips will get one source, and your motors another (higher) voltage source.

Install the PICAXE Program Editor on a PC (follow the manual for proper installation and to get your USB, serial, whatever, connection set up). Insert the batteries into your (still "headless") robot, insert the computer programming cable into your robot. Enter the Program Editor on your PC and write:

```
servo 0, 150
```

Press F5

Wait for the program to transfer to the bot, and your servo to give a little twitch.

If something goes wrong here, check troubleshooting in your manuals, check comm ports, etc., until no errors are reported, and all seems to be working. If need be, contact me via this project's Instructable page.

To test the bot further, write:

```
servo 0, 200
```

Press F5

The servo's servo horn disk should spin a little and stop. To turn the servo back, write:

```
servo 0, 150
```

Press F5

Now your robot's "neck" is facing forward, and you can stick on the Sharp IR sensor (or SRF05 if you went for that).

17. Heads up and go!

You're done with the basics. You've just built yourself a programmable robot. Now the fun starts. You can program your bot to do almost anything, and attach anything to it, expand it in all sorts of creative ways. I'm sure you're already full of ideas.

A basic control program, code for making your robot sound off, and cool ideas for other ways your robot can be programmed can be found at this project's Instructables page and at letsmakerobots.com/node/254.

Frits Lyneborg (aka "fritsl") was born in Denmark in 1970. He is a freelance inventor and developer of toys and robots and is world-famous for projects like "Robot Wall Racers" and the "Yellow Drum Machine" series. Visit fritsl online at letsmakerobots.com/user/4 or via email at fritslyneborg@gmail.com.

How to Make an OAWR (Obstacle Avoiding Walking Robot)

Build a no-solder robot that can navigate a space without the need of a computer brain

By Clement Fletcher

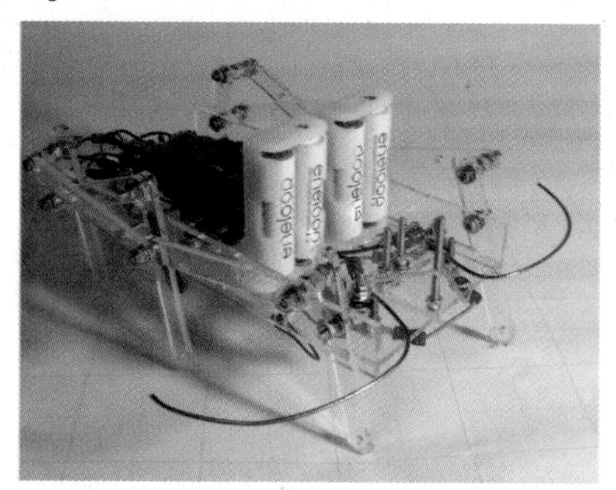

Figure A: The finished OAWR about to explore its environs

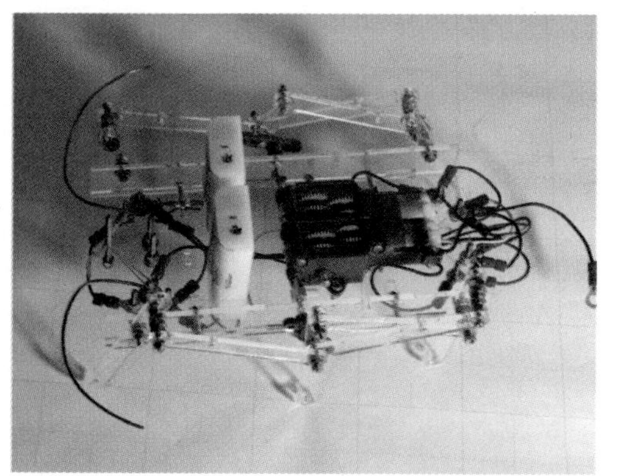

Figure B: A top view with a good look at the Tamiya Twin Motor Gearbox Kit

This Instructable shows you how to make a little walking robot that avoids obstacles (much like many commercially available bots). But what's the fun in buying a pre-made toy, when you can start with a motor, a sheet of plastic, and a pile of bolts and wires and proceed to make your own!

Features:

- Not difficult to source parts (no switches, relays, or computer chips, everything but the motor is available at Home Depot)
- No soldering
- Has a Mechano for grown-ups feel
- Choice of options for cutting out the pieces (scroll saw and drill, access to a laser cutter, purchasing parts online from Ponoko—http://tinyurl.com/6ehwsg)

Note: The project page for this Instructable is required to complete this project. It includes videos of the assembly and the finished robot in action. There are also numerous, very high-quality downloadable documents (PDFs) of the parts list, wiring diagrams, cutting patterns, and a detailed assembly guide. It is highly recommended that you download and study these before you start the project.

1. Parts and tools

All parts, with the exception of the motor, can be found at Home Depot. The motor can be ordered from a number of online stores for about $10.

Parts list

Nuts and bolts:

- 3mm x 15mm bolts (20)
- 3mm x 20mm bolts (2)
- 3mm x 30mm bolts (9)
- 3mm washer (48)
- 3mm nuts (45)
- 4mm nuts (26)
- 5mm washers (2) (12mm OD)

Electrical:

- Crimp wire terminals (18) (red 5mm ring)
- 2-battery AA box (2)

For additional information, discussion, and more, please visit the Instructables project page:

- A motor (Tamiya Twin Motor Gearbox #70097, available from many online sources)
- Various colors of electrical wire

Miscellaneous:

- Acrylic (150mm x 300mm x 3mm thick)
- Whisker Wire (260mm x 1.6mm) or two large paper clips
- Elastic band

Tool list

Required:

- Printer
- 5.5mm wrench (2)
- Screwdriver
- Pliers
- Crimp terminal crimpers
- Hot glue gun

Additional Tools Depending on Choice of Sourcing Acrylic Parts

Option 1: Scroll saw and drill

- Glue stick
- Scroll saw
- Drill
- Drill bits (3.2mm, 12.5mm, 16mm)

(I was going to use this option, however, I snagged a free shipping coupon from Ponoko so I had my pieces laser cut instead.)

Option 2: Ponoko.com

A Ponoko account (this is the option I used)

Option 3: Access to a laser cutter

Access to a laser cutter

2. Cutting pieces

Please choose which steps to follow based on the cutting option you have selected.

Option 1: Scroll saw and drill

Download and print the PDF pattern included with the Instructable page for this project (please choose the file corresponding to your paper size).

- A4 size paper ('31A-(OAWR)-Scrollsaw Pattern(A4).pdf')
- Letter-size paper ('31B-(OAWR)-Scrollsaw Pattern(Letter).pdf') (It is important to not scale the drawing while printing.)

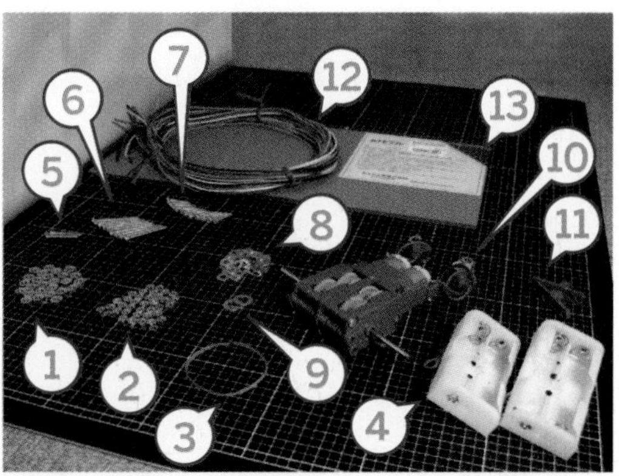

Figure C: The parts—1. 4mm nut (26) 2. 3mm nut (45) 3. Elastic band 4. 2 x AA battery holder (2) 5. 3mm x 20mm bolt (2) 6. 3mm x 30mm bolt (9) 7. 3mm x 15mm bolt (20) 8. 3mm washer (48) 9. 5mm washer (2) 10. Motor (Tamiya Twin Motor Gearbox Kit (#70097) 11. Crank (2) (included with motor) 12. Multicolored wire 13. 150mm x 300m x 3mm acrylic sheet. Not pictured: 260mm x 1.6mm solid wire, crimp wire terminal (red 4mm ring) (18)

Measure the ruler on the printout against a ruler you trust. If they do not match, the pattern has been scaled and you need to look at your printer settings before reprinting. If they do match up, onwards!

- Glue the pattern to the acrylic sheet
- Drill holes
- Cut out pieces using a scroll saw

Option 2: Online digital manufacturing through a service like Ponoko.com

- Get a Ponoko account (www.ponoko.com)
- Order the pieces I've created here (http://tinyurl.com/6ehwsg). They are priced at cost.

Option 3: Access to a Laser Cutter

- Download the laser cutter optimized pattern found on this project's Instructable page. The pieces are placed side-by-side and duplicate lines are removed. The file is called 32-(OAWR)-Laser Cutter Outline.eps, in encapsulated postscript (.eps) format.
- Cut the file on your laser cutter.

3. Whiskers

The last step before we start putting it all together is to add the whiskers. Bending the whiskers is quite straightforward. Use pliers and a 130mm length of 1.6mm wire (a large paper clip will also work), using the pattern on the PDF found on the project page

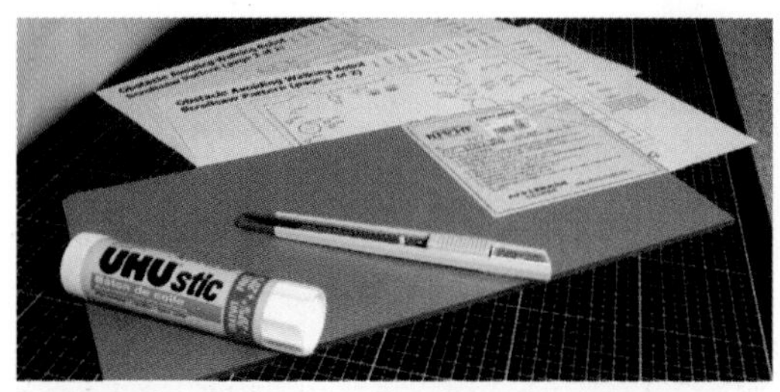

Figure D: Getting ready to glue the patterns onto an acrylic sheet

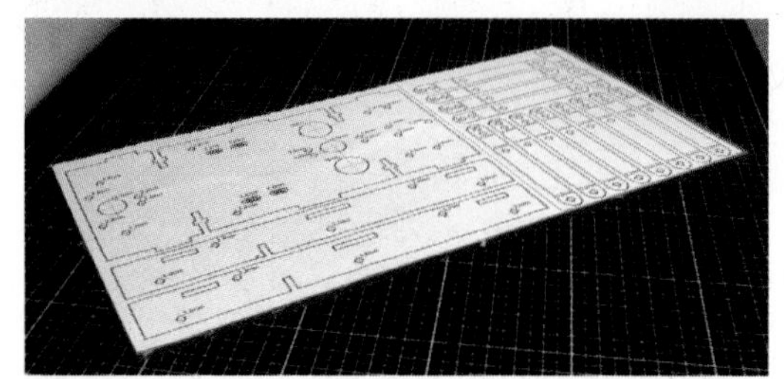

Figure E: The acrylic sheet with pattern glued and ready to cut/drill

Figure F: The pieces as they arrived from Ponoko. Burn marks are on protective sheet, not parts themselves

(41-(OAWR)-Whisker Bending Guide.pdf).

Note: While initially designing this robot, I experimented with different shapes of whiskers. The pattern I created is the one I found to work best, however it is quite interesting to experiment with different shapes. I was surprised how even small changes can drastically alter the navigational behavior of the robot.

4. Assembling

I tried to make assembling all the pieces together as straight-forward as possible. To this end, I've included a Lego-style assembly guide (51-(OAWR)-Assembly Guide.pdf).

A step before you begin:

Assemble the Tamiya twin motor gearbox (I used

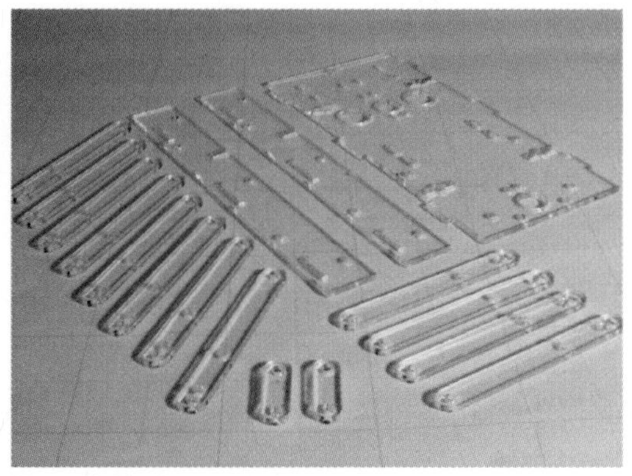

Figure G: The laser cut pieces once the protective film has been removed

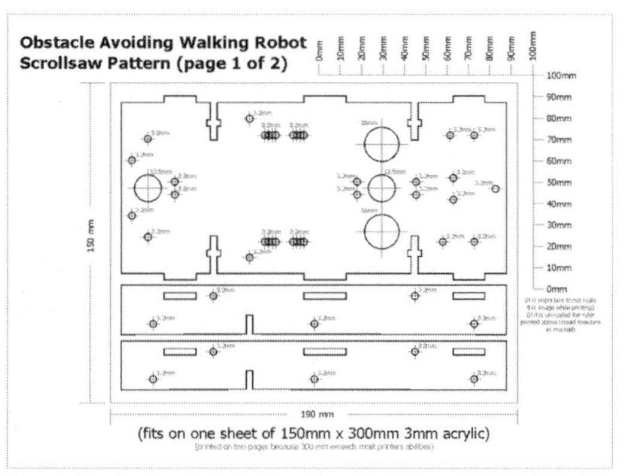

Figure H: Scroll saw pattern (and laser cutting pattern) available for download on project's Instructable page

the 58:1 ratio with the output shaft exiting at hole "A," however battery life on this setting is not great. Mounting holes have been included to allow for using the 203:1 ratio with the output shaft exiting at hole "C". If you prefer a slower, longer battery life version.)

A step after you finish:

Add shoes to the feet of your robot (the rounded acrylic feet don't grip surfaces very well). I applied a bead of hot glue to the bottom edge of each leg and performance was greatly enhanced. (But if you have access to six miniature-sized running shoes, by all means, use them.)

5. Wiring

With the big pieces all fit together and our bot starting to look pretty nice, it's time to add the copper veins that will give it life. A first look at the wiring

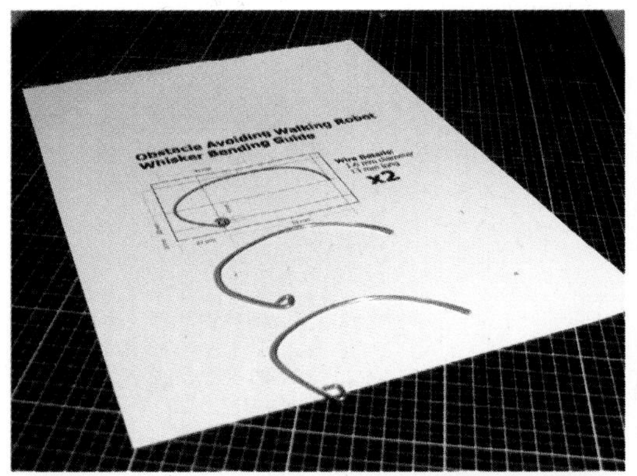

Figure I: Bending the whisker. Whisker Bending Guide available on website.

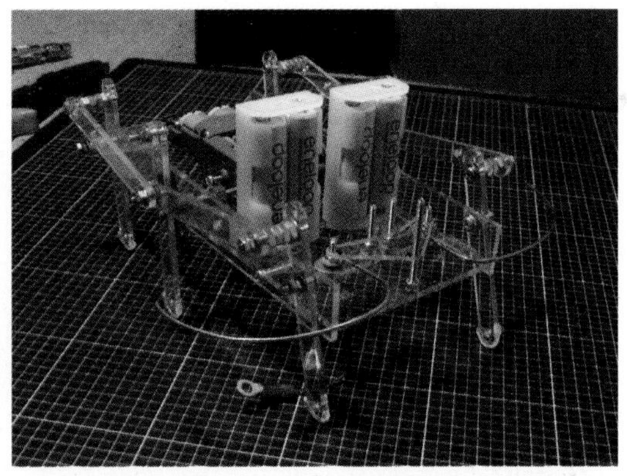

Figure J: All the parts assembled and ready for wiring.

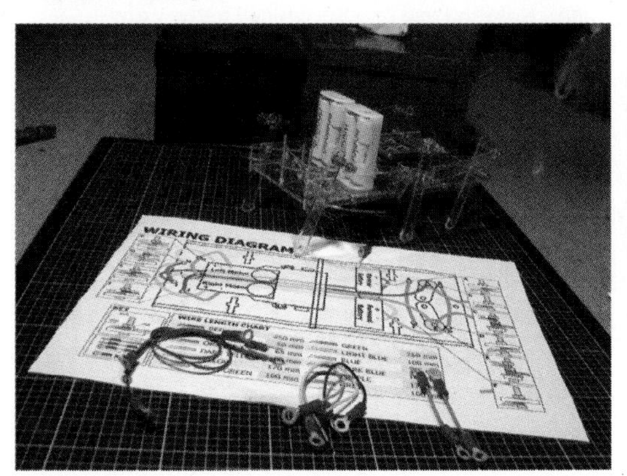

Figure K: Using the printable Wiring Diagram to easily wire up the walker

Figure L: Head-on shot of the finished walker

diagram (61-(OAWR)-Wiring Diagram.pdf) can be intimidating, but if you tackle each wire individually, it's really quite straight forward.

Four tips to (hopefully) help you out:

- Each wire-end that connects to a connection point should have a crimp wire terminal (red 4mm ring) affixed to it (there are 18 of these points).
- The exploded view linked to each connection point illustrates whether the wire is meant to attach above or below the acrylic sheet.
- Any connection point that does not already have a bolt in it uses a 3mm x 15mm bolt and a matching 3mm nut.
- Most of all, don't worry. The next step is fully devoted to troubleshooting so have a go and if it's not working properly chances are you'll find your answer there.

6. Troubleshooting

If you've made it this far and your robot is walking and avoiding obstacles, then you may skip right over this step. However, if it isn't quite working, or is not working at all, check out the Troubleshooting section on this Instructable's webpage and also check the Comments section where other builders share their questions and show off their successful builds.

7. Finished

Congrats! I hope you've reached this point without too much frustration and that you are happy with the results. If you have any tips or suggestions on how the design or Instructable could be improved, I'd love to hear them. Also, if you have finished, it would be lovely if you could post a photo to the Comments section or perhaps send me one so it can be added to this stage.

Clement Fletcher says he's "having a lovely time being the best Clement he can be."

Balancing Robot

A simple robot that uses a switch as a sensor and stands on only two wheels

By Mohammad Yousefi

Figures A-C: Three views of our ingenious Balancing Robot. Note the use of a button cell battery and momentary switch as the balance sensor.

This is a simple robot that uses a basic switch as a sensor and stands on two wheels by means of an inverted pendulum mechanism. When the robot is going to fall, the motor turns on and moves the bot in the direction of the fall. The robot's motor generates restoring torque, and when the restoring torque is equal to the falling torque, the robot becomes balanced.

1. You will need the following parts and tools

- A small electric motor
- An axle
- Wheels (2)
- AA battery holders (2) and batteries (4)
- A button cell battery (can be a dead one)
- A SPDT (single pole double throw) switch with a metal lever (aka a momentary switch)
- A toggle switch for the on/off power switch
- Some plastic strips to make robot body parts
- Some gears (or a motor with a gearbox)
- A small nail/brad (used as an attachment pin)
- Hook-up wire
- Soldering iron and solder
- Glue/epoxy

2. Motor, gears, shaft, and wheels

In this step, you must make a drive system to move your robot. You can make it easily by adding some gears to a small DC motor, then connect it to an axle and two wheels (Figure D). You could also use a readymade motor with gearbox. For my gear train, I used a 12.16 gear ratio (32*38:10*10).

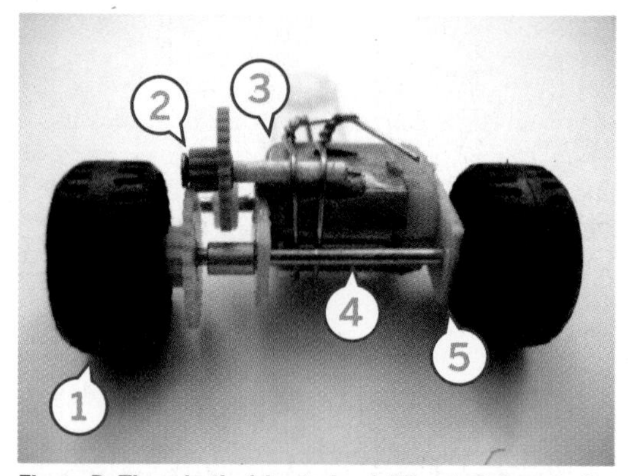

Figure D: The robot's drive train—1. Wheels 2. Gear train 3. Small DC motor 4. Axle 5. Plastic axle brackets glued to motor

Figure E: Attaching the robot's "neck" to the drive train with glue and wire

For additional information, discussion, and more, please visit the Instructables project page:

Figure F: Attaching the two AA battery holders, with tape and glue, to either side of the plastic "neck"

Figure G: The basic bot assembly. The tape is used only to hold parts in place while glue is drying.

Figure H: The momentary switch with button cell soldered to lever to use as touch-trigger

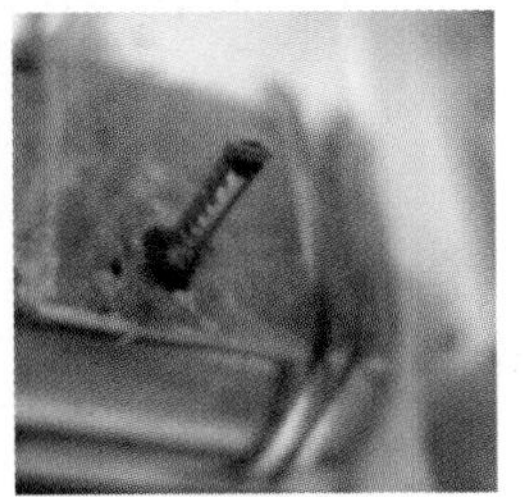

Figure I: Heated nail pressed into plastic to make mounting-pin for sensor switch

Figure J: Sensor switch installed

Figure K: Wiring the battery packs together (positive to negative)

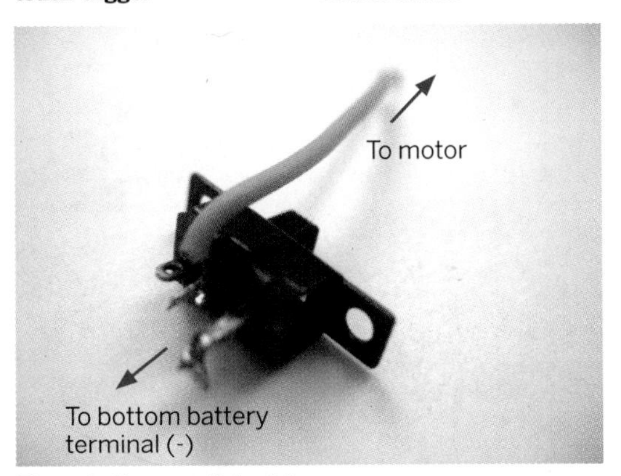

Figure L: Power switch wiring

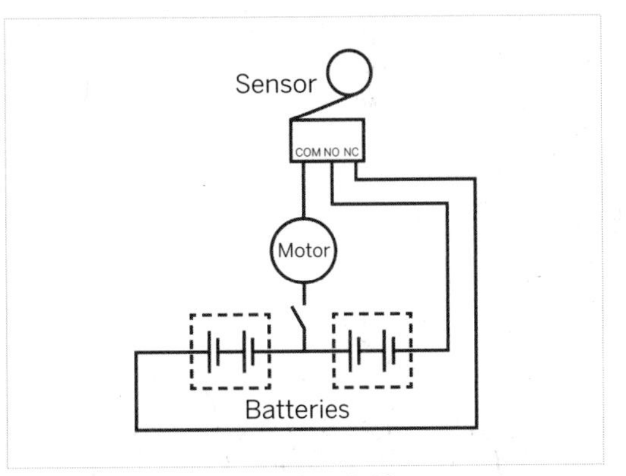

Figure M: Balancing Robot Circuit Diagram

3. Attach the robot's neck and head

Use glue to attach a strip of plastic (approx. 2-3/4" long) to the motor. Then put some glue on one side of the battery holders and attach them so that they are flush with the top of the plastic material. The battery holders act as a "head" for the robot.

4. Making the sensor

Solder a button cell battery to the SPDT momentary switch. This is the kind of switch you can scavenge from an old computer mouse (Figure H). The battery is only used as a weight/trigger, so it can be a dead one. Heat the nail head over a flame until it's hot and press it into the plastic sheet glued to the

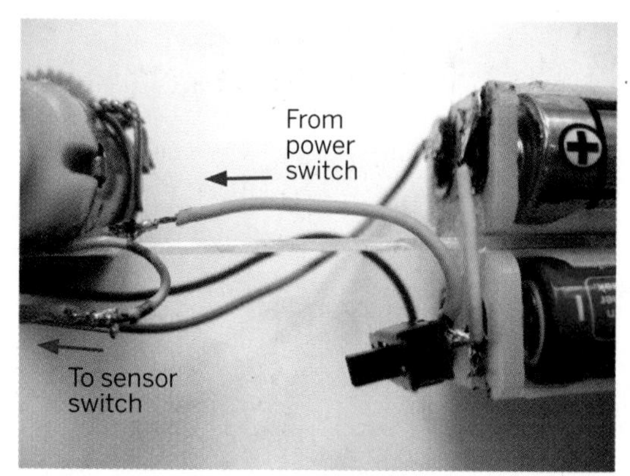

Figure N: Wiring arrangement (one side)

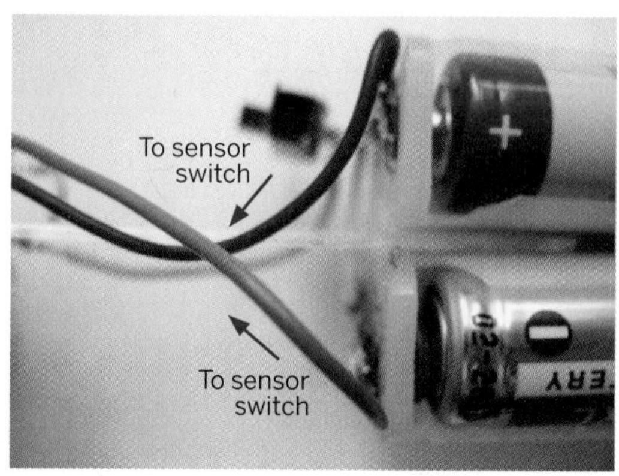

Figure O: Wiring arrangement (other side)

Figure P: Another view of the finished bot

motor in a position so that, when the robot stands vertically, the button cell will touch the ground (when the sensor switch assembly is attached to this nail pin you've made). Cut any excess length on the nail/pin (Figure I). Then attach the sensor switch to the pin with glue (Figure J).

5. Connecting the switch

Solder a wire from the positive pole of one of the battery holders to the negative pole of the other holder and attach the toggle switch to it (Figure K). Next, attach the other side of the switch to the motor (Figure L).

6. Wiring

Now it's time to solder up the robot's circuit wires. Note that you must solder the wires in such a way that the robot moves toward the direction where it naturally wants to fall. The photos of the wiring may be a bit confusing, but if you follow the circuit diagram, you should be okay.

7. Testing

The robot is now completed and it's time to test it. Put four batteries into the holders and turn on the switch. Adjust the position of the sensor switch so that the robot balances better. If the robot operates in reverse (moves away from the direction it wants to fall), swap the wires on the sensor or on the battery holders to reverse the rotation of the motor.

Note: A video of this robot in action is available on the project page for this Instructable.

Mohammad Yousefi was born in Shahroud, Iran, and is studying mechanical engineering. He likes making things and is interested in robotics, computer programming, electronics, mechanical design, and aeronautics.

Give the Gift of Robot Invasion

These adorable little bots are easy to build and make great gifts By Seth Munki

Figure A: Merry Xmas!

Figure B: ET phone home

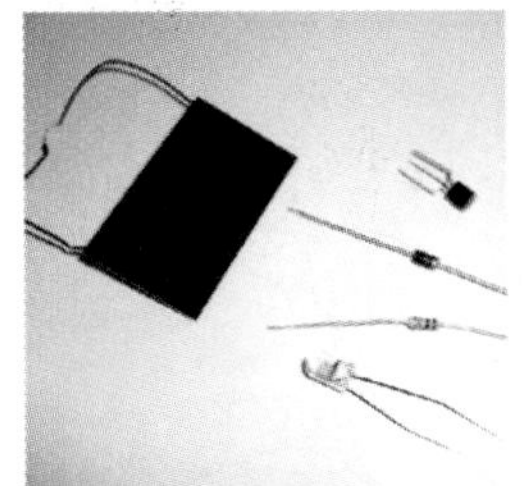

Figure C: The electronics parts you need

Figure D: The parts to make the body

This solar-powered robot ornament wishes you holiday greetings during the day, but when the lights go out, it radios home for reinforcements!

1. What you need

Electronics:

- A 2N3904 transistor
- A rectifier diode
- A 1M resistor
- A blinking LED
- Rechargeable 1.5V batteries (2)
- A solar panel

Miscellaneous Parts:

- Copper plumbing parts
- Grommets
- Rivets
- Brass and copper wire
- LED holder
- Copper mesh or beer can
- Spring
- Epoxy
- Hot glue

2. Battery

I used two cells from a rechargeable 9V for some of the bots; for others, I used rechargeable button cells. Be careful when soldering to batteries.

3. Circuit

I found the circuit I used here: http://tinyurl.com/6ayyxf.

I really like the simplicity of this circuit.

4. Robot shell

I decided to use copper plumbing parts because I had them on hand, but this part could really be anything that has enough room to put the electronics in or on.

Drill holes in the end cap for the eyes and in the

Figure E: Batteries taken from a rechargeable 9V

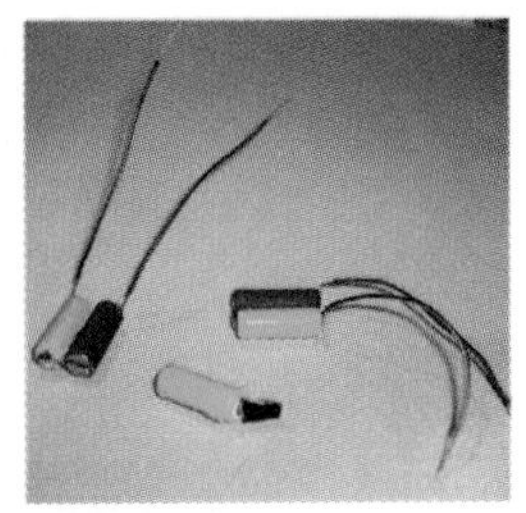

Figure F: The batteries wired up

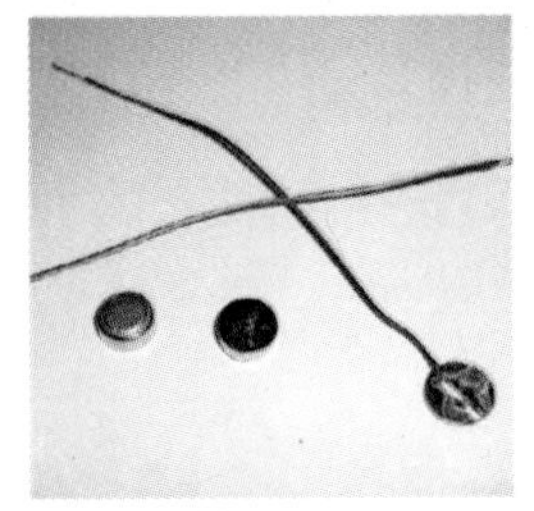

Figure G: You can use button cells, too

Figure H: Button cells taped together

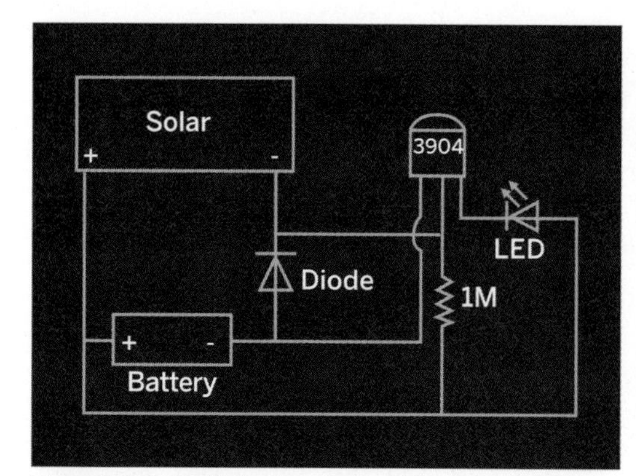

Figure I: Schematic for the simple "dark activated" circuit

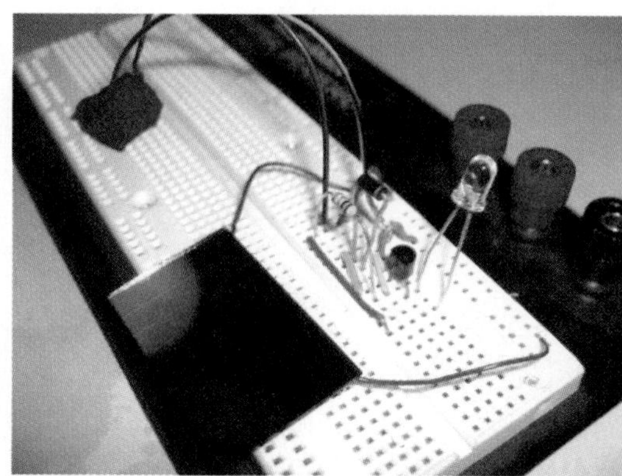

Figure J: Breadboard the circuit

Figure K: Drilling holes for the eyes

Figure L: Little robot face and body emerges

Figure M: Use rivets and grommets for eyes, armholes, and belly buttons

Figure N: Fitting the arms into the body

body for arms, belly button, satellite pole, and wires. I use a hand drill and go slowly, starting with a small bit and stepping up bit size incrementally.

I made the mouth with a Dremel and a cutoff wheel, but a hacksaw would work well, too.

If you're feeling fancy and/or have them on hand, finish the holes with aluminum grommets. On some I just used rivets for eyes; on others it's a grommet and rivet mix.

5. Soldering arms

I used a small hobby torch to solder the wire arms onto the body. You will have a much stronger joint if you spend some time fitting the wire so it makes good contact with the inside of the body. I also sand both pieces right before I solder them.

I went with a four-finger-hand approach because it was easier to use small Vs of wire.

I like to pretty much get the arms and fingers into position at this point; if you didn't do a very good job soldering, they will break off. It's easier to fix them now, before the electronics are inside.

6. Satellite

This could be anything that kind of resembles a dish. I happened to have some fancy copper mesh/fabric, so I went with that. If this part seems like a pain, you can cut out the bottom of a beer can and use it as is.

If you have the fancy copper, you will still need a beer can. Turn the can over and gently work the copper mesh into the bottom; I used the handle of a screwdriver. Once you have it shaped, cut the excess off with scissors.

Punch the hole in the center after you form the dish; if you do it before, the hole will warp and spread.

I also solder some wires onto the LED at this point. The LED holders I used have a rubber cork that has to be on the LED before soldering.

7. Get ready

Turn the body over and put a piece of tape at the bottom of the neck; this is important because at the end we will be pouring epoxy into the neck.

Figure O: Soldering the arms inside the body

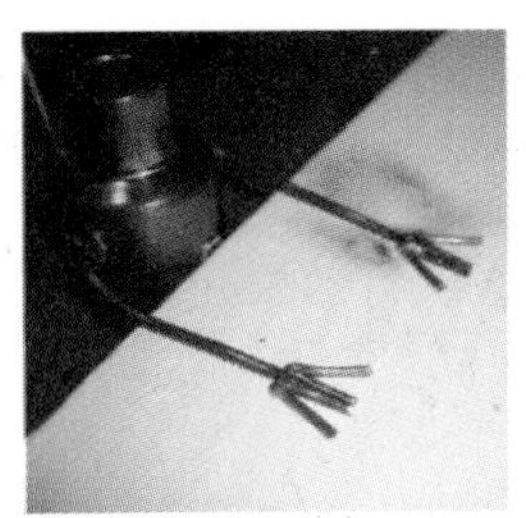

Figure P: Wee little robot hands!

Figure Q: Shape hands so they're ready to hold the solar panel

Figure R: Support wire for future satellite tower

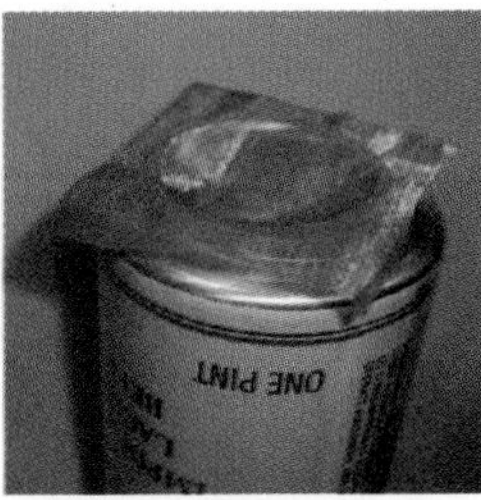

Figure S: Shaping the satellite dish

Figure T: Cut the center hole after you shape the dish

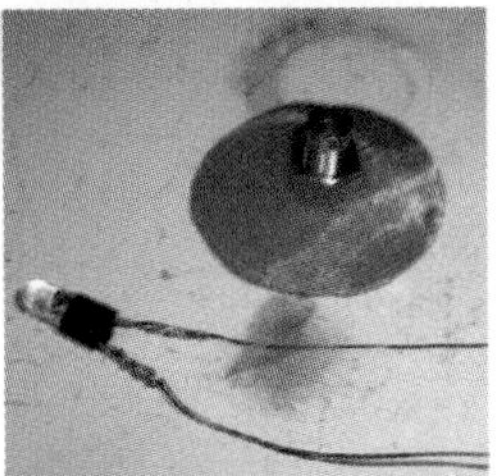

Figure U: Solder wires onto the LED

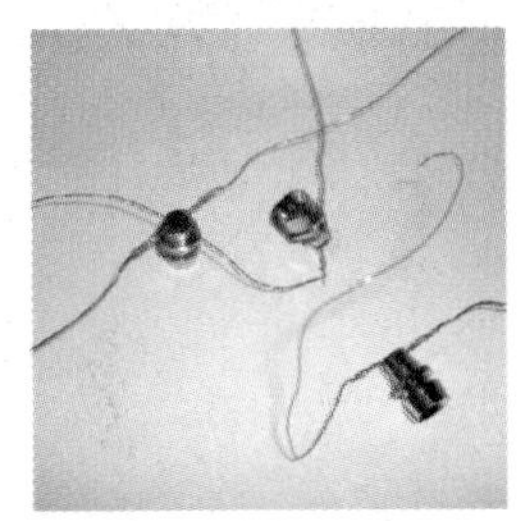

Figure V: LED holders, wired and ready to install

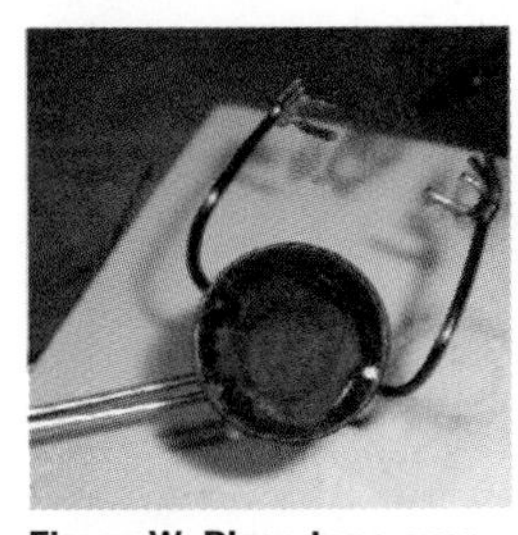

Figure W: Place tape over the neck hole in the body

Figure X: Starting to wire things up

Figure Y: Solar cell leads go through the belly button

Figure Z: 9V cells are like a backpack

When I used the 9V cells, I kept the batteries on the outside like a backpack; with the button cells, I just stuffed them inside. Either way we need to get those leads ready, either through a hole in the neck if they will be on the outside, or out on the desk if they will be inside.

The LED wires need to be strung through into the interior. The leads for the solar cell need to be put through the belly button; better too long than too short.

8. Solder the circuit

Note: Diode terminology is confusing; I'm going to call the end of the diode with the white stripe the negative (-) terminal. Looking at these pictures, it may be hard to figure out what attaches to what, but it should be clearer when you're holding the components in your hands. Solder the components together (Figures ZB–ZE).

STEP 1:
+ Solar
+ Battery
Resistor
+ LED

STEP 2:
PIN 1
- Battery
+ Diode

STEP 3:
PIN 2
- Solar
- Diode

STEP 4:
PIN 3
- LED
Resistor

Figure ZA: Button cells tucked inside

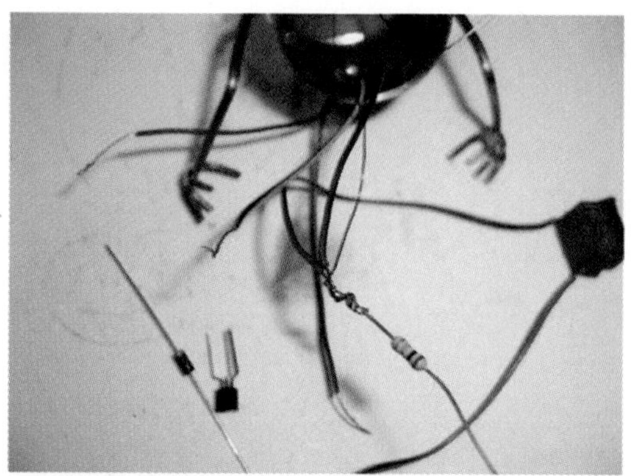

Figure ZB: Soldering STEP 1

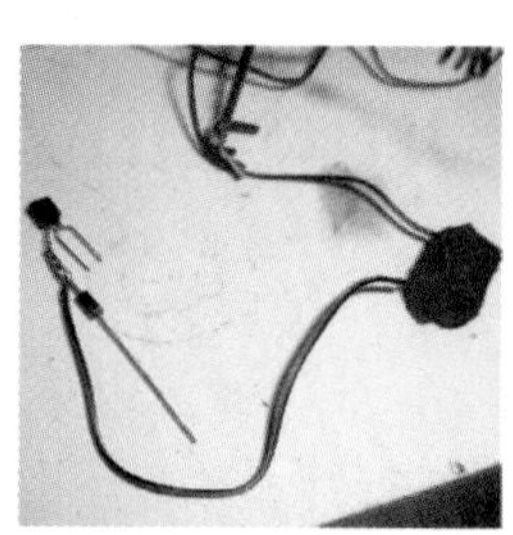

Figure ZC: Soldering STEP 2

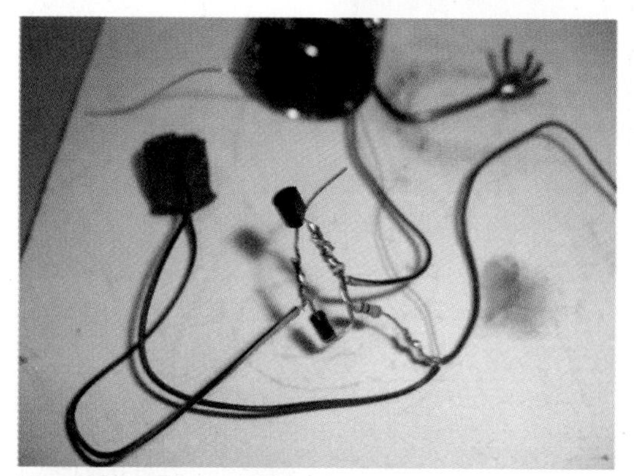

Figure ZD: Soldering STEP 3

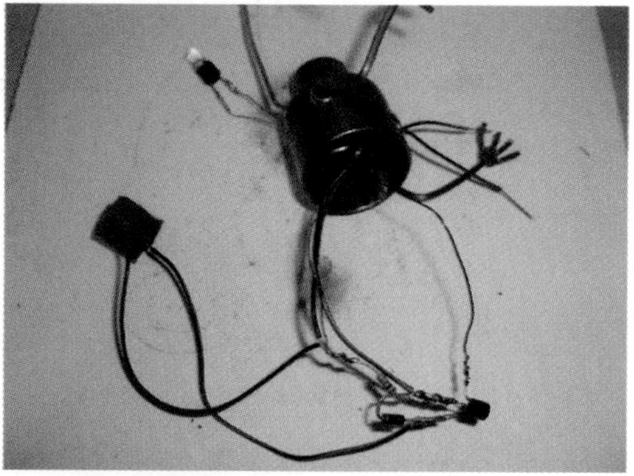

Figure ZE: Soldering STEP 4

Figure ZF: Position parts on tape so they don't touch

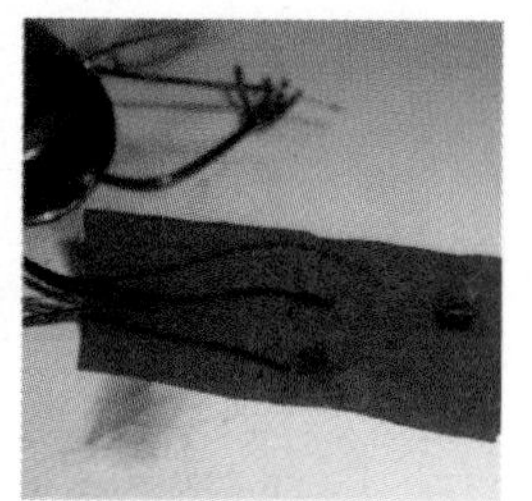

Figure ZG: Secure with more tape

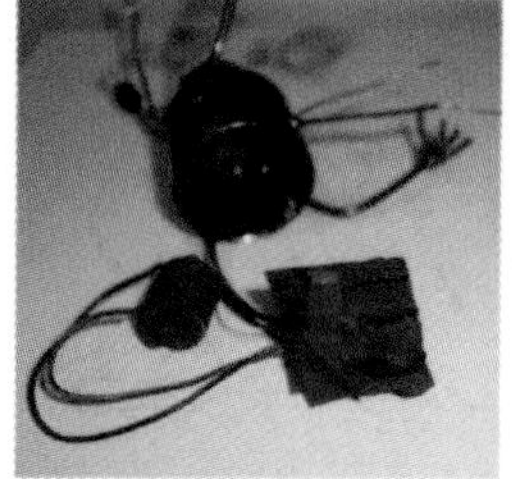

Figure ZH: Fold it so it will fit inside the body

Figure ZI: Solder on the solar panel

After you have it soldered together, cover the solar panel and make sure the LED starts to blink. Tape everything up so the leads don't touch, and fold it into a shape that is going to fit inside.

9. Attach solar panel

Solder the solar panel to the leads. Carefully position the panel into the hands. Flip over and hot glue the panel in place. I also use hot glue to hold the electronics inside.

Attach satellite dish to pole and position at a rakish angle.

10. Put your head on

Use epoxy to glue the spring inside the head. After it cures mix another batch of epoxy and pour into the neck; hold the head in place until the epoxy cures. This creates a nice "bobblehead" effect.

Figure ZJ: Hold with hot glue

Figure ZK: Add the satellite dish and position it

Figure ZL: Put the LED in the holder on the dish

Figure ZM: Mix up your epoxy

Figure ZN: Epoxy the spring inside the head

Figure ZO: Adding more epoxy into the neck

Figure ZP: Your legion of robots is almost ready to do your bidding!

Figure ZQ: Write the message of your choice on the solar panel

Figure ZR: Leave most of the solar panel exposed

Figure ZS: Packaged for Christmas

User Notes

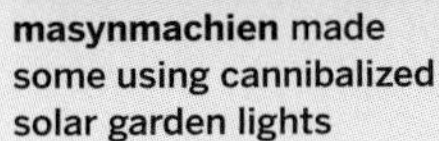

masynmachien made some using cannibalized solar garden lights

Mutant solar-powered garden light robots

11. Write your message

If you're anal, or just like to make life easier for yourself, you should probably do this step before you attach the solar panel.

I used a semi-fine point silver paint pen. Leave most of the solar panel exposed so the batteries can charge.

12. Start the invasion

Box these up and give them away. I packaged mine a couple of days before Christmas and they were still blinking when they were unwrapped.

You can play with the resistor value to determine how sensitive the solar panel is. With the 1M resistor it has to be pretty dark before the LED starts blinking. With a 100k resistor it will blink in a lit room, but turn off in direct sunlight. It all depends on how sneaky you want your robot to be.

Seth Munki lives in Oakland, California, with his insanely talented wife Maggie, and is a founding member of "The Dragon Club."

Mousebot Revisited

A crazed light-seeking robot you can make from an old computer mouse By Jacob McKenzie

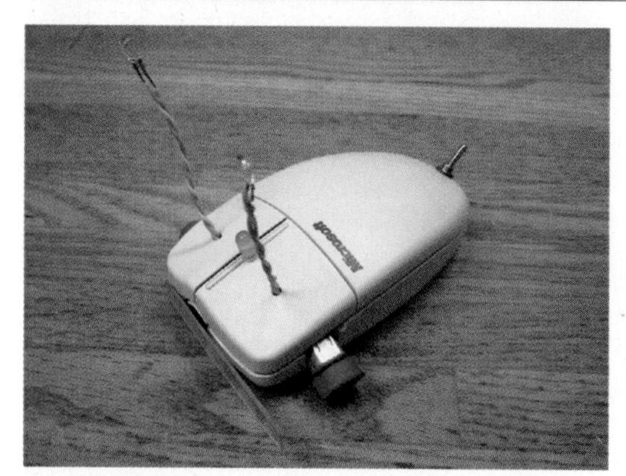

Figure A: Mousey the Junkbot, ready to terrify the cat!

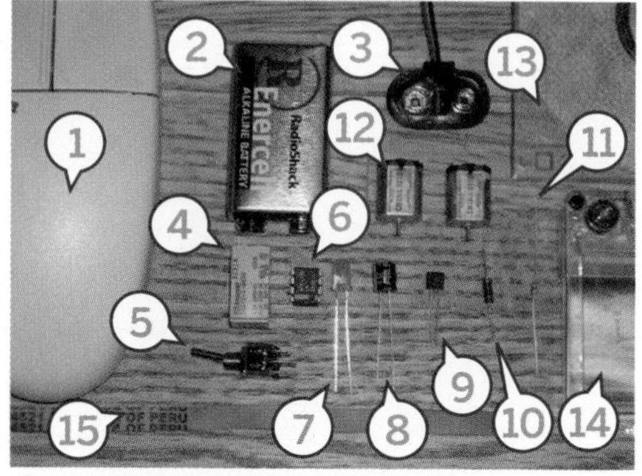

Figure B: The materials list—1. Old computer mouse (for case, switch, and infrared emitters) 2. 9v battery 3. Battery snap 4. 5V relay 5. Toggle switch 6. LM386 chip 7. LED 8. 100uF capacitor 9. 2N3904/PN2222 NPN transistor 10. 1KΩ resistor 11. 10Ω resistor 12. Two small DC motors 13. Bumper plastic material 14. Cassette tape wheels 15. Rubber band for tires

Gareth Branwyn's "Mousey the Junkbot," from MAKE Volume 02, is a fun introduction to robotics. So fun, that I've created this expanded documentation of a Mousey build from start to finish, with a few extra tips and tricks you won't find in the magazine project. This how-to is best understood after reading the original article, found on page 100 of Volume 02 (or in *The Best of MAKE* collection), however it is probably not required.

Mousey is a simple bot that uses two "eyes" to sense light and then turns towards the light it detects. A single large "whisker" is mounted on the front to detect collisions. A collision with a wall will cause the bot to reverse and turn, then take off in another direction. This project is pretty cheap if you have a mouse to use, the other parts can be obtained for less than ten dollars.

1. Gather the materials and tools you'll need

Materials:

- An analog mouse
- Small DC motors (2)
- A toggle switch
- A DPDT 5v relay (Aromat DS2YE-S-DC5V works)
- An LM386 op-amp
- A 2N3904 or PN2222 NPN transistor
- An LED (any color)
- A 1KΩ resistor
- A 10KΩ resistor
- A 100uF Electrolytic capacitor
- An audio cassette tape (you know, from the '80s)
- A CD-ROM or floppy disk (for the bumper)
- A 9V battery snap
- A 9V battery
- Wide rubber bands (2 or 3)
- 22 or 24 gauge wire (some stranded, some solid core)

Tools:

- Multimeter
- Phillips screwdriver
- Dremel
- Small pliers
- Wire cutter/stripper
- Razor knife
- Soldering iron
- Desoldering tool of choice
- Superglue or epoxy

For additional information, discussion, and more, please visit the Instructables project page:

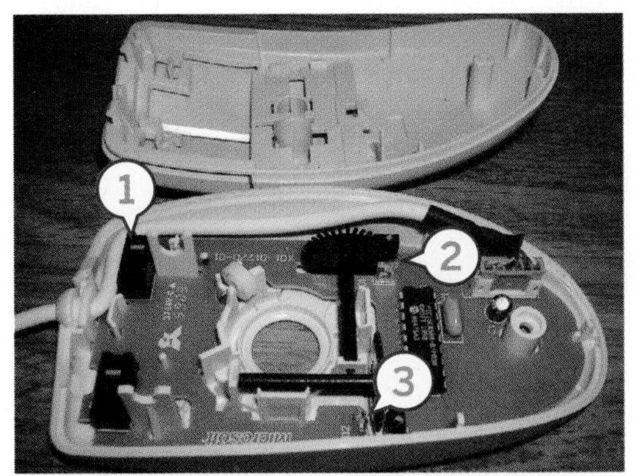

Figure C: The components scavenged from a mouse: 1. Momentary switch 2. & 3. Infrared emitters

Figure D: An infrared emitter close-up

Figure E: The mouse case before the carnage begins

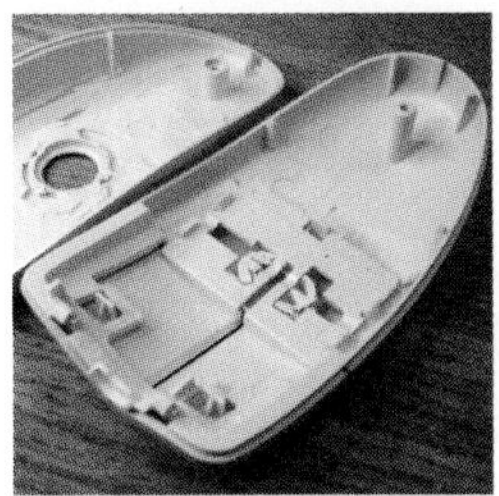
Figure F: The bottom case with everything removed except the screw post

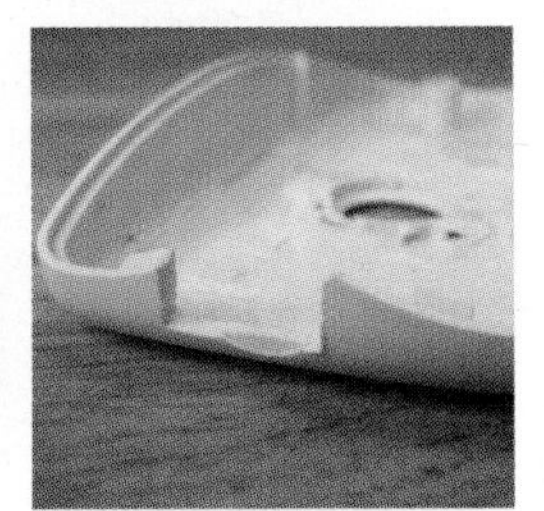
Figure G: The notch cut for the bumper switch

Figure H: The bumper switch installed

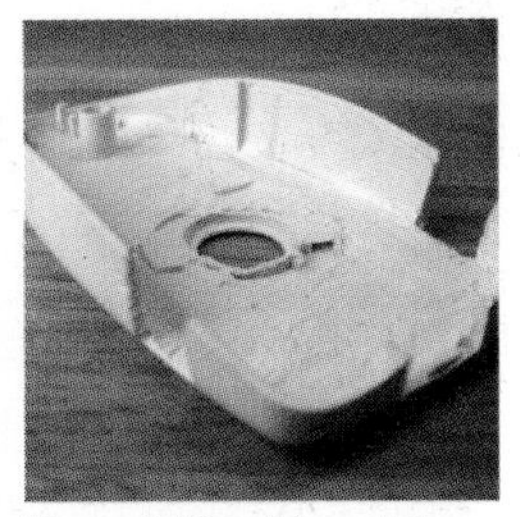
Figure I: The side notches for the motors cut

- Hot glue gun and glue
- Hacksaw

2. Scavenge some parts

Mousey requires several parts that we can conveniently borrow from a donor mouse, its eyes and its whisker.

Open up the mouse and locate the components that we'll be harvesting, the momentary switch and the infrared emitters. The emitters are the components in the clear package (Figure C).

Remove the PCB and desolder the push switch and both infrared emitters.

3. Prepare the case

Next we need to give Mousey's insides a cozy place to live, so break out your Dremel and remove all of the internal plastic structure from the top and bottom of the mouse. If your mouse is small, you may have to remove the screw posts that hold the mouse together. Now use the Dremel to cut openings for the bump switch in the front of the mouse and motors on the sides.

The best Dremel bit to use for this is the short cylindrical one, it will cut a good right angle if the Dremel is held vertically.

4. Make the wheels

The axles on these motors are pretty small and if we want Mousey to be stable at the blazing mouse speeds he travels, we need to make some rims. Cassette tapes have a rim that's the perfect size in the bottom right and left corners (if the open side is down). It might take a couple tries opening different brands of tapes before you get ones that fit your axles perfectly (Figure J). Once you find some rims you're happy with, super glue them to the axles (Figure K).

Cut the rubber band and super glue it to the rim then wrap it around three times, adding super glue every half-turn or so to keep it together (Figure L). Cut off the leftover rubber.

Next, glue another rubber band back to the rubber band that you just wrapped. Complete one revolution and cut off the extra (Figure M). Don't forget to add enough glue to keep the outside

Figure J: The transport wheels from a cassette tape

Figure K: Our tape wheel hub fitted to a motor

Figure L: The "inner tube" of our wheel

Figure M: The outer tire finishes it off. Traction, baby!

Figure N: The component layout on this mouse. Mice are different, so you'll have to carefully think out where everything goes.

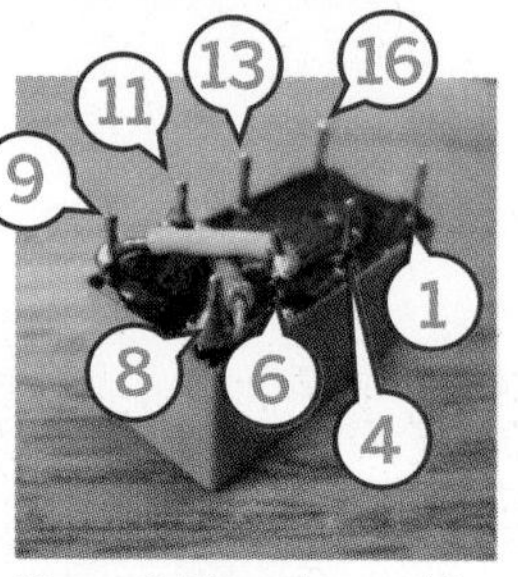

Figure O: The relay upside down with pins 8 to 11 and 6 to 9 soldered

Figure P: Pins 1 and 8 connected and stranded wire soldered to pins 8 and 9

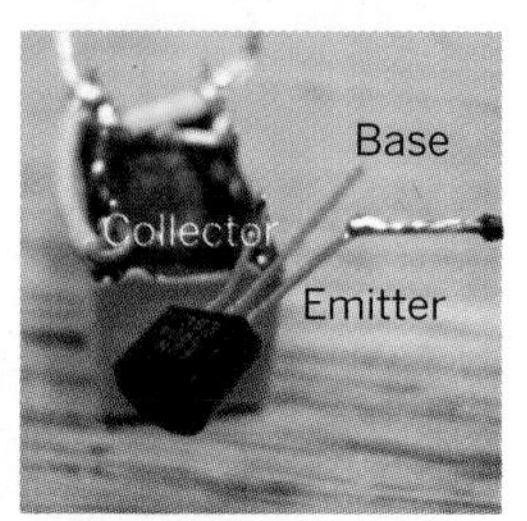

Figure Q: The transistor soldered to the relay. Note location of collector, base, and emitter pins in relation to flat side of the transistor.

rubber band on. Repeat the same process for the other wheel.

5. Layout the design and install the relay

There are quite a few good component layouts for Mousey. The best layout is probably the one pictured on the top of page 100 of MAKE Volume 02. However, this alternate setup pictured in Figure N works better on certain mice. I chose to use the standard layout. The circuitry of the mouse will be free-formed since there is not much extra room for a PCB.

Note: Max Headroom! Even with the tightest wire management, this little bot has a lot of wires inside. Be mindful while you're building it that you have to be able to fit the top onto it when you're done. Keep your wire runs as short as humanly possible.

Once we know where everything will go, it's time to get to the real work. Set down the relay and solder wires in an "X" connecting pins 8 to 11 and 6 to 9 (see the pin labels in Figure O).

Connect pins 1 and 8 with a wire along the side and add stranded wire to leads 8 and 9 (Figure P).

Solder the collector of the transistor (right pin looking at the flat side) to pin 16 and clip the lead short. Then connect the wire we soldered to pin 9 to the emitter (left pin looking at the flat side) leaving a little bit of slack (Figure Q).

Now glue the relay into the mouse case. I have added two clipped leads to act as positive and negative voltage rails that will get rid of some clutter in the motor area. Use your razor knife to strip the shielding from the wire connecting pin 9 and the emitter and solder it to the negative (-) voltage rail. Then connect pin 8 to the positive (+) voltage rail.

Note: In Figure R (and Figure T), ignore the blue wire. It was going to be the one connected to the transistor, but I replaced it with the black wire you see in the photo. Also note the power rails marked (+) and (-).

6. Install the bump switch

Now let's give the Mousebot his whisker. Make it by soldering the positive lead of the capacitor and the 10kΩ resistor to the end pin on the switch that is Normally Open (NO). You can check which side is the open side of the push switch by using the Continuity Check feature of your multimeter. There should be no connection between the middle and Normally Open pin until the switch is pressed. Once

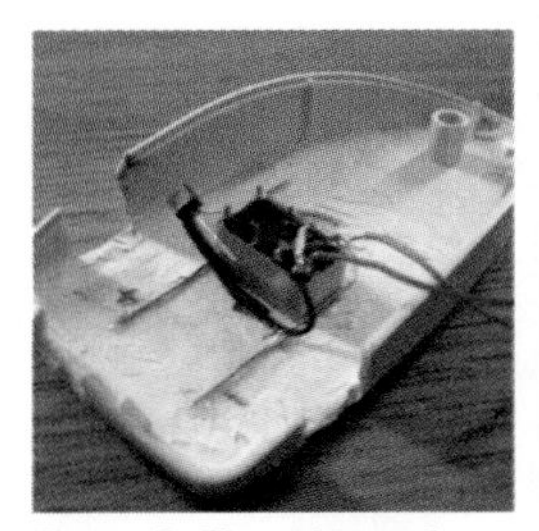
Figure R: The relay and transistor glued in place and our "power rails" installed

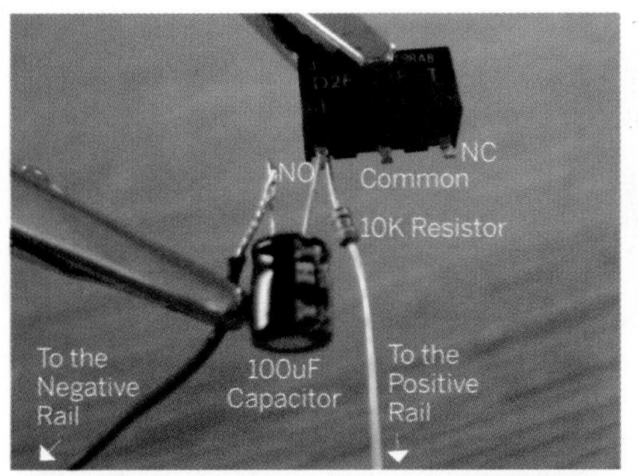

Figure S: Resistor/capacitor circuit attached to bump switch

Figure T: This mouse does not have enough room in the back so the chip should go in the front where there is more free real estate.

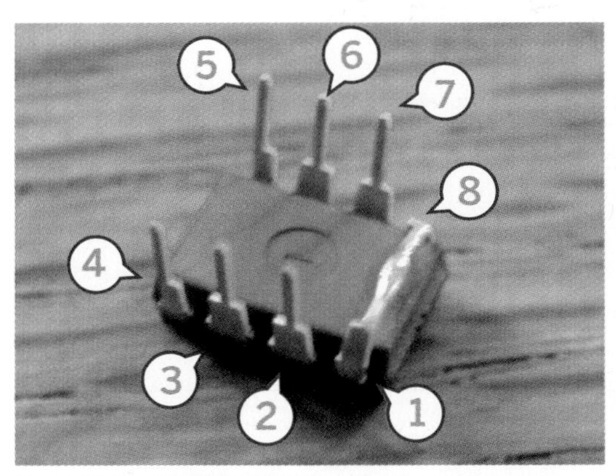

Figure U: The LM386 with pins 1 and 8 connected

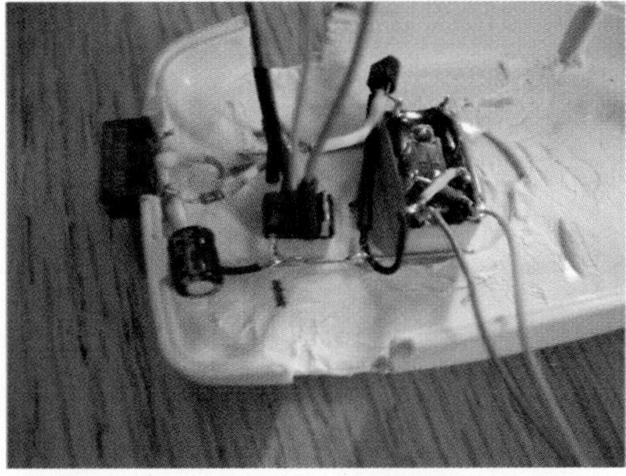
Figure V: Pin 4 of the 386 connected to the (-) rail, pin 6 to (+) and stranded wire added to pins 2, 3, 5

this is done, add stranded wire to the ground lead (-) of the capacitor and the middle pin of the switch.

Connect the resistor on the switch to the base (middle pin) of the transistor and the wire from the ground side (-) of the capacitor to the (-) voltage rail. Then connect the middle pin to the (+) voltage rail. To make your joints a bit more secure, you can use heat-shrink tubing to insulate the connections and bend the capacitor to the side to free up a little more space (Figure T).

7. Build Mousey's brain

Mousey's brain is the LM386 op-amp chip, flip it onto its back (pins up) and bend pins 1 and 8 so that they are touching and add some solder to connect them (Figure U).

Note: There are much larger versions of all these images on the Instructables page for this project. On the webpage, just click on the [i] symbol in the upper left corner of each image to view an enlargement.

Now place the LM386 into position and connect pin 4 to the (-) rail, pin 6 to the (+) rail, and add stranded wire to pins 2, 3, and 5 (Figure V).

We are almost ready to connect the motors, so solder some stranded wire to pins 4 and 13 of the relay. At this point your mousebot should look something like the one in Figure W.

8. Construct Mousey's top half

First drill three small holes at the front of the mouse for the two eyestalks and sensitivity-boosting LED (Figure X). Then drill a larger hole for your toggle switch at the back of the mouse and install the switch to form the bot's on/off tail switch (Figure Y).

To create Mousey's eyestalks, twist two pieces of solid-core wire together and solder the infrared emitter to the leads on one end (Figure Z). Place the

Figure W: Most of the bottom half of the bot is in place and wired up

LED in the middle hole and connect the (+) lead to the 1kΩ resistor (Figure ZA).

Next use the Diode Check feature of your multimeter to find the (-) leads of the infrared emitters and connect them to the (-) lead of the LED.

9. Glue down the components

Use hot glue or epoxy to secure the bump switch and the motors to the mouse chassis. I used a combination of Super Glue and hot glue to hold in the bump switch and hot glue on the motors. Make sure the angle of the motors are roughly equal and extend down far enough to raise the front of the mouse slightly off of the ground.

10. Finish up the connections

Connect pin 13 of the relay to the left motor and pin 4 of the relay to the right motor. Now connect pin 5 of the IC (brown wire in Figure ZF) to the ground node of both motors. If you aren't sure which side is (+) and which is (-) on your motor, connect it to a battery and observe the direction of the spin. The right motor should spin clockwise if you are looking at the wheel and the left should spin counter-clockwise.

Locate the wire coming from IC pin 2 (green) to the (+) lead of the left eyestalk and IC pin 3 (blue) to the (+) lead of the right eyestalk. Then wire the 1kΩ resistor to the (+) voltage rail.

Hook up the battery by soldering the black wire (-) on the battery cap to the negative voltage rail. Connect the red wire (+) on the battery cap to the switch and then connect the switch back to the (+) voltage rail.

Replace the cover of the mouse and cut a thin strip of your bumper material with a saw or knife.

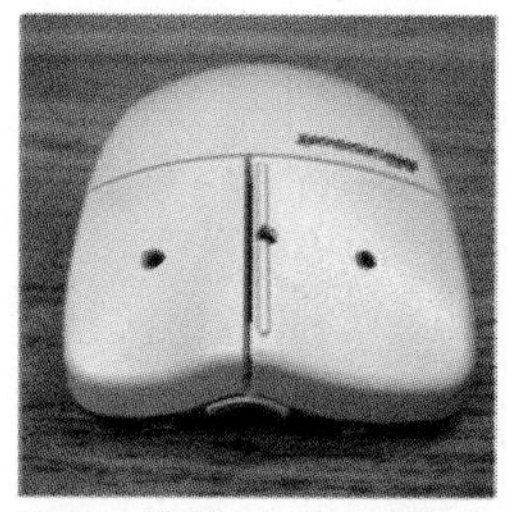

Figure X: The holes drilled for the eyestalks and the LED lamp

Figure Y: The hole drilled for the tail toggle switch

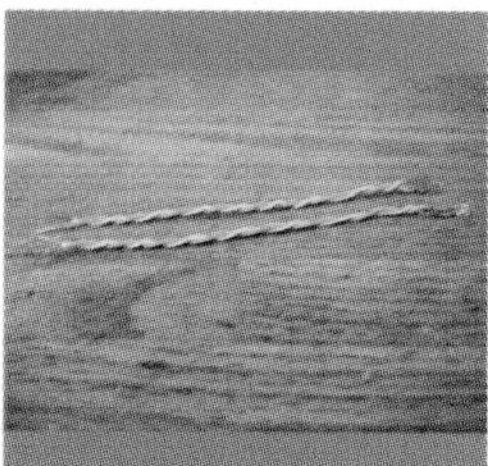

Figure Z: The two infrared eyestalks

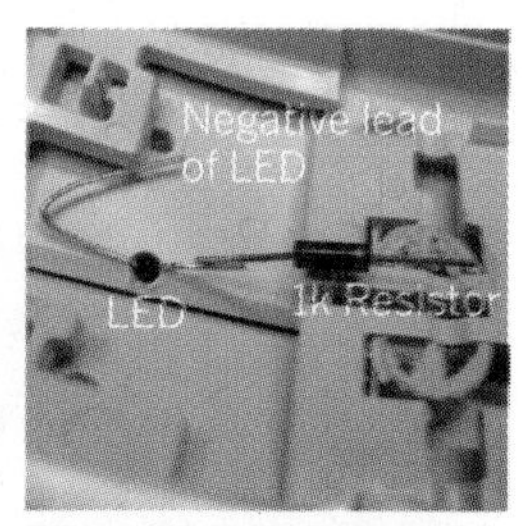

Figure ZA: Installation of the LED sensitivity circuit

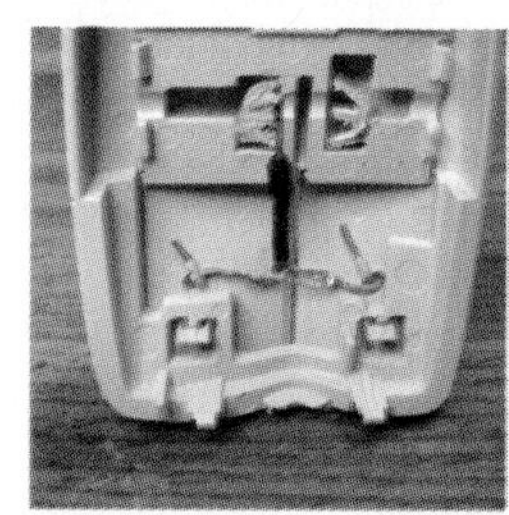

Figure ZB: The eyestalks installed and connected to the LED

Figure ZC: The top half of Mousey completed

Figure ZD: The bump/ whisker switch glued in place

Figure ZE: The DC motors glued in place

Attach the strip with epoxy or hot glue on one side so that wherever you apply pressure the button clicks. Once you have the strip attached, give yourself a pat on the back, you're done!

Flip the switch and enjoy your crazy mousebot!

Jacob McKenzie is a mechanical engineering student at UC Berkeley and former MAKE magazine intern. He thinks bicycles and Lagrangian mechanics are both awesome.

Figure ZF: Making the final wire connections—connecting the motors

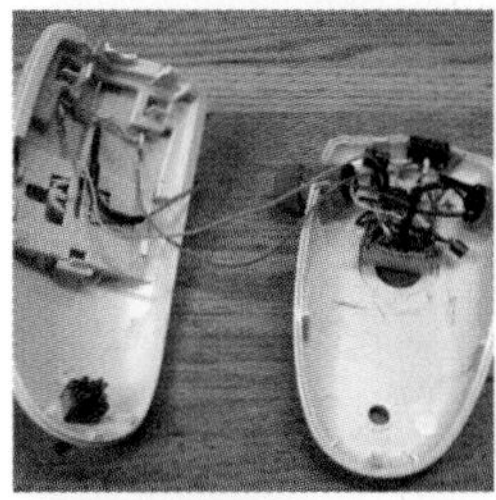

Figure ZG: Connecting the eyestalks to the LM386

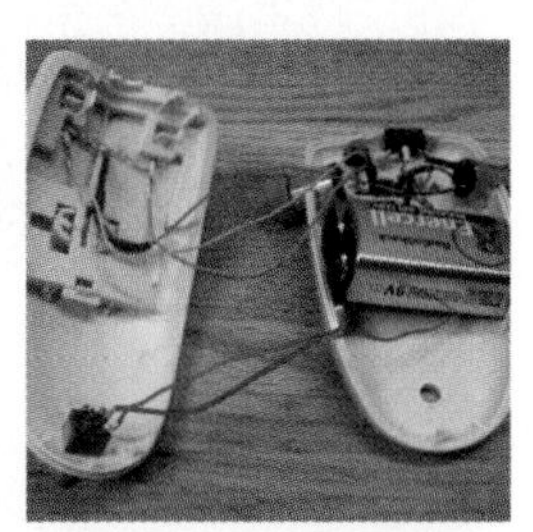

Figure ZH: Connecting the power switch to the circuit

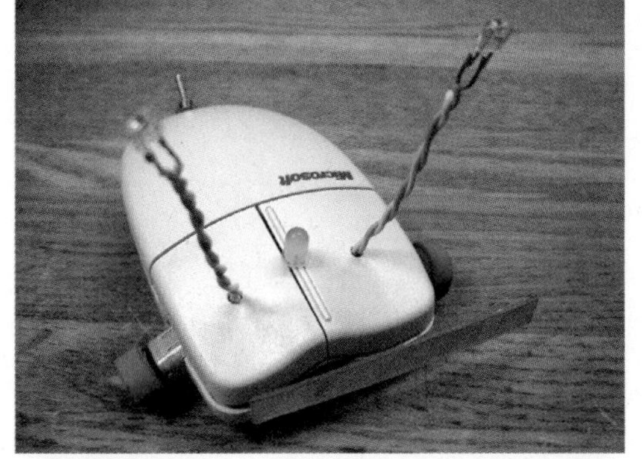

Figure ZI: Attach the bumper and you're ready for your mouse to roar

User Notes

Gareth Branwyn says: Builders of this project should be aware of the fact that the wiring for the LED light sensor sensitivity subcircuit, as detailed in the MAKE article, in my book *Absolute Beginner's Guide to Building Robots*, AND in Dave Hrynkiw's *Junkbots* book, is incorrect. You can find the corrected circuit diagram at streettech.com/robotbook/circuitMousey.html. It's only a minor change in how the circuit is wired. The bot works fine as seen here, but the eyes will be a bit more sensitive with the corrected wiring.

There are many resources related to Mousey the Junkbot that can be found, starting with the Mousey the Junkbot FAQ (http://tinyurl.com/5pv6uw).

A complete, free PDF of the original project in MAKE Volume 02 can be downloaded here: cachefly.oreilly.com/make/mousey.pdf.

Solarbotics.com, purveyors of fine robots and robot accessories, sells a Mousey the Junkbot parts bundle. It comes with everything you need to build this project. It even includes a pair of infrared optical eyes and a SPDT switch with a lever on it (for easily attaching the bumper), parts you'd normally have to get from inside the mouse. With this bundle, all you need to provide is the mouse case itself (or anything else you'd care to house your bot in).

Mousey funfact: Mousey the Junkbot was a guest, along with MAKE Editor in Chief Mark Frauenfelder, on the Colbert Report. In the midst of the segment, Stephen Colbert turned Mousey on, it tore off of the table, onto the floor, and smashed to pieces. Colbert seemed genuinely embarrassed. You can see the segment here: http://tinyurl.com/585u2f. Mark also shows off the Marshmallow Shooter also featured in this book.

The Best of Instructables Ride

My Grandma used to say "If I'd a-knowed I could a-rode, I would-a went!" A vehicle gives you the awesome power to escape. You can transform your surroundings by simply moving from one place to another. Fed up with humans hassling you? Instructables to the rescue! Humans are mostly land animals and don't seem to care too much what you do on the water. Get a free yacht (details inside) and head for the liquid horizon! Tired of working so hard and still being broke? There are Instructables on various styles of vehicle dwelling. There's plenty of free land (parking) available, even in urban areas—experience the freedom of the pioneer wagon trains today!

Remember when they "killed the electric car?" Fight those petro-retro powers! Make your own electric bike or car from plans found on the site. And, of course, bicycles! A bicycle makes a human into the most efficiently mobile animal on the planet. Instructables offers an awesome variety of bike projects.

Want your ride to be more of an expression of you? Want something *really* specific? Then, you've got to do it yourself! Companies have to be cautious and make products with mass appeal. They can't rapidly change their product lines. And their marketing departments will cover the thing with logos, brand names, and unnecessary "features." It won't be exactly what you want. You know what you want, and thanks to sites like Instructables, how to get it.

I used to build projects for my own satisfaction. When I started sharing them on Instructables, I got a *lot more* satisfaction. The best thing is when people build your project, or one inspired by it, and post it to the Comments section. You can feel the waves of goodness spreading over the world. Got a project of your own that turned out well? Or failed in a hilarious way? Share it with the group!

—Tim Anderson (TimAnderson)

contest WINNER!

Turn Signal Biking Jacket

Safety with style

By Leah Buechley

Figure A: The Turn Signal Biking Jacket

This tutorial will show you how to build a jacket with turn signals that will let people know where you're headed when you're on your bike. We'll use conductive thread and sewable electronics so your jacket will be soft, wearable, and washable when you're done. Enjoy!

A version of this tutorial is also on my website at www.makezine.com/go/turnsignaljacket.

1. Supplies

Get the following (Figure B) from SparkFun Electronics (www.sparkfun.com):

- LilyPad Arduino main board, part number DEV-08465
- LilyPad USB Link, part number DEV-08604
- Mini USB cable, part number CAB-00598
- LilyPad power supply, part number DEV-08466
- 16 LilyPad LEDs, part number DEV-08735
- 2 LilyPad switches, part number DEV-08776
- A spool of 4-ply conductive thread, part number DEV-08549
- A digital multimeter with a beeping continuity tester. I use RadioShack's 42-Range Digital Multimeter with Electric Field Detection (catalog number 22-811).
- A garment or a piece of fabric to work on
- A needle or two, a fabric marker or piece of chalk, puffy fabric paint, a bottle of fabric glue, and a ruler (available at your local fabric shop or Jo-Ann Stores: www.joann.com)
- A pair of scissors
- Double-sided tape (optional)
- A sewing machine (optional)

Note: Disclosure: I designed the LilyPad, so I'll make some $ if you buy one.

2. Design

Plan the aesthetic and electrical layout of your piece: Decide where each component is going to go and figure out how you will sew them together with as few thread crossings as possible. Make a sketch of your design that you can refer to as you work. Figure C shows the sketch for my jacket. Stitching for power (+) is shown in red, ground (-) in black, LEDs in green, and switch inputs in purple.

Important note about the power supply: As you design, plan to keep your power supply and LilyPad main board close to each other. If they are too far apart, you are likely to have problems with your LilyPad resetting or just not working at all.

Why? Conductive thread has non-trivial resistance. (The 4-ply silver-coated thread from SparkFun has about 14Ω/foot.) Depending on what modules you're using in your construction, your LilyPad can draw up to 50 milliamps (mA) of current, or .05 amps. Ohm's Law says that the voltage drop across a conductive material—the amount of voltage that you lose as electricity moves through the material—is equal to the resistance of the conductive material times the amount of current that is flowing through it.

For example, if your LilyPad is a foot away from the power supply, the total resistance of the conductive material that attaches your LilyPad to your power supply is about 28Ω. (14Ω in the conductive thread that leads from the power supply's negative terminal to the LilyPad's negative petal and 14Ω in the conductive thread that ties the positive terminals together.) So we can expect a drop of 1.4 volts

(28Ω * .05 amps). This means that while 5 volts is coming out of the power supply, the LilyPad will only be getting 3.6 volts (5 volts – 1.4 volts). Once the voltage at the LilyPad drops below about 3.3 volts, it will reset. The resistance of the traces from + on the power supply to + on the LilyPad and - on the power supply to - on the LilyPad should be at most 10Ω. Plan the distance accordingly.

If all of this was confusing, don't worry! Just keep the LilyPad and power supply close to each other in your design.

Transfer the sketch to your garment: Use chalk or some other non-permanent marker to transfer your design to the garment (Figure D). If you want, use a ruler to make sure everything is straight and symmetrical.

Use double-sided tape to temporarily attach LilyPad pieces to your garment. This will give you a good sense of what your final piece will look like. It will also keep everything in place and, as long as the tape sticks, make your sewing easier.

3. Sew your power supply and LilyPad to your jacket

Trim the leads off the back of the power supply: Get out your LilyPad power supply piece and trim the metal parts that are sticking out the back of it. Small clippers like the ones shown in the photo work well, but you can also use scissors (Figure F).

Stabilize your battery on the fabric: Generally, you want to do everything you can to keep the power supply from moving around on the fabric. I recommend gluing or sewing the battery down before starting on the rest of the project. You may

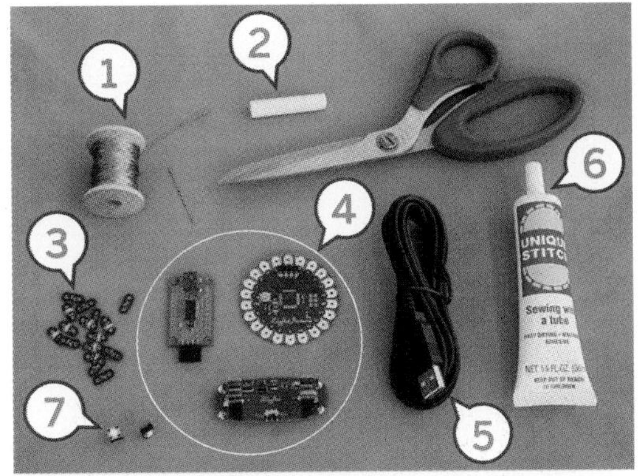

Figure B: 1. Conductive thread and needle 2. Chalk for drawing on fabric 3. LilyPad LEDs 4. LilyPad Arduino main board, power supply, and USB link 5. Mini USB cable 6. Fabric glue 7. Switches

also want to glue or sew something underneath the power supply to help prevent it from pulling on the fabric and bouncing around as you move.

If you are working on a thin or stretch piece of fabric—first of all, reconsider this choice! It's much easier to work on a heavy piece of non-stretchy fabric. If you are determined to forge ahead with a delicate fabric, choose the location for your power supply wisely. It's the heaviest electronic module, so put it somewhere where it will not distort the fabric too badly. Definitely glue or sew something underneath the power supply.

Sew the power supply's positive (+) petal to your garment: If you are new to sewing, check out the "How to Sew" Instructable (www.instructables.com/id/How-to-Sew.) before you start for info on how to thread a needle, tie knots, and make stitches. Cut a 3-4 foot length of conductive thread. Thread

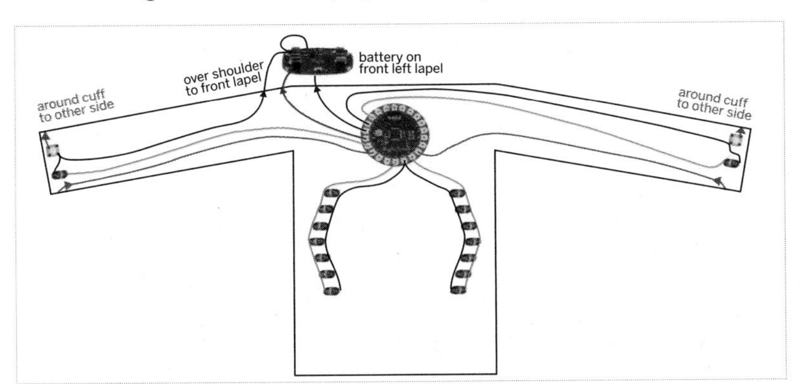

Figure C: Here's the design for what you'll be sewing

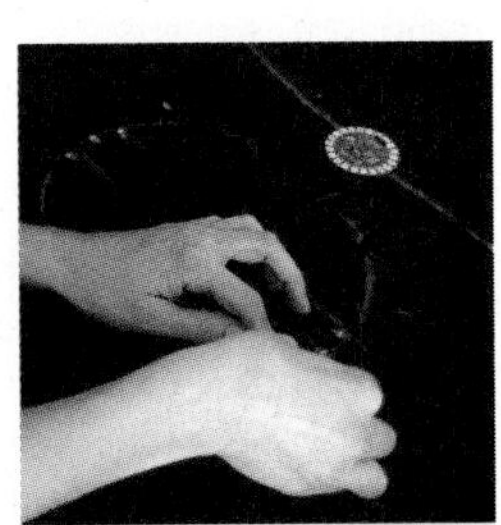

Figure D: Chalk it out

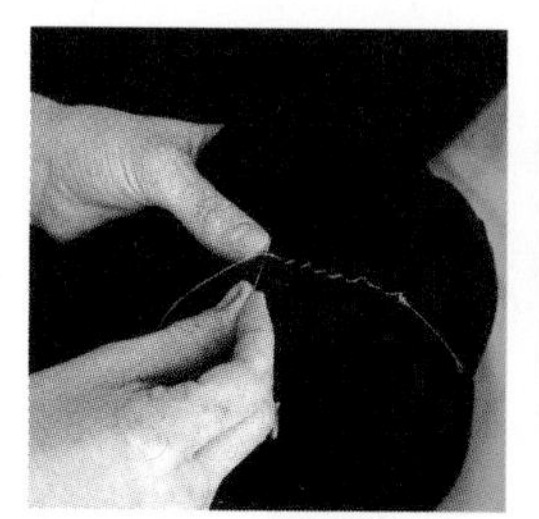

Figure E: Sewing with conductive thread

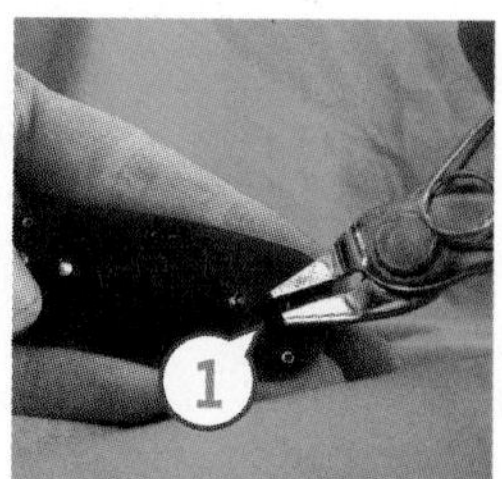

Figure F: 1. Trim the battery posts off the power supply

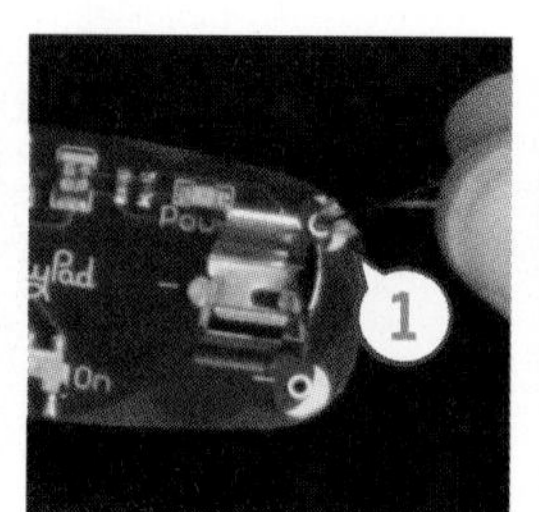

Figure G: 1. Sewing on the positive (+) petal of the power supply

Figure H: Sewing connections

your needle, pulling enough of the thread through the needle that it will not fall out easily. Tie a knot at the end of the longer length of thread. Do not cut the thread too close to the knot or it will quickly unravel (Figure G).

Coming from the back of the fabric to the front, poke the needle into the fabric right next to the + petal on the power supply and then, from the front of the fabric, pull it through. The knot at the end of the thread will keep the thread from pulling out of the fabric. Make a stitch going into the hole in the + petal on the power supply. Do this several more times, looping around from the back of the fabric to the front, going through the + petal each time.

Pay special attention to this stitching. It is the most important connection that you'll sew in your project. You want to make sure you get excellent contact between the petals on the power supply and your conductive thread. Go through the hole several times (at least 5) with your stitching. Keep sewing until you can't get your needle through anymore. Do not cut your thread, just proceed to the next step.

Sew from the battery to the LilyPad: Once you've sewn the + petal of the battery down, make small neat stitches to the + petal of your LilyPad. I used a jacket with a fleece lining and stitched only through the inner fleece lining so that no stitches were visible on the outside of the jacket.

Sew the + petal of your LilyPad down, finishing the connection: When you reach the LilyPad, sew the + petal down to the fabric with the conductive thread. Just like you were with the battery petal, you want to be extra careful to get a robust connection here. This stitching is making the electrical connection between your power supply and LilyPad.

When you are done with this attachment, sew away from the LilyPad about an inch along your stitching, tie a knot, and cut your thread about an inch away from the knot so that your knot won't come untied.

Connect the negative (-) petals: Sew the - petal from your battery to the LilyPad's - petal.

Put fabric glue on each of your knots to keep them from unraveling: Once the glue dries, trim the thread close to each knot (Figure L).

4. Test your stitching

Measure the resistance of your stitching: Get out your multimeter and put it on the resistance measuring setting. Measure from power supply + to LilyPad + and power supply - to LilyPad -. If the resistance of either of these traces is greater than 10Ω, reinforce your stitching with more conductive thread. If you're not sure how to measure resistance, check out this tutorial: http://tinyurl.com/46eooy (Figure M).

Put a AAA battery into the power supply and flip the power supply switch to the on position. The red light on the power supply should turn on. If it doesn't and you're sure you flipped the switch, quickly remove the battery and check for a short between your + and - stitches. (Most likely there is a piece of thread that's touching both the - and + stitching somewhere.) You can test for a short between + and - by using the continuity tester on your multimeter. See the tutorial at http://tinyurl.com/4l8fvq for information on how to use the continuity tester.

Also check the resistance between the + and - stitching. If the resistance is less than 10KΩ or so, you've got a mini-short (probably a fine conductive

Figure I: Connecting the petals

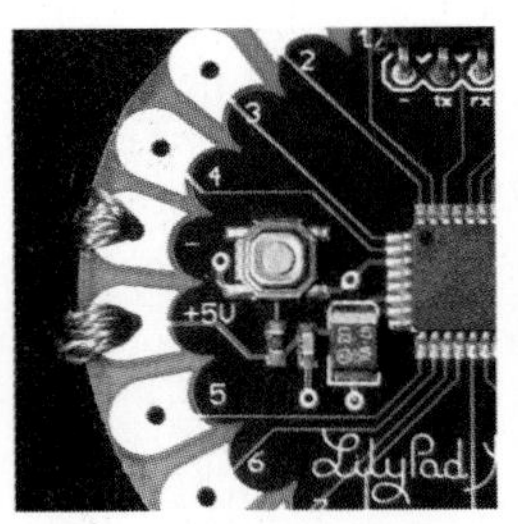

Figure J: Notice how dense my stitching is here. This is what your stitches should look like.

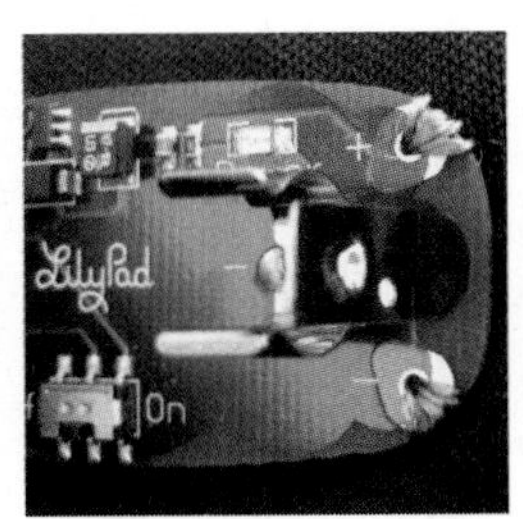

Figure K: Connecting the power supply

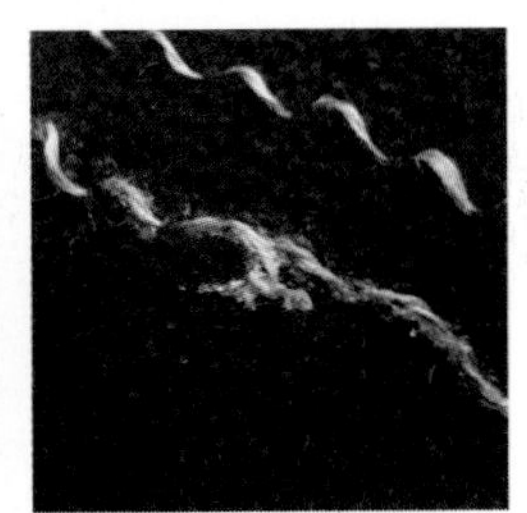
Figure L: A glued and trimmed knot. Knots without glue will come unraveled quickly.

thread hair that is touching both + and -) that you need to find and correct.

If the power supply does turn on, look at your LilyPad. It should blink quickly each time you press its switch. Once these connections are working, turn off the power supply and remove the battery.

Insulate your power and ground stitching: So, your jacket is now full of uninsulated conductive stitches. This is fine when a body is inside of it. A body will prevent sewn traces from contacting each other. But when the jacket is off of a person and you bend or fold it, traces will touch each other and short out. To fix this problem, cover your traces with puffy fabric paint (or another insulator like a satin stitch in regular thread).

Note: you don't want to cover traces until you're sure that everything works! So, use good judgment deciding when to coat traces.

5. Sew on your turn signal LEDs

Sew in your left and right signals: Using the same techniques you used to sew the power supply to the LilyPad, attach all of the + petals of the lights for the left turn signal together and to a petal on the LilyPad (petal 9 for me) and all of the + petals for the right signal together and to another LilyPad petal (11 for me). Attach all of the - petals of the lights together and then to either the - petal on the LilyPad or another LilyPad petal (petal 10 for me). Refer back to my design sketches if any of this is confusing (Figures N-Q).

Seal each of your knots with fabric glue to keep them from unraveling. Be careful to avoid shorts; don't let one sewn trace touch another. In this case, the - traces for the LEDs are all connected, but you want to make sure that the + traces for the left and right signals do not touch the - trace or each other.

Test your turn signals: Load a program onto your LilyPad that blinks each turn signal to make sure all of your sewing is correct.

Note: If you don't know how to program the LilyPad, work through a few of these introductory tutorials before proceeding: www.makezine.com/go/lilypadhome.

Here's my test program:

```
// The LED on the LilyPad
int ledPin = 13;

// My left turn signal is
// attached to petal 9
int leftSignal = 9;
// My right turn signal is
// attached  to petal 11
int rightSignal = 11;
// the - sides of my signals
// are attached to petal 10
int signalLow = 10;

void setup()
{
 // set ledPin to output
 pinMode(ledPin, OUTPUT);
 // set leftSignal petal to output
 pinMode(leftSignal, OUTPUT);
 // set rightSignal petal to output
 pinMode(rightSignal, OUTPUT);
 // set signalLow petal to output
 pinMode(signalLow, OUTPUT);
 // set signalLow petal to LOW (-)
 digitalWrite(signalLow, LOW);
}
void loop() // run over and over
{
```

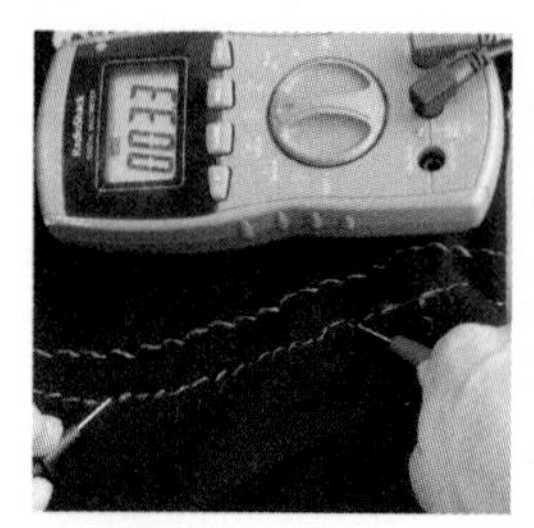

Figure M: Measuring resistance

Figure N: Sewing on an LED

Figure O: Stitching in process, outside view—3 positive petals are sewn together.

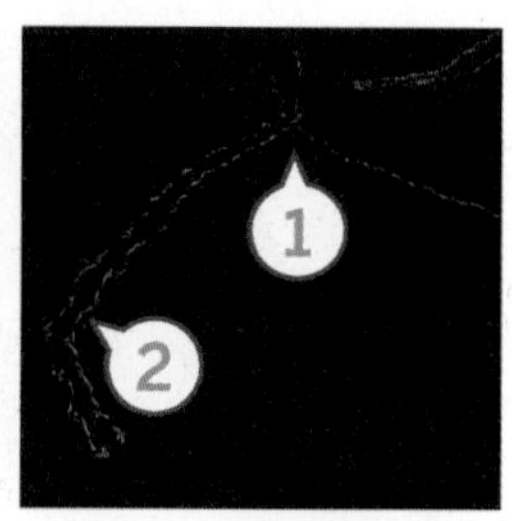

**Figure P: 1. Negative (-) traces for turn signal LEDs attached to petal 10
2. Positive (+) traces for right turn signal LEDs attached to petal 11**

```
  delay(1000); // wait 1 sec

  // turn the left signal off
  digitalWrite(leftSignal, LOW);
  delay(1000); // wait 1 sec

  // turn the right signal on
  digitalWrite(rightSignal, HIGH);
  delay(1000); // wait 1 sec

  // turn the right signal off
  digitalWrite(rightSignal, LOW);
  delay(1000); // wait 1 sec
}
```

If your layout is the same as mine, you can copy and paste this program into your Arduino window.

If your turn signals don't work, use your multimeter (and the instructions from the last step) to test for shorts or bad connections and make sure that your program matches your physical layout.

Insulate your turn signal stitches: Cover your traces with puffy fabric paint. Remember, *don't cover your traces until you're sure everything works!*

6. Sew in your control switches

Place your switches: Find a spot for your switches where they'll be easy to press when you're riding your bike. I mounted mine on the underside of my wrists. I found a good spot by trying out different places. Check out the photos to see what I mean (Figures R-S).

Once you've found a good position, push the legs of the switch through the fabric and bend them over on the inside of the fabric.

Sew in your switches: Sew your switches into the garment. Sew one leg to the switch input petal on the LilyPad and another leg, *one that is diagonally across from the first*, to ground or another LilyPad petal. I used petal 6 for the switch input on the left side and petal 12 for switch input on the right side. I used - for the - connection on the left side, but petal 4 for the - connection on the right side. Refer back to my design drawing if any of this is confusing (Figures T-X).

When you're done sewing, go back and reinforce the switch connections with glue. You don't want your switches to fall out of their stitching.

7. Sew in your indicator LEDs

Sew a single LED onto the sleeve of each arm: These will give you essential feedback about which turn signal is on. They'll flash to tell you what the back of your jacket is doing, so make sure they're in a visible spot. Sew the + petals of each LED to a LilyPad petal and the - petals of each LED to the - side of the switch (the - trace you sewed in the last step). I used petal 5 for the LED + on the left side and petal 3 for the LED + on the right side. Again, refer back to my design drawings if any of this is confusing.

As always, remember to glue and trim knots and be careful not to create any shorts. Once you sew both wrist LEDs, you're done with the sewing phase of the project! Now, on to programming...

8. Program your jacket

Decide on the behavior you want: I wanted the left switch to turn on the left turn signal for 15 seconds or so, and the right switch to do the same thing for the right signal. Pressing a switch when the corresponding turn signal is on should turn the signal off. Pressing both switches at the same time should put the jacket into nighttime flashing mode.

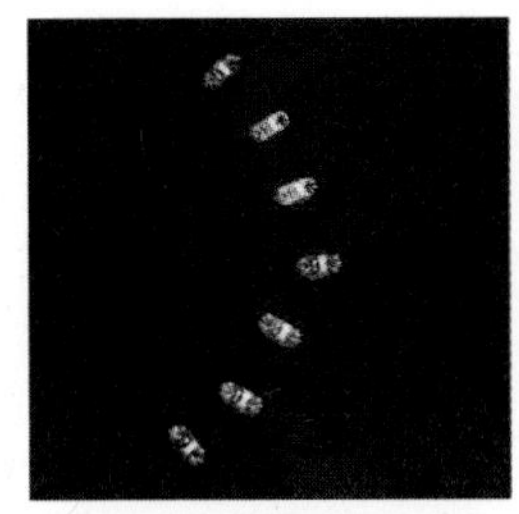

Figure Q: My finished right turn signal. Notice how my stitching doesn't come through to the outside of the garment.

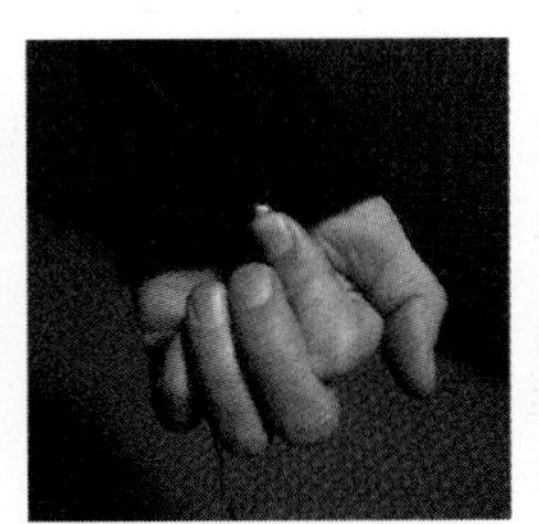

Figure R: One of the push buttons in action

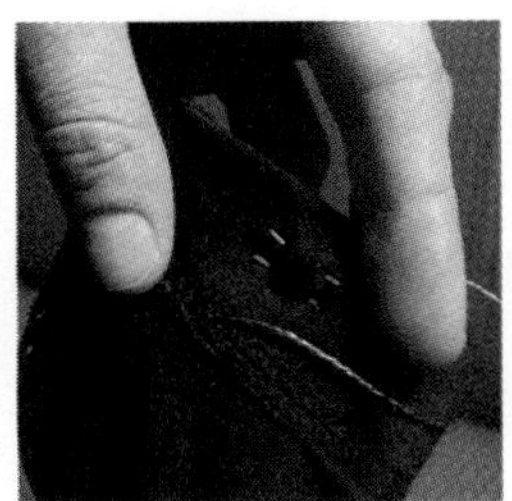

Figure S: The push button in place

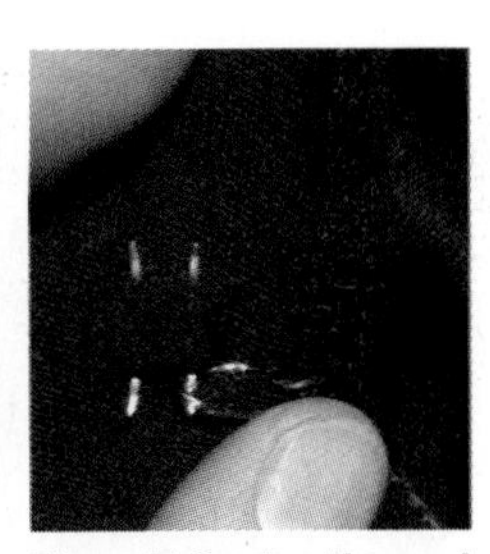

Figure T: Sewing the push button

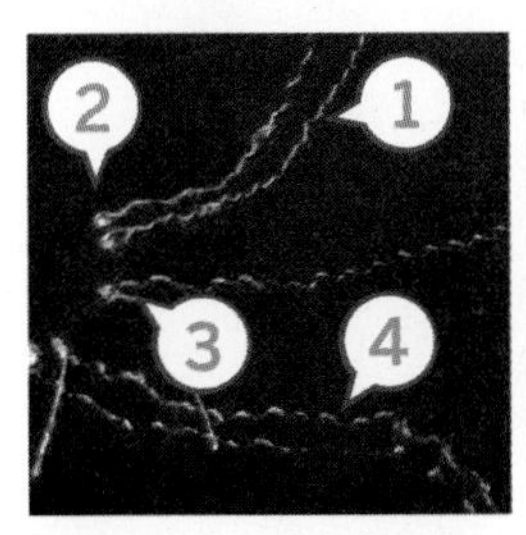

Figure U: 1. and 2. Stitches from power supply to LilyPad 3. First trace from left switch 4. Left turn signal stitches

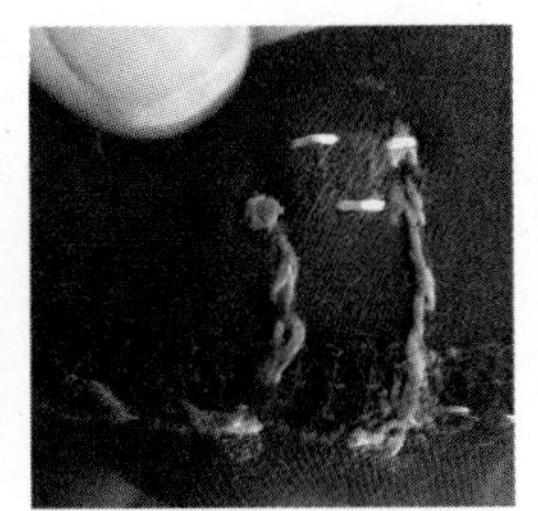

Figure V: The push button sewn in place

Figure W: An indicator LED

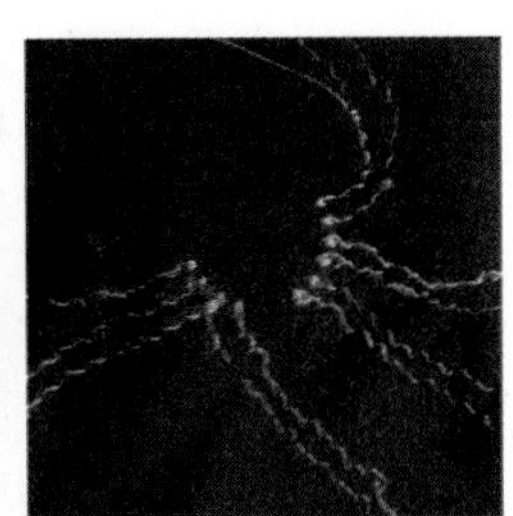

Figure X: The circuit from behind

The wrist-mounted LEDs should provide feedback about the current state of the jacket. Here's the code I wrote to get that behavior: www.makezine.com/go/lilypadcode.

Program your jacket: To program your garment, copy and paste my code into an Arduino window and load it onto the LilyPad. You may have to make some small adjustments first depending on where you attached lights and switches. Play with delays to customize your blinking patterns. Follow my LilyPad introduction instructions at www.makezine.com/go/lilypadhome if you need more information on how to program the LilyPad or how to make sense of my code.

Plug your battery back in and see if it works and... go biking! Take it out for a spin and make sure it all works OK.

Insulate the rest of your traces: Cover the rest of your traces with puffy fabric paint. Again, don't coat anything until you're sure it works.

Figure Y: Connecting the USB programmer

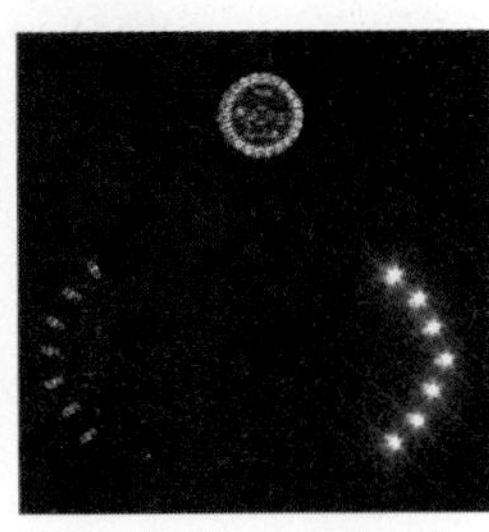

Figure Z: The completed jacket

Note: About washing—your creation is washable. *Remove the battery* and wash the garment by hand with a gentle detergent.

Note: Silver coated threads will corrode over time and their resistance will gradually increase with washing and wear. To limit the effects of corrosion, insulate and protect your traces with puffy fabric paint or some other insulator. You can also revive exposed corroded traces with silver polish. Try this on a non-visible area first to see what it does to your fabric!

Leah Buechley is an assistant professor of media arts and sciences at the Massachusetts Institute of Technology's Media Lab where she directs the High-Low Tech research

Visor-Cleaning Wiper Glove

Keep your vision clear without getting cold hands!

By Mike Warren

Figure A: Amazing wiper glove

Last year I was hanging out with a friend while he was checking out snowboards and I saw a glove that had a tiny wiper blade attached to it for cleaning your visor when it gets wet. While I don't snowboard, I do ride a motorbike. I also live on the rainy west coast of Canada, which means even in the spring and summer I can expect to get rained on. I had just bought new gloves for this season a few months back and decided that this would be a neat project to try out. Enough talk, let's make something!

1. Acquire materials

I'd say go out and buy a wiper blade, but that's just crazy talk. Wipers are expensive if bought new and we only need about 1/10th of just one of the blades, so instead of buying one just to throw away, why not just walk down the street and take one off that pesky neighbor who decides that mowing the lawn first thing on a Saturday morning after a heavy night out is a good idea, or you can just happen upon one in the recycling bin outside your apartment. Either way works.

2. Break apart the wiper

Yup, it's my favorite part. Time to break stuff! Wipers come in a variety of types and makes, but they are all going to follow the basic structure of assembly, so taking them apart may be different than what I've shown. Don't worry, we need just the rubberized part, so destroying the outer housing isn't an issue. Take a sharp knife and carefully cut into the rubber until you hit the silver metal shanks that hold the rubber blade in place. Cut off more than you think you need just in case; I cut about 13cm (5"). The rubber part extends beyond the shanks, but your knife will not fit—instead of trying to cut it, just grab the rubber and pull it out sideways along the length of the shank; the small section of rubber still attached will easily break. In Figure C I have also removed the metal shanks, we don't need them. I thought about having a system that would make the blades removable but it was too much effort and not worth the trouble. Mechanically fastening them to the glove is far easier.

Figure B: Check the recycle bin.

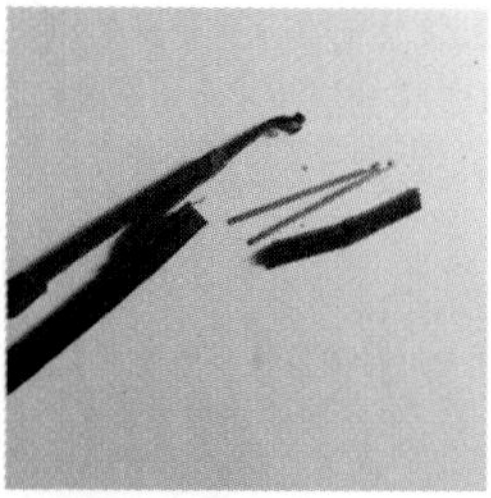

Figure C: Taking apart the wiper.

Figure D: Trimming the blade.

3. Rubber surgery

From the rough-cut piece of the blade, we're going to need to trim that profile down a bit (Figure D). Grab a sharp knife and start at one end and simply slice down the length of the blade to trim off the

For additional information, discussion, and more, please visit the Instructables project page:

Figure E: Careful!

Figure F: Blade after trimming.

Figure G: Heavy needle, heavy thread.

Figure H: Sewing on the blade.

smaller underside portion of the blade (Figure E). We're doing this so the wiper sits flat on the glove and won't wobble around so much. Again, every wiper will be a little different, but should all be roughly the same, so if yours doesn't look just like this don't worry, just trim down what you can so you have a nice solid base for your wiper to sit flat on your glove.

4. Needle and thread

Now I'm not sure if there are Instructables about how to thread a needle and simple sewing stitches, but there should be. I'll leave that in the realm of members Smexy Dead and threadbanger. The type of needle you use should be a fairly thick gauge so it doesn't snap and poke out your eyeball, and the thread we're using here is a heavy nylon with a gloss coat. Really, anything will work, just make sure you're using some heavy thread as you really don't want this to come off.

5. Putting it all together

The original I saw had the wiper on the thumb, but for this application it wasn't the ideal location since it would possibly rub against the part where my bike grips meet the clutch reservoir. I elected to place the wiper where the thumb joins with the hand as shown in the picture. I positioned the blade and attached the ends first, then made a few more simple stitches in the middle. It's that easy!

6. Done!

If all goes according to plan then you should be done. It's a fairly simple project and it works great, especially since the alternative is a smeared visor, which isn't very much fun.

7. Final thoughts

I had originally designed this to be a slide using the metal shanks from the blade body, that way the blades could be interchangeable. However, it was too difficult to implement for such an easy project, and sewing it directly on the glove is idiot-proof. Also, after a few days riding, I thought an alternate position for the blade might be under the meaty part of the thumb, which would still allow you to wipe your visor and keep it out of the way of any bike controls. I'd be interested to see any modifications of this cheap and easy build, such as if anyone finds a way to make the blades replaceable and fix it to the glove without scratching the visor. Good luck, and please post your results, it's a fun build!

---

michaelsaurus, aka Mike Warren, stays mostly hidden these days caring for his unicorn and eating dragon wings, emerging to see if any spaceships have come close enough to take him away from this planet and into the stars. Until then he creates small wonders, biding his time until he unleashes his next heinous experiment on the world.

Atomic Zombie's ChopWork Orange Chopper Bicycle

Re-make found bicycles into stylin' road hogs

By Brad Graham & Kathy McGowan

Kids' bikes with 20" wheels are abundant at the dump and yard sales, especially the cheap steel frame units. These bikes not only take a beating, but they are outgrown in a year or two, so you will probably find a lot of them at your favorite scrap yard or city dump. These frames are easy to chop, and 20" bikes make great choppers, but unless you are only 4-1/2 feet tall, there won't be much leg room on one of these bikes.

The problem of size is compounded even more if the head tube angle is taken back to add more rake, as this pushes the handlebars even closer to the seat. At this point, your only option is to move the seat higher or farther back, creating either a goofy looking chop, or a flying death trap that pulls uncontrollable wheelies on so much as a sneeze.

To get a little more leg room on a chopper made from a kid's 20" frame, two frames will be joined together in order to move the bottom bracket further up. The head tube will also be pushed forward, allowing for a nice long set of forks to be installed without creating a super tall wheelie machine.

More about bike building, robotics, and electronics can be found at Atomic Zombie Extreme Machines! (www.atomiczombie.com) Check out the Support Forums, Builder's Gallery, Blog, Downloads, Games, Videos, and more!

1. Find a donor bike

The sacrificial lamb is a typical steel girl's frame 5-speed bike, fresh from the garbage heap at the local dump (Figure B). For this project, you will need two 20" frames, and the components to make one complete bicycle. Depending on how you join the frames, the condition of the front half of one frame and the rear half of the other may not be important, as you will soon see.

Figure A: The finished chopper

2. Find a second donor frame

The second frame is another small 20" steel kid's bicycle, most likely another 5 speed or possibly a BMX wannabe. When joining two frames together to create a Frankenstein chopper, it really doesn't matter too much how similar the frames are, just make sure the rear triangle to be used fits whatever rear wheel you end up using. Since I planned to use 20" wheels all around, the two donor frames are just perfect (Figure C).

3. Inventory all necessary parts

It's always good to take everything apart in order to assess what will be usable and what will be tossed. Cracked bearing rings, rusty bearings, and bent pedals should all be replaced (Figure D).

4. Cut the frame tubes

There are many ways to join two frames together in order to create one longer frame, and depending on the condition and size of each frame, you will have to decide what goes where. I cut the top tube and

Figure B: The sacrificial lamb is a typical steel girl's frame 5-speed bike

Figure C: The second frame is another small 20" steel kid's bicycle

Figure D: Here are two donor frames and enough guts to assemble one complete bicycle

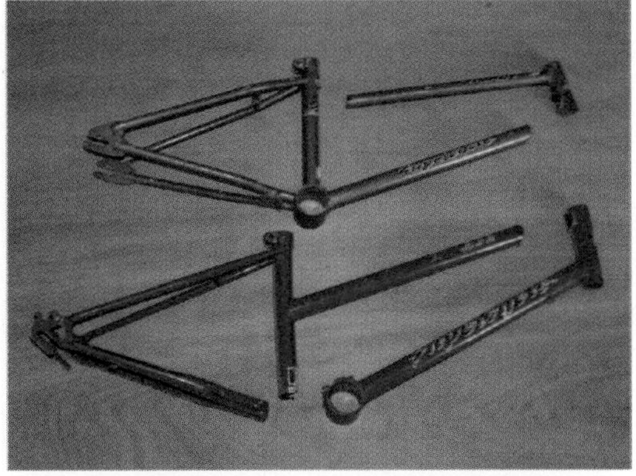

Figure E: Cut the top tube and head tube from the first frame, and the down tube, head tube, and bottom bracket from the other

head tube from the first frame, and the down tube, head tube, and bottom bracket from the other. I used the rear part of the first frame and the bottom bracket, down tube, and head tube section from the second frame (Figure E).

5. Grind and clean the bottom bracket

As soon as any part is cut, it's a good idea to grind away any leftover metal, as the part is easily handled on the workbench at this point. Once you start welding, it may be difficult (if not impossible) to get the grinder disk into the area to be ground (Figure F).

6. Tack weld the two frames

I admit there was no real plan here, just the idea of making a longer, taller frame for a chopper with extended forks. I decided to lay both bottom brackets on the ground and see where the head tube on the front frame would end up. The resulting layout was perfect! The bottom bracket was farther ahead, the head tube was nice and high, the rake was increased, and the distance between the head tube and seat was longer. I promptly tack welded the two frames together right where they sat, making sure vertical alignment was correct (Figure G).

Safety Warning: Welding fumes and coatings on some metals contain hazardous particles. Identify the metal and coatings prior to welding. Some coatings contain lead and other chemicals that can cause serious health problems. Always weld in a well-ventilated area, away from combustibles. Always wear safety equipment while welding.

7. Fill gaps and make seat stays

With the basic frame tack welded together, the next step was to fill in the gaps using whatever scrap was

Figure F: Bottom bracket, cleaned and ready for welding

Figure G: Tack weld the two frames

Figure H: Cut a seat stay in half to separate the two small tubes

Figure I: Cut a small piece of tubing from the leftover frame to fit between the two bottom brackets

cut from the other frames. Since the frame we tack welded would not be anywhere near strong enough to hold up to a rider's weight, some tubing was needed to create a solid shape. Cut a seat stay in half to separate the two small tubes. Let's see where I can find room to weld these on the frame (Figure H).

8. Make triangles for strength

Two lengths of tubing were just long enough to form two triangles from the head tube to the top of the seat tube. Since the triangle is the strongest shape you can form with tubing, this is a good thing. The two bottom brackets still needed to be joined, so another small section of tubing was cut from the leftover frame to fit between them (Figure I).

9. Frame is taking shape

The tube running between both bottom brackets almost completed the frame. In fact, I would imagine the frame would be strong enough right at this point, but something looked missing—just not enough going on in there yet. Besides, I had a lot of leftover scrap from the two frames. At this point, the welds were completed and ground clean (Figure J).

10. Add more tubes

I wanted a tube that would form another triangle in the frame, and since the tubing was becoming gradually smaller in diameter from the bottom to the top, I found an even smaller bit of steel rod (from an old fridge rack) to install. Now the frame was made of many triangles, and looked completed (Figure K). What else could I weld onto this thing?

11. Front fork construction

Before adding any more, I decided to work on the front forks. The front forks will be a typical set of round tube BMX forks, cut and extended to some length using 1" thin-walled electrical conduit. The first thing to do was cut the dropouts from the fork legs, sparing as much of the metal as you can since they will be put back on the new forks (Figure L).

12. Cut the fork legs

The fork legs were then cut so that only the vertical portion is removed. Imagine drawing a line from the inside of each leg and continuing it up past the head tube. This is the line that will be cut. This is done so that new fork legs can be installed later (Figure M).

Figure J: Welds are completed and ground clean

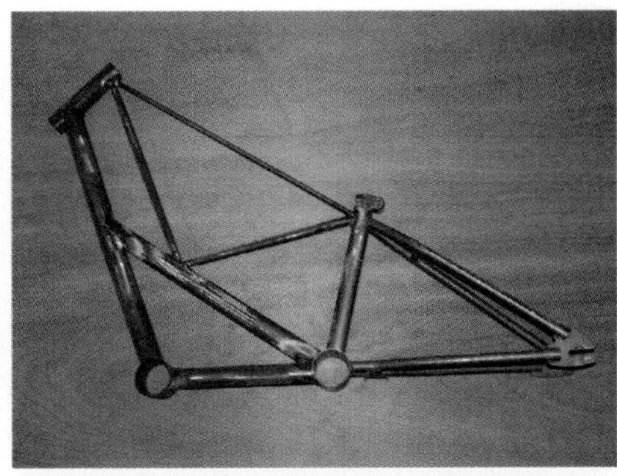

Figure K: Add another piece of rod to make another triangle to add strength

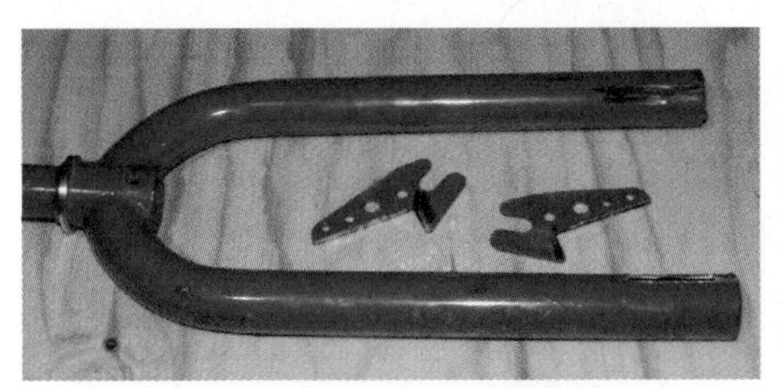

Figure L: Cut the dropouts from the fork legs

Figure M: Cut the fork legs so only the vertical portion is removed

Figure N: Extend the forks by welding two lengths of 1" conduit where the original legs were cut

Figure O: This leftover chainring fit nicely in the frame

13. Extend the forks

The forks are extended by welding two lengths of 1" conduit where the original legs were cut. You will have to grind a little away from the original fork material in order to make a proper joint with the conduit for welding. When I welded the fork extensions to the original fork stem, the front dropouts were already welded to the other ends of the fork legs, and a wheel was installed to hold it all in place. The two extension legs are also laid on a flat board to help alignment (Figure N).

14. Use leftovers for pizzazz

This leftover chainring fit nicely in the frame. It was at this stage that the bike was given the name ChopWork Orange, because of all the gears, and yes, it would indeed be painted orange (Figure O).

15. Assemble the chopper

The chopper was assembled in order to make sure everything was going together correctly. A banana seat and some wide handlebars were installed to give the bike an old school cruiser look. A fork length was chosen that put the two bottom brackets in approximately the same position they were on the

Figure P: Assemble the chopper

Figure R: Show off!

original bike, ensuring that the pedals had adequate ground clearance (Figure P).

16. Make a ghost ring

Rather than just leaving the rear bottom bracket empty and unused, I decided to salvage the original crank set to create a ghost ring. This secondary chainring does nothing but spin with the front one, but it will add to the ChopWork theme of this bike. The arms are cut from the crank set, leaving only the center axle. It is ground smooth as well (Figure Q).

17. Paint the chopper

Once completed, the chopper was hand painted with a brush using some spare orange paint that was hanging around the garage. The chainrings were painted black to accent the bike, and the chrome was polished up with some steel wool. The completed chopper turned out quite well considering it only took a few hours and started life without any plan.

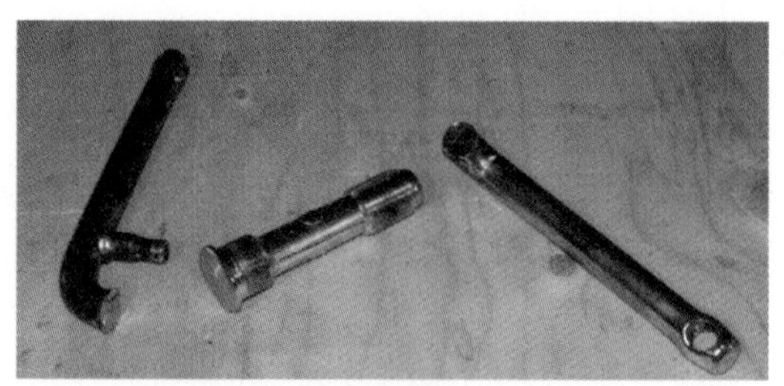

Figure Q: The arms are cut from the crank set; use the center axle to create a ghost ring

18. Start showing off

The bike looks cool with the dual chainrings. I tell people that this doubles your top speed, allowing the chopper to keep up with city traffic. It, of course, does nothing more than look cool!

19. ChopWork Orange Choppers are cool!

Once the hard work is done, it's all about looking cool, you know! The chopper is very comfortable and easy to ride, even for the chopper newbie. Banana seats also let a variety of riders of different heights ride the same bike, just move to a comfortable spot on the seat. Christina takes the bike out for a cruise and it rides like a dream (Figure R).

20. Enjoy your ChopWork Orange Chopper!

We hope that you enjoyed this Instructable. Make sure that you submit your completed chops to the Atomic Zombie Extreme Machines Builder's Gallery at www.atomiczombie.com.

You will also find:

- Builder's Support Forums
- Videos
- Builder's Gallery
- Blog
- Games
- Links

Hope to see you there!

Brad Graham and Kathy McGowan are busy building as many trikes, recumbents, choppers, and electric bikes as possible this summer before winter returns to Northern Ontario. These garage hackers have little "spare time," but always make time to connect with other bike builders worldwide. "No fate but what we make."

Drainage Luge

Turn a cement drainage ditch into a downhill racecourse By Phillip and Mars Shoemaker

Figure A: Four trucks

Figure B: Clamp the strips

Figure C: Sand the surface

Figure D: Secure the parts

Figure E: Rope handholds

Figure F: Shoot the pipe!

Have you ever seen those cement drainage pipes running down hillsides? They're meant to let water flow without eroding the landscape. With their flowing, organic lines, we decided to turn them into a new kind of ride with the Drainage Luge!

1. Get the parts: Use the drainage pipe width to determine how big your 3/4" plywood needs to be. And you'll need four skateboard trucks with bolts big enough to go through your wood (Figure A). Also rope to make hand-holds.

2. Affix reinforcements: Determine where the trucks go, length-wise, on your board. Cut truck reinforcement strips from 3/4" ply and glue and clamp these to the luge (Figure B). Let dry. We spaced ours about 4" from each end.

3. Prepping the board: It isn't necessary to add padding, but it would behoove you to sand the board surface and to remove any sharp edges... and splinters! (Figure C)

4. Install the trucks: Depending on the width of your board, you want to make sure the outside wheels are roughly 30" apart—from outer wheel to outer wheel. Also make sure they're the proper distance from the front of the board. Mark the drill locations for each truck hole, remove the truck, and drill. Once all the trucks holes are drilled, add trucks, bolts, washers, lock nuts, and tighten (Figure D).

5. Handholds: As an afterthought, we added rope handholds to help with staying on the board, and to aid steering (Figure E). At roughly the middle of the board (front-to-back), we drilled 1/2" holes, 4" apart on either side, roughly 2" from the edge.

6. Test Your Ride! Now for the fun part. Riding. First, scout the terrain. You don't want to shoot down the pipe only to find that it ends at a cement wall, steel pipe, or off a cliff! Remove any obstacles, and note where you need to stop. Wear boots, helmet, gloves, and pads. Mount and give yourself a few luge-like pushes. Before you know it, you'll achieve incredible speeds (Figure F). To stop, use your feet, Fred Flintstone-style, on the dirt.

For more detailed instructions and tips, check out the project page for this Instructable.

Phillip and Mars Shoemaker live in San Jose, CA, with the rest of their family. They love to build and invent things, including new sports, robots, and things that attack.

contest WINNER!

Solar-Powered Trike

Travel for free with the power of the sun

By David Pearce

The purpose of this project is to build a vehicle that:

- Provides free, "green" transportation for short distances (~ 10 miles), thus it must never plug into a wall socket, or emit any pollutants
- Charges while at work
- Is cheap, simple, and low maintenance
- Draws attention to the practical application of green energies and promotes fossil fuel alternatives
- Reduces excess automobile wear and pollution from cold driving and short, in-town trips.

This is a project for Dr. Reza Toosi's "Energy and the Environment, a global perspective" class at California State University, Long Beach. We look at the sources, technologies, and impacts of energy on our environment.

1. Acquire a vehicle

Find a lightweight vehicle with low rolling resistance. A two-, three-, or four-wheeler will do, depending on how much work you want to do, but the concept is the same. Four-wheeled vehicles may be regulated under different laws. Of course, the best vehicle is one that you already have, if you happen to have a three- or four-wheeled pedal-powered vehicle. In the interest of simplicity, a three wheeler was chosen for this project. This Schwinn Meridian Trike was $250 new, readily available locally, and the basket provides a convenient location for batteries and solar panels with minimal fabrication.

The first thing to do was to completely disassemble the trike and paint it a bright "fern" green. This step may not be necessary, but I felt that it was in my case since this is a school project that is supposed to grab attention, and let you know that it is a true green vehicle. It is a vehicle that does not use gas, and does not plug into a wall socket, which would defeat the purpose since electricity from the grid likely comes from a non-renewable energy source. It runs on pure solar energy.

Figure A: The Solar Trike. It's a cool green machine.

Before painting the frame, I used this stage as an opportunity to reinforce the frame where the batteries were going to mount. Lead-acid batteries are heavy, but they are relatively cheap. One tube was welded in to distribute the load over four points on the axle carrier instead of two. It also ties the rear subframe together, which makes the tube the load bearer rather than the weld beads, which may eventually fatigue and fail. High pressure (65psi) tubes were equipped and the Trike was meticulously assembled in order to minimize rolling resistance.

While the welder was out, a battery mount was fabricated, and bolts welded to the basket to be used as battery mount studs making removal easier. 12V LEDs were put in the reflectors and wired as brake lights through the brake levers that cut the motor when you brake. They are wired through only one of the three 12V batteries.

2. Drive train/running gear

The drive train consists of your electrical system and electric motor. The Electric Hub Motor Kit was purchased from GoldenMotor.com, costs $259, and consists of a front wheel with an integrated brushless 36V electric motor as part of the hub, along with the necessary components such as a twist grip throttle, brake levers that are wired to cut power to

For additional information, discussion, and more, please visit the Instructables project page:

Figure B: Reinforcing the frame where the batteries will mount

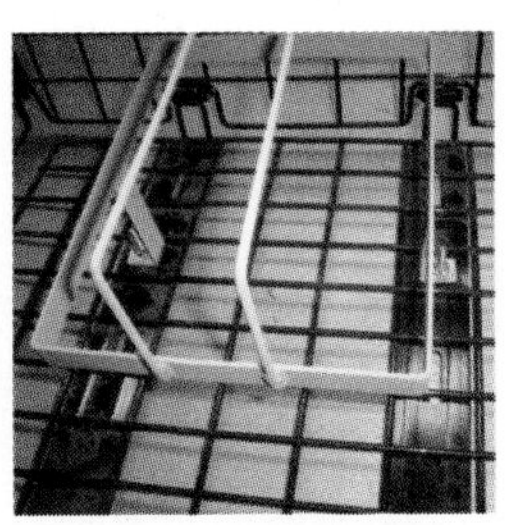

Figure C: The fabricated battery mount installed

Figure D: The drive train system on the front wheel

Figure E: The battery in its basket mount

the motor, battery level indicator, and the motor-speed controller, 36V battery charger, and a battery pack connector. Not sure if the kit is still available, but they still sell everything needed. The customer service is basically an owners' forum, which did prove useful in diagnosing a bent pin on one of the electrical connections.

The motor installation requires a simple front wheel change, and routing the wires back to the controller which will be mounted under the rear basket (Figure D). Slack must be left in the wires around the steering tube/fork juncture so they will not be in tension even at the maximum steering angle. The grips and brake levers are replaced with the new ones, and their wires also routed back to the controller.

Choosing the right battery is a compromise between price, weight, and range vs. charge time. Lots of money can be spent on batteries, but since I was on a budget, I had to take what I could get. I took a multimeter to a local industrial liquidation warehouse and found three batteries for $20 each, and they have worked well so far. Three 12V, 20 Amp/hour batteries are run in series to make 36 volts. 20A/hr provides long range, with the trade-off being a longer charge time. A battery cut-off switch was added so the rider doesn't have to unplug the battery pack to shut the electrical system off.

3. Charging system/solar panels

The solar panels need to be as large as possible to maximize the available wattage, but they also must provide the right voltage. Solar panels produce a range of voltages, which peak and drop, but the nominal voltage of the panel is what matters for selecting the right charge controller. I purchased three Q-cell brand mono-crystalline solar panels that I found on eBay for $110 each. They produce 21.8V peak and 17V nominal, at about 1.2 amps nominal. With the three panels wired in series, this makes around 66V peak and 51V nominal, which is plenty over the 42V needed to charge the batteries. A basket was added in the front to accommodate the third solar panel.

From Ohm's Law, Power (P) is equal to Voltage (V) times current (I), (P=V*I), so the panels produce ((17V*3)*1.2 Amps)= 61.2 Watts nominal, and over 80 Watts peak. A Maximum Power Point Tracking (MPPT) charge controller tricks the panels by hiding the battery load from them and allowing them to operate at their peak power when conditions allow.

A charge controller was purchased from solarsellers.com. The controller basically takes the varying voltage/amperage input from the solar panel array and converts it into a constant voltage (42V) or current, to optimize charging the 36V source. Maximum input voltage to the controller is 100V, so the peak of 66V will not harm the controller. The controller is a MPPT type, which charges faster as more sun is available, rather than at a set rate as most controllers do.

In order to charge the batteries in a practical amount of time, they need to charge about as fast or faster than the provided 110V wall socket to 36V charger/converter, which charges at a rate of 1.5 amps. At 1.2 amps the panels do not quite achieve this, but with the MPPT controller, it takes right around the same amount of time for a charge. The bike is stored in a location that gets a few hours of sun every day (where I live the sun is pretty reliable), which keeps the batteries topped off and ready to go whenever needed.

And for those of you wondering, the electric motor draws up to 20 Amps, and the 1.2+ Amps added by the solar panels do not make it go faster, since the 1.2 amps are routed through the controller and only serve to charge the batteries. The motor speed controller does not see this extra amperage, and

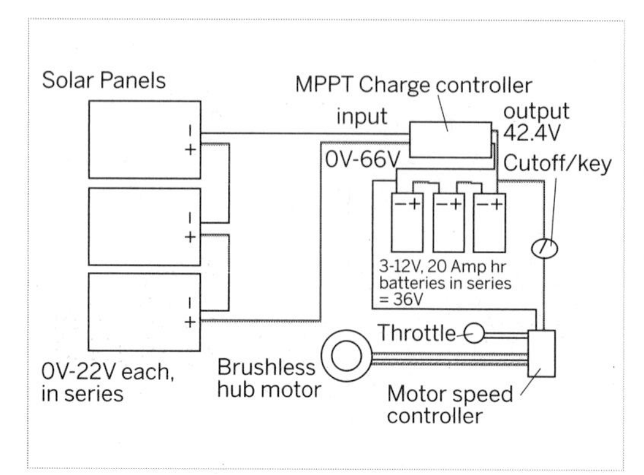

Figure F: The Solar Trike wiring diagram

Figure G: The solar panel hinging

outputs just the same as without panels, except the batteries will stay charged slightly longer, (extending your range) with the net drain being (20-1.2) A= 18.8A rather than 20A without the panels. The motor only pulls 20A when taking off though, so the draw is much less when at cruising speed. The motor speed controller cuts the voltage off at 32V to keep the batteries from going below 10.5V, but I monitor the voltage and try not to discharge the batteries below 36V. Consult the wiring diagram in Figure F to see how the system was put together.

4. Solar panel mounts

Now you have to figure out how you're going to mount the panels on your vehicle. Hinges were welded on the baskets to mount the panels and allow them to tilt for access to the basket, with rubber hold-downs on the other side to keep them from opening while riding (Figure G).

Once your wires are all routed and zip tied, your batteries and panels held securely down, double check every thing and you are ready to go.

Figure H: Feel the power of free energy

Performance:

This solar-powered trike does about 15-18mph, depending on the weight of the rider. The furthest I've gone is a little over ten miles with small hills and little pedaling, and the battery meter still read full (green) at the end of the trips.

At ten miles, the voltage drops to around 36V, safely above the controller's cut-off voltage. If the batteries are kept from discharging too low, the panels take about the same amount of time as the plug-in charger, since both the plug-in charger and the solar charge controller charge with constant wattage. With constant wattage charging, Power (P), and Ohm's Law again (P=V*I), the charging current goes down as the voltage goes up, as the batteries near their fully charged state.

What this means is if you keep the voltage from dropping too low, the panels provide adequate current to match the charging speed of the plug-in charger, but if it drops below a certain point the panels are slower at charging. This is easily avoided since my typical trip range is around three miles or less, semidaily at most, so low voltage is not an issue, but on longer trips I bring the multimeter.

The Trike cost a little over $910 to build:

Schwinn Meridian Trike	$250.00
Q-cell Mono-crystalline solar panels	$330.00
Charge controller	$95.00
Electric Hub Motor Kit	$260.00
12V batteries (3)	$ 60.00
High pressure tubes	$ 15.00

David Pearce is a mechanical engineering student at California State University, Long Beach. His interests include all forms of human propulsion, surfing, skateboarding, nature, and traveling.

How To Get a Free Yacht

Sound impossible? It's not...

By Tim Anderson

Figure A: Our first free yacht. Note the diesel tank made from a beer keg.

1. Get the free yacht

This is the easy part. *Wooden Boat* magazine has a "Free Boats" section in every issue. Any harbormaster can show you some free boats. They're especially plentiful in the northeast in the fall. Divorce season, whenever that is, produces lots of project boats that "must be removed from my yard before such-and-such a date."

My friend Patrick and I got the beauty seen in Figure A from the "free" section of craigslist.com.

2. Rent a marina slip

That's where you'll put your new boat. Our slip costs $200 a month because it's a 30-foot slip. Our bowsprit is a lot of that length. We could saw it off to save some rent, but it's a great thing to ride on. Some marinas don't allow wooden boats because they don't like beauty or suffering. Or boats older than ten years, or worth less than 40 grand, or some-such because they worship Moloch. We found the Emeryville Marina, which is a righteous place.

On the project page to this Instructable, you'll find the application we filled out to get into the marina. It requires insurance, which in turn asks what marina the boat is in: Catch-22.

3. Get insurance

To rent a marina slip, you need insurance. Get whatever the marina requires or recommends. The following companies wouldn't sell us a policy because the boat was too old: Geico, West Marine, BoatUS, United Marine Underwriters. Progressive Insurance (progressivedirect.com) did it online, with no hassles. Yay! It costs about $300 a year. I'm told hagerty.com is also good for insuring weird old boats, cars, etc.

4. Get it ready to move

We swam around the boat with spatulas and brushes, scraping off all the seafood that was thriving on the bottom. When we were done, the boat sat an inch or so higher in the water. The diesel engine runs great, but the RUBBER exhaust hose leaked and had to be replaced. There's a pump that sprays water into the exhaust pipe, cooling the hot gas just as it hits the rubber hose going out the back of the boat. That's apparently the usual way of doing it. We went to West Marine and bought a new one. The sticker said "$7.71" and I thought that was very reasonable. We also got some flares that the Coast Guard, the DMV, or some-such requires. The price on the hose turned out to be PER FOOT. $85.00 got us a rubber hose and four highway flares.

Definition of boat: "A hole in the water into which you pour money."

5. To be continued...

Continued at: Maiden Voyage of the Free Yacht (www.instructables.com/id/Maiden-Voyage-of-the-Free-Yacht).

Tim Anderson co-founded zcorp.com, manufacturer of 3D printers that are computer-controlled machines that build sculptures. He travels looking for minimum-consumption technologies developed by poor people. He writes the "Heirloom Technology" column for MAKE magazine and has written 150+ Instructables.

This Instructable was only the beginning of a grand adventure, and eight Instructables, chapters in an amazing saga of high-sea adventure, many repairs and improvements, a fire, how to give away a free yacht, how to get another free one, and lots more. There's a good book in here, which may be why Tim has come to calling each of these Instructables "chapters." Links to each successive Instructable can be found at the end of the previous one.

Home & Garden | Food | Photography | Science | Computers | Electronics

The Best of Instructables Craft

For years, I was under the impression that "crafting" was an activity for bored grandmothers smelling of potpourri, or New Age hippies selling home-made hand soap. Not to disparage perfumed grandmothers or soap-hocking hippies, but crafting seemed as foreign to me as the surface of the moon.

Somewhere along the line, my thinking changed, probably around the time that people started viewing my Instructables and describing them as "Craft." At first, I was outraged. I wasn't crocheting doilies; I wasn't making soap. I wasn't even using natural fibers!

I had inadvertently stumbled into an emerging craft movement that had directly spawned from the DIY (do it yourself) ethos of previous decades. This new wave of crafters didn't lean on tradition so much as modify, mash up, subvert, and transcend standard conventions.

Shunning store-bought items for handmade, this maverick community of crafters believes in making things themselves, by whatever means necessary. Frequently, hefty ideas are constrained by thin wallets, limited tool sets, and modest materials, leading to a resourcefulness and flexibility that can rival a highly trained Navy SEAL.

With equal parts independent thinking, unconventional materials, and highly unorthodox methodologies, crafters on Instructables create objects that are both useful and personally meaningful. It's this tendency towards extreme personalization that makes a trip to Instructables such a treat.

Say you need a wallet. Go to a store and wallets might be made of leather, or vinyl, or canvas, but they'll roughly be of the same style, size, and material. Go to Instructables, and you'll find dozens made out of everything from playing cards to inner tubes, all constructed with more design variance than seems possible. It's this endless variety, spurred on by fierce individualism, that makes the Instructables Craft community like no other.

—Randy Sarafan (randofo)

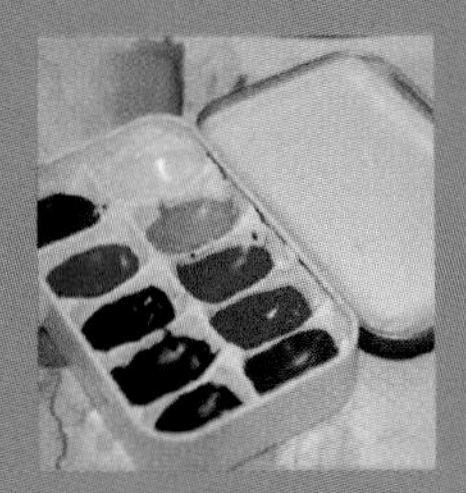

Gallery

Altoids Tin Projects

It's a mint tin modification extravaganza! What *can't* you do with an Altoids tin?

Sour Time

William Craig

/altoids-clock

You'll always know what time it is with this Altoids clock, made from a round Mango Sours tin

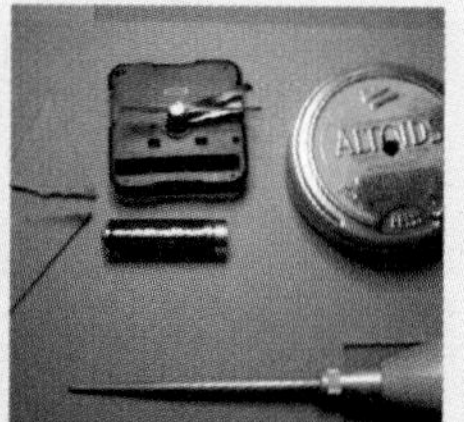

Take the workings out of a cheapo clock and install them in a tin; poke a hole in the face for the hands

Minty Strobe

Martin de Selincourt

/Minty-Strobe

Make a triggerable strobe to capture those high-speed moments, with the innards of a camera flash unit and some switches

Catch the action—look, levitating pennies

Guitar Unplugged

bumpus

/Altoid-Guitar

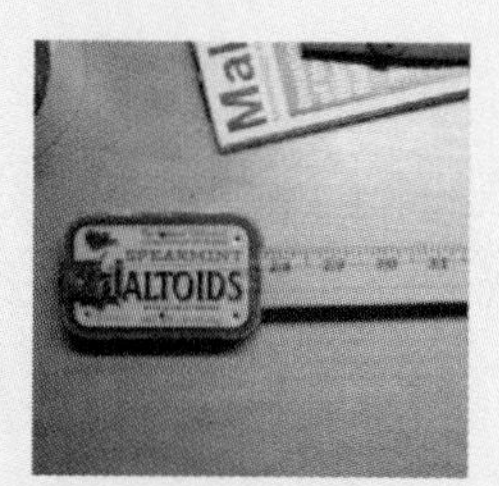

Add a ruler, guitar strings, some eye screws for tuners, and you have a tiny guitar

Tin Guitar—Electric Version

Jason and Ian Wilson

/Altoids-Tin-Guitar

Turn your tiny guitar up to 11 by adding a piezo pickup and an amp jack

To view these Instructables, and read comments and suggestions from other Altoids hackers, type in www.instructables.com/id followed by the rest of the address given here. Example: www.instructables.com/id/Altoids-Tin-Guitar

Altoids Case From Old iPod

Courtney Sexton

/Altoids-case-made-from-old-iPod-Shuffle

Don't know what to do with all the mints you're kicking out of their tins? How about this iPod Altoids case?

Ultimate iPod Case

David and Chuk Industries

/The-Ultimate-Ipod-Altoids-Case

Go EXTREME with this Ultimate iPod Altoids case—whoa, dude!

Speaker

Matthew Poage

/Altoids-Tin-Speaker

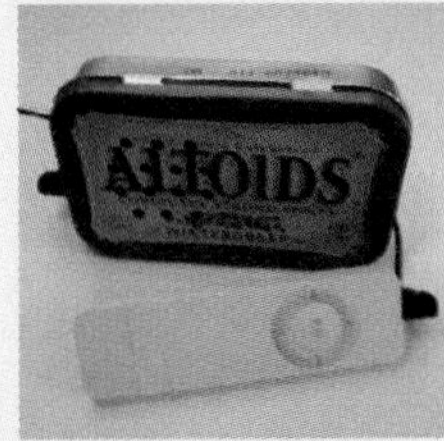

Plug your iPod into this Altoids tin speaker, inspired by MAKE magazine's Crackerbox Amp

You need some experience making printed circuit boards to build this one

Survival Kit

Daniel Matthews

/Altoids-Survival-Kit_2

Our Altoids tins will help us survive if we're out in the wild... at least they will if we pack all these essentials in them.

Bicycle Survival Kit

=SMART=

/Bycicle-Survival-Kit-

For a bike ride, all this fits into a tin

Write important numbers in the lid

DIY BBQ in Action

Alex Fleming

/DIY_bbq_in_action

Be prepared to barbeque at a moment's notice with this foldable pocket BBQ grill. Lilliputian grilling utensils sold separately.

Gallery

Altoids Tin Projects
continued

Minty Boost

Limor Fried

/MintyBoost!---Small-battery-powered-USB-charger

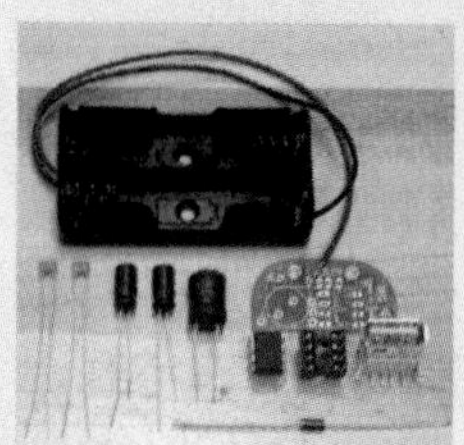

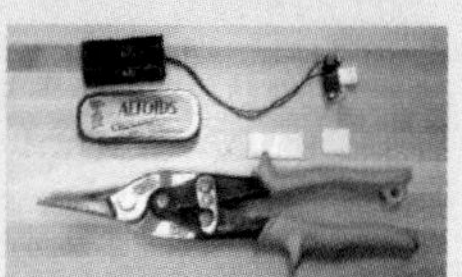

The now-classic Minty Boost is a portable battery-powered USB charger. Beginners can get a build kit through the adafruit webshop (www.adafruit.com).

Pocket-sized Watercolor Box

JPMartineau

/Altoids-Tin-Pocket-Sized-Watercolor-Box

Use Fimo to make little wells to hold your paste watercolors. Let the paint dry, and you have a portable art studio.

Emergency Candle

EAKLONDON

/Altoids-Tin-Emergency-Candle

Never be left in the dark with this multiwicked emergency candle—store matches in the lid

To view these Instructables, and read comments and suggestions from other Altoids hackers, type in www.instructables.com/id followed by the rest of the address given here.
Example: www.instructables.com/id/Altoids-Tin-Guitar

The Blow Brush

A funky-fine airbrush that you power by blowing into it

By Marek Soliński

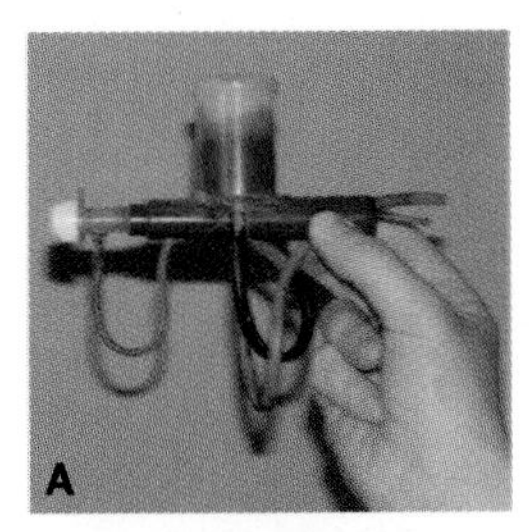

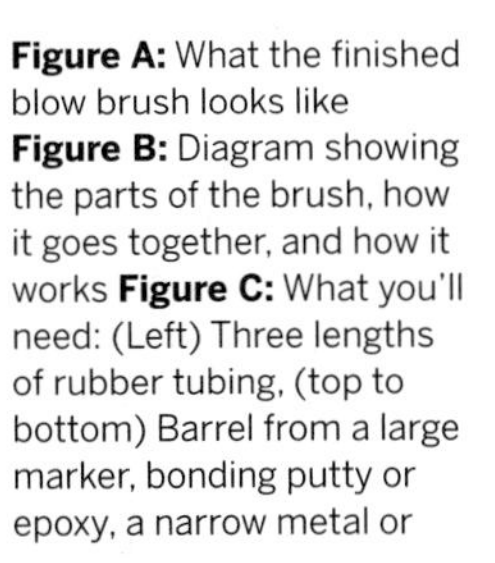

Figure A: What the finished blow brush looks like **Figure B:** Diagram showing the parts of the brush, how it goes together, and how it works **Figure C:** What you'll need: (Left) Three lengths of rubber tubing, (top to bottom) Barrel from a large marker, bonding putty or epoxy, a narrow metal or

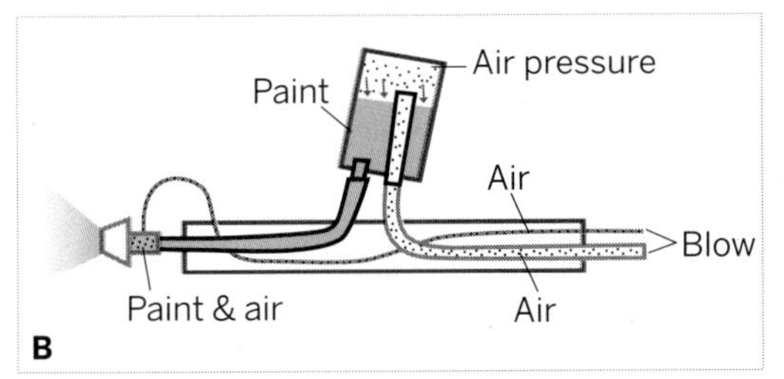

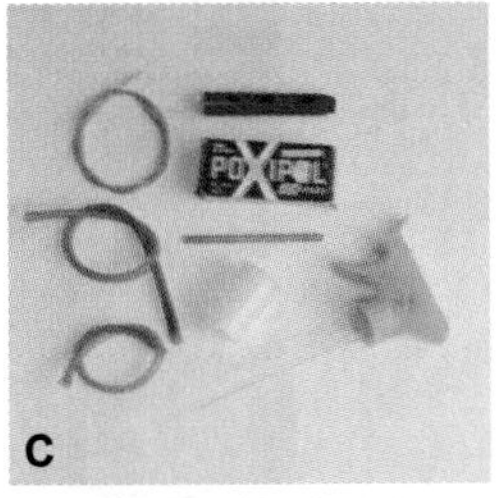

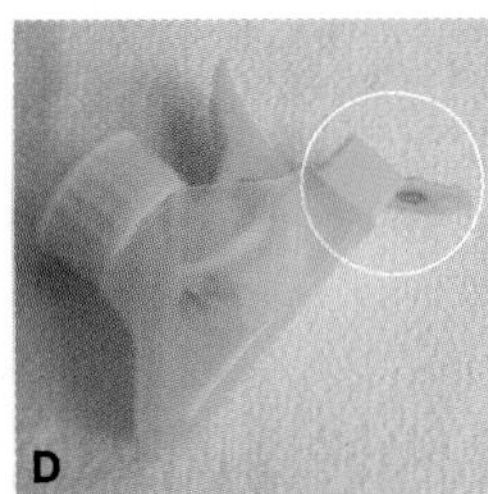

Figure C (continued): plastic tube (such as from an ink pen), a 35mm plastic film canister, an adjustable spray bottle nozzle **Figure D:** Remove the nozzle from the bottle. You may have to take the head apart to get the nozzle off in one piece (retain the throat part behind the nozzle head). **Figure E:** The adjustable nozzle **Figure F:** Create a nozzle connector tube. The tube shown is attached to a different nozzle, but it's the same idea. Drill a hole to accept the short metal or plastic tube (which the rubber tubing needs to fit over) in the throat of the spray nozzle. Use epoxy or other bonding material to glue it in place.

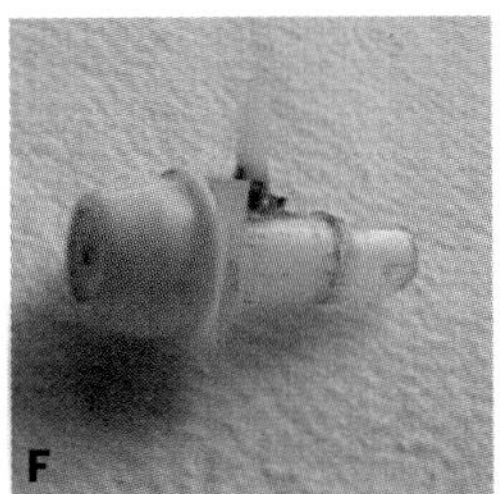

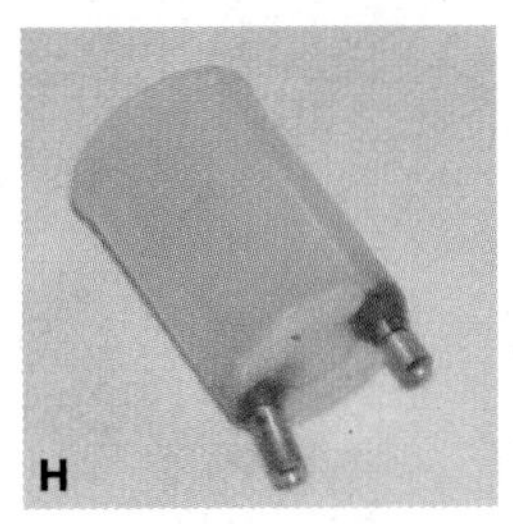

Figure G: Preparing the color cup. Drill two holes in the bottom of the film canister large enough to accept the rigid tubing. One tube length should come about 3/4 of the way up into the canister (the air tube), and one should be near-flush with the bottom (the paint tube). Use the diagram in Figure B for reference. Glue the connector tubes in place. **Figure H:** The bottom view of the color cup, air and paint tubes in place **Figure I:** Prepare the body of the brush. Drill three elongated holes along the length of the marker barrel to accept the hoses. Again, use Figure B for reference.

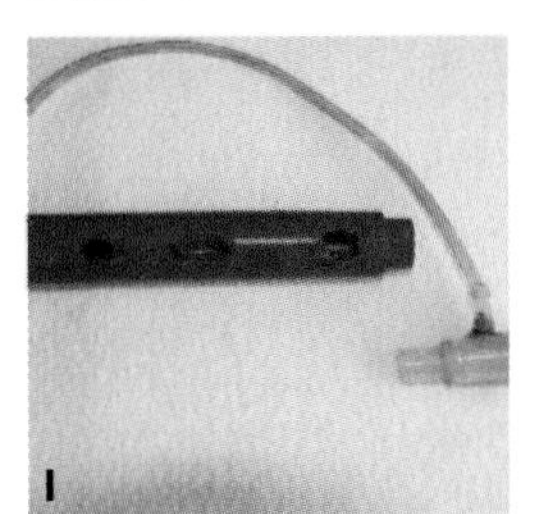

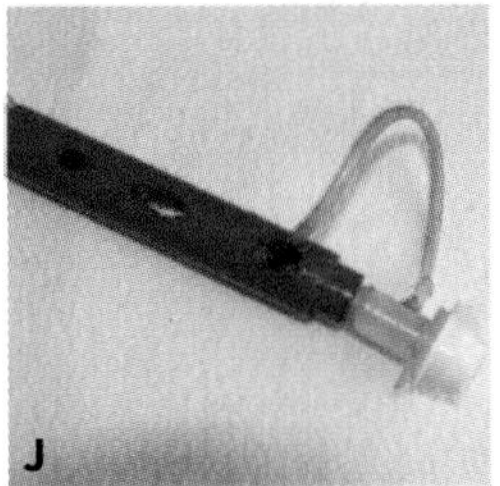

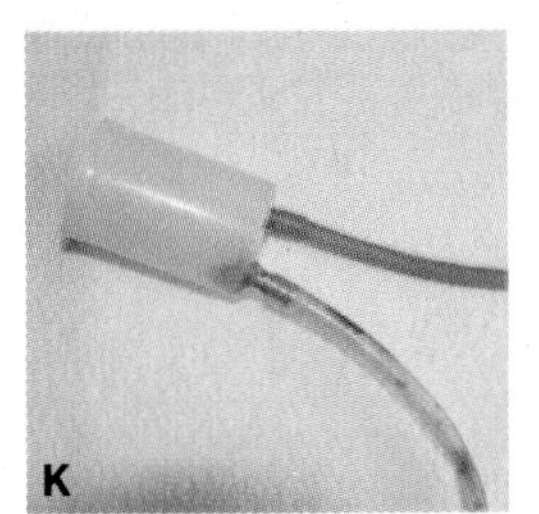

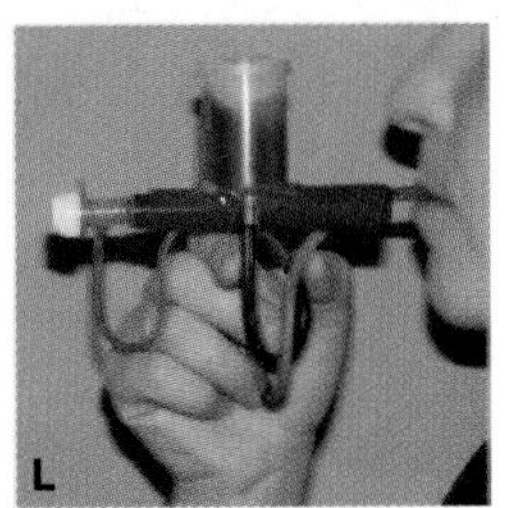

Figure J: Attach all three hoses (consult Figure B) and thread them through the body of your brush through the holes you made. **Figure K:** The air and paint tubes attached to the paint cup. One attaches to the nozzle, the other is threaded through the body to the mouthpiece. **Figure L:** The fully assembled blow brush, filled with paint and ready to fire.
Note: A video showing the assembly process for this project is available on its Instructable page.

Marek Soliński is a tinsmith living in Poland with his wife and son. He works as a car paint shop assistant and loves fishing, PC games, and tinkering.

Electric Umbrella

Turn an ordinary umbrella into something whimsical and magical

By John Kowalski

Figure A: The Electric Umbrella

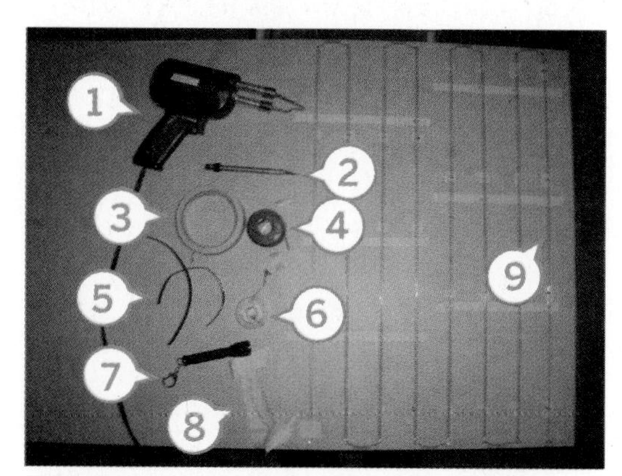

Figure B: 1. Soldering gun 2. X-Acto knife 3. Masking tape 4. Solder 5. SMD LEDs (still in their packing material) 6. Solder wick to remove excess solder 7. Mini flashlight (it's hard to see those tiny LEDs when it's dark!) 8. The ends of the wire—you'll apply power here to see all the LEDs light up! 9. Solder LEDs here!

The Electric Umbrella will glow with many pinpoints of light. Carry the sun and the stars with you at night! Perfect for nighttime strolls through the countryside or just being silly. And it's dimmer-adjustable so you can set how bright you want to be—anywhere from dim ambient light for strolling in the dark to carrying your own portable supernova beacon of light!

1. What you need

The things that you need may be found through some combination of local stores, electronics parts shops, online, and scrounging parts from old electronic junk you may have lying around.

Parts and equipment:

- One umbrella, preferably light colored (I picked yellow), with a straight handle and a hollow shaft so that you may pass wires through it. It is very important that the umbrella be simple—none of that spring-loaded automatic stuff! You want the shaft to be hollow.
- 64 SMD (surface mount) LEDs in your color of choice. The actual size does not matter except that smaller will look more invisible (preferable) but will be more difficult to work with. I used size 805 (2mm wide) 3.5V white LEDs. White, blue, UV, and some greens require 3.5V and won't require additional resistors on each LED, but 1.8V LEDs (red, yellow, green) do (more trouble!).
- A spool of thin single strand, lacquered copper wire. Thin enough to be almost invisible against the umbrella, but thick enough to withstand the occasional stresses/snags. This is what the SMD LEDs will be soldered onto.
- One 3-AA battery holder, preferably compact and arranged in an L shape, as the batteries will have to lie over the umbrella's shaft. 3 AAA batteries would work well too, and are more compact, but won't last as long.
- Normal plastic-coated multistrand copper wire, preferably the kind that will not break easily after repeated flexing.
- One 750Ω variable resistor with built-in on/off switch for dimming and turning the umbrella on and off.
- Needle and thread (of the same color as the umbrella)
- Solder and soldering iron/gun
- Wire cutters, wire strippers, scissors, X-Acto knife

For additional information, discussion, and more, please visit the Instructables project page:

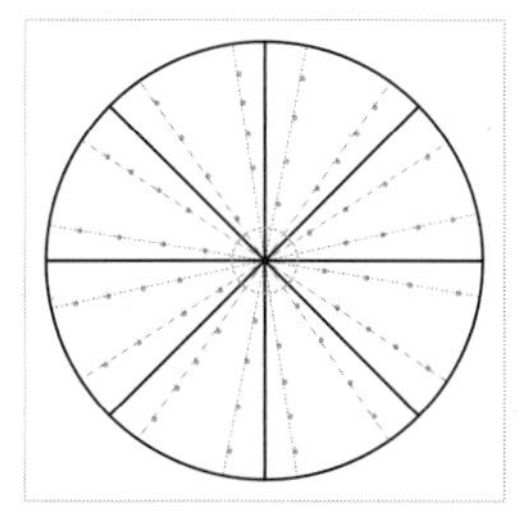

Figure C: Each blue dot represents the placement of an LED within the umbrella

Figure D: Two fine strands of wire running back and forth across the board

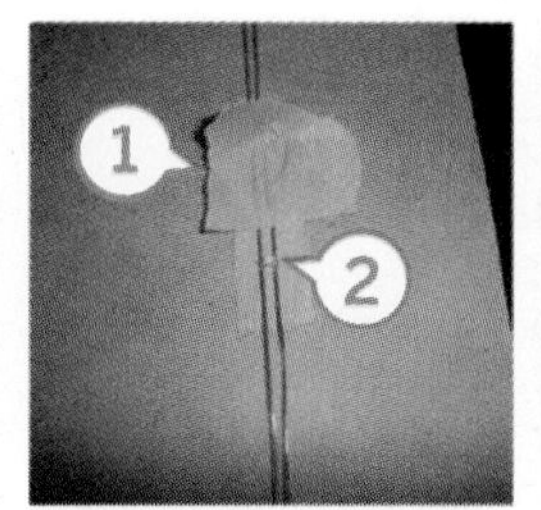

Figure E: 1. This tape keeps the wires pressing against the LED 2. Surface mount LED being soldered onto the wires

Figure F: 1. LEDs before being placed onto the wire 2. Soldered LED 3. The location of an LED from the first wire/LED set

- Drill and drill bits
- Large board and small nails, to be used for laying out the wires and soldering the SMD LEDs onto the wires.
- Masking tape and double sided tape/carpet tape
- Clear epoxy or glue, Super Glue

Figure B shows many of the parts and equipment.

2. Solder the LEDs onto the wires

Be prepared for some long and tedious steps. Carefully soldering 64 individual LEDs, each not much larger than a grain of rice, onto thin and uncooperative wires takes patience.

Before starting, measure out your umbrella and plan where each of the LEDs will go. This umbrella will have 16 spokes radiating out from the center, each spoke having 4 LEDs. I chose to have 4 different sets of LED spacing (8 of each set) to make a pseudo random looking pattern. I set the LED spacing so they're generally closer together towards the outside of the umbrella in an effort to make the LED distribution reasonably even throughout the surface of the umbrella. Figure C shows the distribution.

Get a large board (wider than the radius of your umbrella) and hammer a bunch of nails along the sides so that you can string up/stretch out your single-strand copper wires (two wires to each nail as shown in Figure D). Place masking tape and/or mark off the points at which you will be soldering the LEDs. Leave some extra lengths of wire at each end in case you need some extra length once it comes time to actually install them onto your umbrella.

Place some masking tape under the wires to prevent burning the board (in case you ever want to use it for some other purpose) and more masking tape to hold the wires down in place as you solder. My wire was lacquer coated, so I had to first burn the coating off at the points where I planned to connect the LEDs with my soldering gun and hot solder. (You may try scraping it off instead, or using a wire stripper to strip it off.) Once the wires are tinned with solder, try to wedge an LED between the two. Be careful to place all LEDs in the same polarity!

Time to solder your first LED. I tried to apply some masking tape onto the wires so that they pinch the LED in place—makes it easier to solder the LEDs in if they're not always moving around. With a very quick, light touch, touch both sides of the LED with the hot soldering tip, and the solder coating the wires and tip will flow into the LED contacts. If you're not sure you got it, hook up 3V (two AA batteries) to the wires hanging off the board and see if the LED lights up! Once you get the hang of it, move on and do the rest of the LEDs. I soldered mine in two sets—half of the wire/LEDs on the board at one time (16 'spokes') and the other half after finishing the first.

Check out Figures E and F for a detailed look at this.

After all the LEDs were soldered (Figure G), I applied power to the board/wires to see all the LEDs light up in their glory, as shown in Figure H. This is also a good time to determine how many volts you want your umbrella to run on, and what value variable resistor you want to use for dimming. I decided on 3 AA batteries (4.5V—3.6V if using rechargeable batteries) and a 750Ω variable resistor.

3. Assemble the central hub for the wires

All the LED spokes connect to a center hub near the tip/center of the umbrella. The tricky part is assembling this outside of the umbrella first and

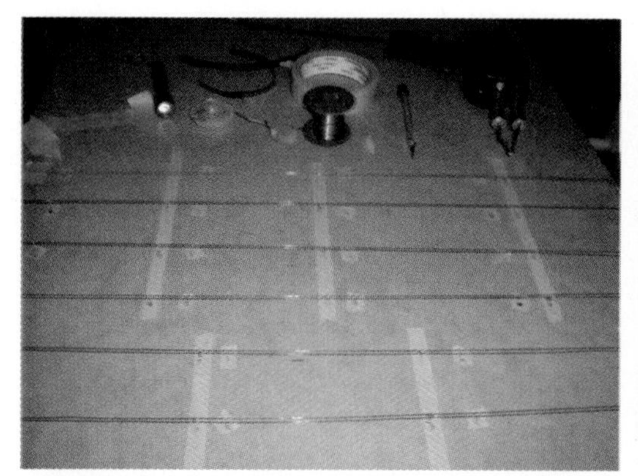

Figure G: Hard to see, but all the LEDs are soldered in place

Figure H: All 64 LEDs placed onto the board. I attached 32 at a time. The second set of 32 is placed loosely over the others.

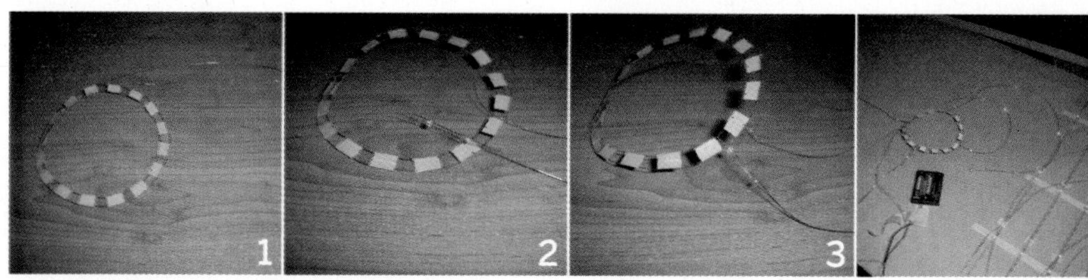

Figure I: 1. The loops of wire are only taped together at this point; I've left some excess wire for later 2. The ends, cut to line up with the two wires in the hub (the tape identifies which of the four different LED sets/spacings this is and also marks the start of the length of wire) 3. First LED string attached and soldered to the hub 4. Testing the LEDs on the hub as I go along

then carefully fitting it into the umbrella between the spokes and the cloth. I assembled it separately because it's hard to reach the inside of an open umbrella. I also didn't want to risk burning holes into the cloth when soldering.

Make two rings of wire. I used masking tape to hold its shape while putting it together. The masking tape also served to mark the spacing where each LED string was to be attached. The exact size of the rings doesn't matter except that you want it to be fairly close to the center of the umbrella. Do not actually solder the wires into a full circle just yet as you will need these to be detached when you fit it into the umbrella later on—just hold the circle together with more masking tape for now and leave lengths of wire long enough at one end to reach the batteries once it's inside the umbrella. Figure I shows the process.

Warning: I used the same single strand copper wire as I used for the LED strings/spokes. This was a mistake! Every time you open and close the umbrella, the wires in this hub will flex and this kind of wire will eventually break from the stresses... bad bad bad. Later on I soldered additional loops of stranded wire onto the hub. These wires are much better at holding up to stresses of repeated flexing (Figure J).

Note: The next umbrella I made used *only* loops of wire for the hub—see Figures K and L.

Cut the lengths of wire off the big board (4 LEDs per length). Measure the ends/length/LED placement and start soldering the strings of wire onto the hub. Make sure you get the wire/power polarity right! You can add power to the wires to see if you're getting it right. After all 16 strings are attached, you will have an interesting glowy mess... might be used to make an interesting headpiece or hat, but we're trying to stay focused on making an electric umbrella. :-)

4. Get the wires and hub into the umbrella

Loosely place the hub and the mess of wires near the center of the umbrella (Figure M), then carefully start sliding the hub under the umbrella's spines so

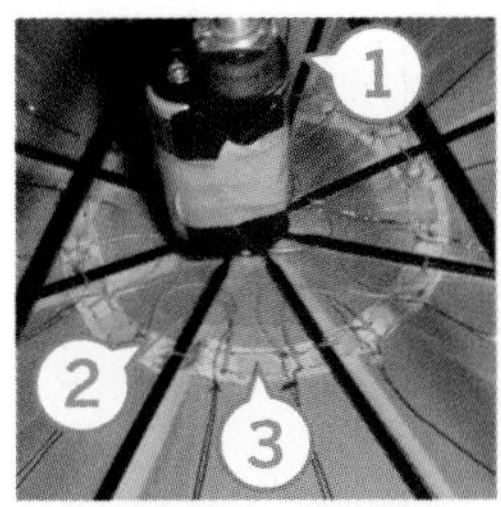

Figure J: 1. The dimmer is here for now—it hasn't been added to the handle yet 2. Loops of flexible wire added to fix breakage problem in the hub 3. Loops of wire added here, too

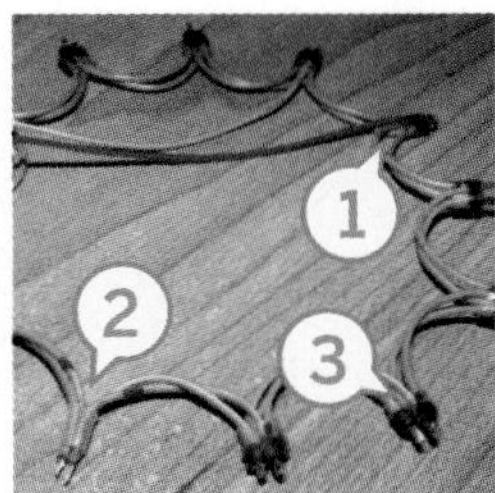

Figure K: 1. These wires go to the power source 2. Leave one connection unsoldered until the assembly is placed in the umbrella. 3. Two sets of wires all the way around, held together with heat-shrink then soldered. Red dots are positive.

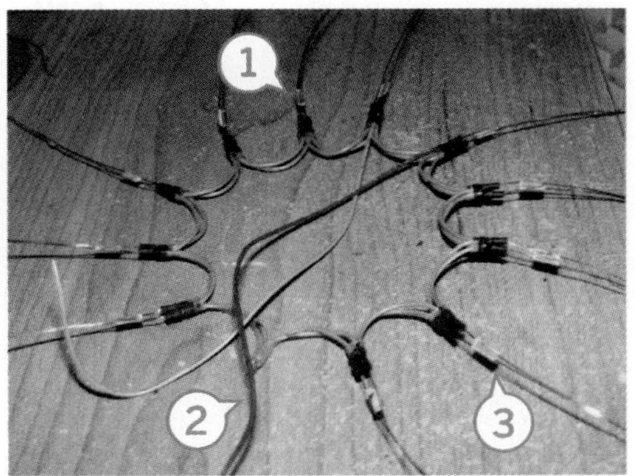

Figure L: 1. I labeled the different strings of LEDs with little bits of colored tape to make sure I didn't mix them up 2. Do not connect a string of LEDs or solder this part together until after **the assembly is placed into the umbrella 3. A string of LEDs is soldered to the hub. Little bits of heatshrink tubing cover the solder points to ensure they won't short out.**

Figure M: The battery clip is only loosely in place at this point, and the LEDs are a glowy mess before they're carefully fitted into the umbrella

Figure N: Things are roughly in place

it runs around the center shaft and rests between the cloth and the spines. Carefully slide the strings of LEDs under the spines until you have two strings in each 1/8th section of the umbrella.

Once everything is roughly in place (Figure N), it's time to attach the open ends of the hub together. Cut, strip, and twist the wire ends together. Once tied together, place some newspaper between the hub wires and the umbrella's cloth so that you won't burn the cloth. Solder the wires together. Once that is done, add tape to that new section on the hub so its shape and spacing matches the rest of the hub. Now you should have two wires coming off the hub. These wires will go to the battery clip/power switch/dimmer assembly.

Cut little bits of double-sided carpet tape and start placing them under the hub to keep it in place. Set it down, centered around the spokes with two strings of LEDs between each spine. Once in place, sew the hub to the spokes and cloth of the umbrella. Figure O shows the steps. Figure P shows the approach I used for another Electric Umbrella.

5. Attach the LED strings to the cloth

Things are finally starting to take shape. Now attach the LED strings to the cloth. Carefully stretch

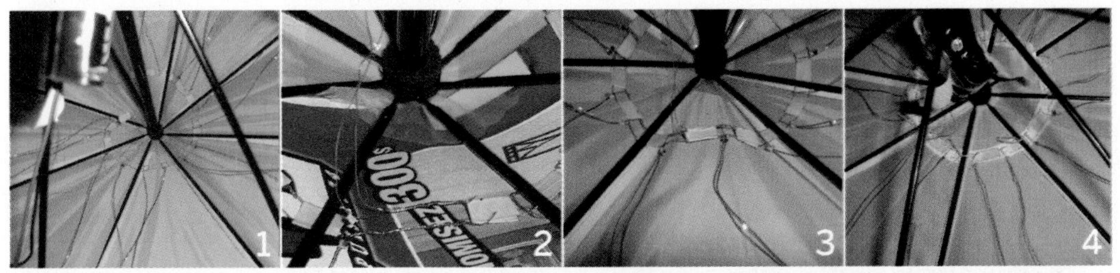

Figure O: 1. Two LED strings per section 2. Two wires going to the batteries/switch/dimmer, and the ends of the hub twisted together 3. The ends of the hub soldered and taped over 4. Sewn in place (the white stuff is double-sided tape)

Figure P: I used a simplified—and more durable—hub on the 2nd umbrella I put together

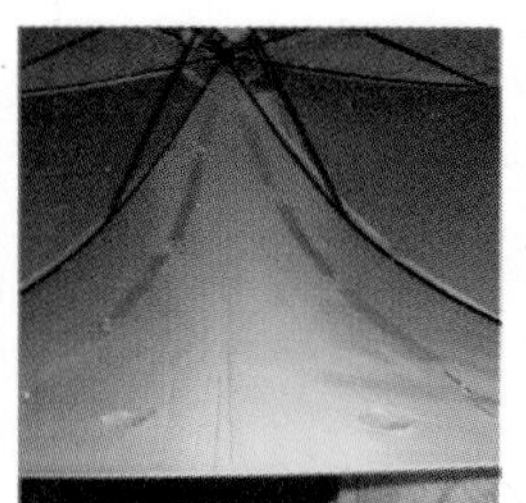

Figure Q: Strings of LEDs are held down with masking tape for now

Figure R: The LEDs in place

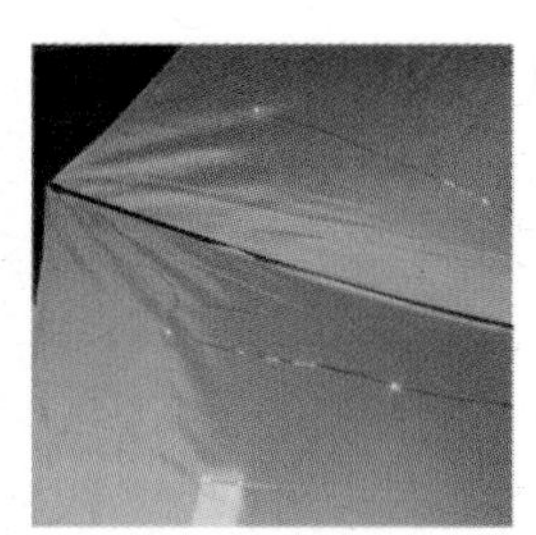

Figure S: Hard to see, but the LEDs are being sewed on here

Figure T: 1. Needle and thread 2. Wires being sewn to the cloth

Figure U: A wire and an LED, both sewn down

out the wires outwards towards the edges of the umbrella. I used masking tape to lie them flat on the cloth (Figure Q).

Once the strings of wire are positioned, you may use a bit of Super Glue under each LED to set them on the cloth, as shown in Figure R. Make sure the wires are not twisted and that all the LEDs are facing up—or rather, facing the person holding the umbrella.

Once they're all set in place, remove the masking tape and cut off the excess wire at the ends/edges. Try adding power to see what the umbrella will look like. It's not finished yet, but this is the point where it finally looks like it is and you can see the lighting effect for the first time.

The Super Glue is not good enough to hold the LEDs in place forever. It's just temporary to keep everything in place while you sew all the wires and LEDs into place. I used small stitches—one on each LED and one on the wire halfway between each LED, as shown in Figures S, T, and U.

6. Add the on/off/dimmer control

In order to add the on/off/dimmer control in the umbrella's handle, you need to drill some holes and run wires down the umbrella's shaft.

Drill one small hole in the shaft at the top of the umbrella—just large enough to let the fine copper wires through, but don't insert wires yet. Next, drill out the umbrella's handle—carefully drill a hole all the way through the handle and into the metal shaft of the umbrella. Be careful to use a drill bit just slightly smaller than the diameter inside the shaft and drill carefully right down the center of the handle until the bit pushes through. You want the hole to be large enough to let the drill shavings that went up the shaft to fall back out again. Try to get them all out.

Next you need to drill a larger hole in the handle, just deep enough for the dimmer switch to rest inside. I used a 3/4 inch bit for this, and then drilled out one side some more for the irregular shape of the dimmer switch. Again, try to get all the shavings out of the umbrella itself.

Now, it's time to run two single strand copper wires down from the top of the umbrella and out the bottom of the handle. This part is really tricky as the wires could get bunched up inside. Pull them out again if they get stuck and try again. If the wires get too bent up, throw them away and try again with new wires.

Once you get them through the bottom of the handle successfully, solder one of the wires at the top to your battery clip, and the other to one of the wires leading to the hub/LEDs. The other wire from the hub goes directly to the second wire on the battery clip. At the bottom (the handle end), solder the wires to the switch and dimmer such that you can disconnect power altogether by turning/clicking the variable resistor all the way counterclockwise, and so that the LEDs grow brighter the more you turn it clockwise.

Once you've tested that it all works, glue the variable resistor in place with epoxy or other glue.

Figure V: 1. 750Ω variable resistor with on/off switch 2. Small hole going all the way through the handle 3. The hole where the dimmer will go 4. The dimmer being fitted into the hole 5. Note the small hole for wires to go in at the top 6. Finished assembly. The batteries slide down a bit as the umbrella is folded closed.

If possible, find a nice decorative knob to put onto the variable resistor. Figure V shows the procedure for adding the dimmer control.

7. Finishing up

Finally, attach the battery clip to the shaft. I left mine tied, but free to move up and down a bit—this way it can move down when you close the umbrella (further from the tip is better as it closes up pretty

Figure W: The finished Electric Umbrella from the top

Figure X: The view from inside

Figure Y: The Electric Umbrella in motion

tight and you don't want to add extra stresses onto the delicate wires on the LEDs), and moves up when you open it (the folding mechanism pushes the battery clip closer to the hub). And now you can take it out for a spin!

It looks amazing (Figures W, X, and Y), but is a bit delicate. Don't take it out in the wind—I don't know if it would survive the umbrella reversing itself in the wind! Also be careful opening, closing, and transporting the umbrella so as not to put too much mechanical stress on the fine wires.

John Kowalski is a software engineer who has a passion for electronics, art, photography, and anything retro. Oh, and MacGyver is his hero.

Etching Brass Plates

How I etched a decorative brass plate for my laptop lid

By Brian Tang

Figure A: Would you like to come over to my desktop and see my etchings?

My brass etching method draws heavily from this instructable (www.instructables.com/id/Make-your-own-Embossed-Business-Cards-using-Acid-E) and this website (steampunkworkshop.com/electroetch.shtml), so I thank the respective authors for their sterling work. There are many different ways of doing this, but when I was researching it, there weren't many thorough tutorials, so this may still be useful for some people. If you're interested, do scour the Internet and you'll find a wealth of information about toner transfer and etching—some good, some bad, some just plain puzzling.

The artwork I used is a piece called Tribal Eagle (http://tinyurl.com/3lh36n) by *xx-trigrhappy-xx and is used with permission.

1. What you need

- Brass plate (I got mine from eBay)
- Fabric iron
- Computer to prepare artwork
- Laser printer and laser OHP acetate (or do what I did and get a print shop to photocopy it onto acetate for you)
- Permanent OHP marker (useful for touching up the mask before etching)
- Insulating (PVC) tape (get at home or hardware stores)
- Masking tape (ditto)
- Scissors
- Non-metal kitchen scouring pad
- Kitchen towel
- Acetone (optional—but useful. Nail polish remover is mostly acetone and will do nicely.)
- (Hydrated) ferric chloride crystals
- Fine wet and dry sandpaper
- Some form of sanding block (I just used a scrap of wood I had lying around)
- Cheap plastic container (you're not going to be able to reuse this for food!)
- Rubber gloves (essential)
- Goggles (essential for safety)
- Dust mask (optional if you're careful and do the painting outdoors—there's not much painting involved)
- Spray paint (I used black enamel satin-finish from Wilkinson)
- Super Glue

2. Preparing the artwork

The first thing is to get your B&W artwork prepared and printed to the right size onto OHP acetate. This can be done directly by a laser printer, or indirectly, by printing onto normal paper and then photocopying onto acetate. The process of toner transfer does effectively reflect the image though, so if you have text, make sure you mirror the image before printing. A word to the wise though: this has to be done with either a laser printer or a photocopier and *not* an inkjet! Only toner will mask off the areas of the brass that we don't want to mask—ink won't work.

Remember, any areas that are black will be masked off, not etched, and will end up brass-colored, whereas any areas that are white will be etched and end up black on the finished plate.

Figure B: 1. Jar of acetone 2. Brass plate for etching 3. Image photocopied onto OHP acetate 4. Image printed four times to allow for screwing up!

Figure C: Lining up the mask. Note that masking tape only goes over one edge of plate. This allows the acetate to be pulled away from the plate easily.

Figure D: Image is mirrored by the toner transfer

3. Prepare the plate

Cut the plate down to the correct size and shape and finish off the edges—it's easier to do this now than after etching. Most importantly though, clean the plate! Use the kitchen scourer to scrub the front of the plate until it's shiny (but it doesn't have to be polished) and free of any tarnish or dirt. Then, use the kitchen towel dipped in acetone to thoroughly clean the surface to remove any oil or grease. Try not to touch the face of the plate now—the cleaner the plate is, the better the toner will transfer later.

Warning: acetone is extremely flammable. If you get any on your hands, it'll dry out your skin, so use some moisturizer on it afterwards.

4. Line up the mask

Align the plate with the artwork on the OHP film and use the masking tape to stick it down carefully. Make sure the tape is only on one side of the plate, as this will make removing the film later much easier. Also make sure the tape is taut, as otherwise, the artwork will move and you'll transfer the toner in the wrong place (Figure C).

5. Toner transfer

Place the brass/acetate sandwich on a heatproof surface (I used some scrap cardboard) and heat up the iron. You want the iron as hot as possible (no steam). Start with the brass-side up and press down on the back of the metal with the iron. This will preheat the brass and help the toner melt. After 20 seconds or so, flip the brass/acetate over *without touching the brass*, which will by now be extremely hot. Place a piece of paper over the acetate, and once again, press down on the brass (this time through the paper and acetate) with the iron, keeping the iron moving. Be careful not to melt the acetate. The actual toner transfer happens very quickly, and I found it needed very little heating with this method (around 5-10 seconds). Take the iron away and immediately peel the acetate back carefully. This must be done while the brass and toner are still hot (which is difficult to do if you're using paper to transfer it). If you let the toner cool first, it will come away with the acetate rather than remaining on the brass. You'll also find that the adhesive on the masking tape will melt with the heat, so it's easier to peel this off while the brass is hot (Figure D). Don't burn yourself!

6. Check and finish the mask

This is the time to check that you're happy with the toner transfer. If you've messed up and need to start over, the acetone will dissolve the toner from the plate easily and you can try again (with a fresh sheet of OHP acetate). If the toner layer is a bit thin, you can carefully align another printed sheet of acetate with your already partially coated brass and try the toner transfer step again. This is one of the major benefits of using OHP film rather than the other options that people have said work (e.g., magazine paper, inkjet glossy paper, press 'n' peel toner transfer paper, etc.). Make sure that the brass plate is free from bits of melted acetate, which can happen if you heat the acetate too much.

If you're happy with it, use the insulating tape to

Figure E: The border is created with tape and the sides of the plate are masked off as well as the back

Figure F: Look closely at the thick part of the wing (compared to Figure E). I've touched it up with black permanent marker.

cover the back, sides, borders (if you want them), and any other large areas that you don't want to etch (Figure E). Touch up any bits of the mask that look slightly thin on toner with the permanent OHP marker (Figure F).

7. Etch

Time to etch! Put on your rubber gloves and goggles. Make up the ferric chloride solution according to the back of the packet in your plastic container and immerse your plate in the solution. The etchant will work much faster if it's warmed—I think the packet recommends 40-50˚C. I placed the container with the etching solution in a larger basin half-filled with hot water to achieve this (Figure H). It will work at room temperature, but take longer.

What I found much more important was to agitate the solution while etching. This ensures a fresh supply of etchant to the plate and will speed up the process tremendously. I just donned my gloves and picked up the plate every 10-15 minutes and used that to swirl the solution around before putting the plate back down and leaving it again. Time-consuming, but it works well and has the benefit of simplicity.

My etching took about three hours until I decided it was done. You can monitor the progress by gently feeling the surface of the etch with your (gloved) fingers. You probably won't see any depth to the etch until it's nearly done. However, keep an eye on the mask, as eventually, it will start to flake off—when this happens, you definitely need to remove the plate, as you'll start to etch in places you don't want to.

Ferric chloride is nasty stuff, so do this outdoors, and don't get it on your skin. Don't breathe the fumes, and beware: it stains anything and everything indelibly.

8. The etched plate

No, we're not quite done yet. But Figure I shows the plate fresh out of the etchant bath. When you're happy with it, remove it from the bath and rinse thoroughly with water. You'll see that my plate is covered with verdigris. This didn't happen on the first two plates I did, but did on this one. I'm not entirely sure why...

9. Disposing of the etchant solution

First: don't bother! You can keep the etchant for the next plate you do (it will gradually take longer to etch the plate as the solution becomes weaker over time). Second: when you do need to dispose of the etchant, do so responsibly. DO NOT put it down the drain! The ferric chloride is corrosive to lots of metals, but more importantly, it now contains dissolved copper that is harmful to wildlife. What you need to do is neutralize the solution with an alkali (something like washing soda) that will precipitate out the copper as a solid. Then, you can dilute the liquid with lots of water and put that down the drain, as long as you filter/decant off the solid copper and dispose of that safely. Check your local rules and regulations for safe disposal of environmental contaminants.

10. Painting the plate

Time to finish it off! Once again, clean and dry is the name of the game here. Get a hold of your spray paint and shake it to mix it. Do some test spraying

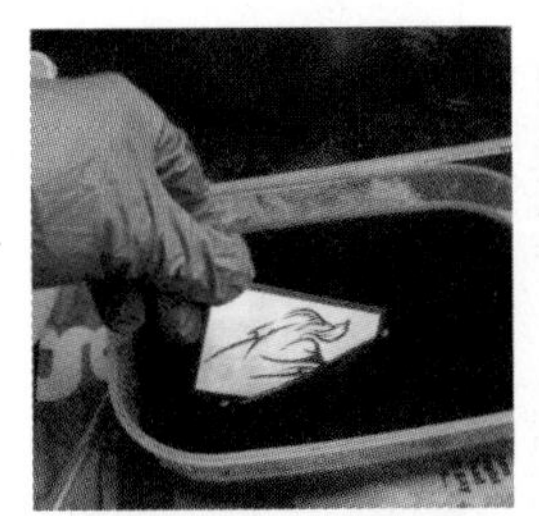

Figure G: Placing the plate in the etchant

Figure H: The hot water bath used to warm the etchant. It's brown because the ferric chloride solution gets everywhere!

Figure I: The plate fresh from the etchant bath after rinsing with water

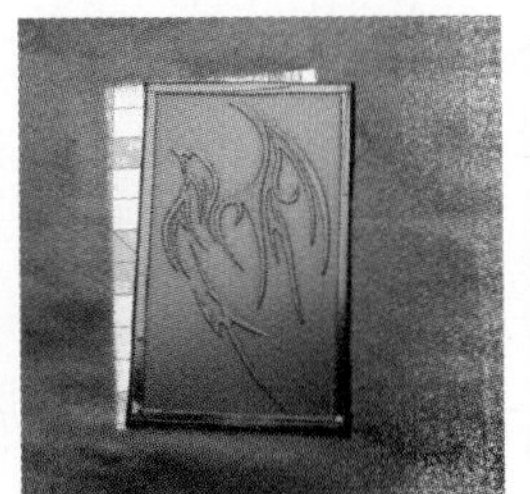

Figure J: Spraying the plate with cheap black enamel paint

Figure K: The finished plate!

onto old newspaper until you're happy with your technique. Remember: light, even passes larger than the object being sprayed, and thin coats. Once you're happy, stick the plate down and paint it! I sprayed mine with three light coats of the paint, letting it dry between each coat. Remember to spray from different angles to ensure that you catch all edges of the etched plate. Ensure the paint is thoroughly dry before continuing. If you're impatient like me, use a hairdryer!

Remove all the tape from the edges and back, and you'll probably find that it's left all sorts of gunk. I'd clean this off carefully by scrubbing gently. Rinse the plate and dry with a kitchen towel.

11. Finishing off the plate

Nearly there. Grab your wet and dry sandpaper and wrap it around your sanding block (the block is just there to keep the paper flat) and sand the surface of the plate. This will expose the brass on any areas that were not etched. Don't worry about the way it scratches the paint surface—it gives it a nice pseudo-aged look.

Once you've cleaned down to the brass all over the unetched areas, add a bit of water to the plate (this makes the wet and dry sandpaper effectively a finer abrasive) and scrub with small circular motions all over the plate to ensure a nice finish. If the paint has been scratched unevenly, then carefully use the paper without the sanding block on the unscratched areas and even it out. When you're happy with the finish, rinse the plate and dry.

Don't forget, if you completely screw up this stage, all is not lost—just clean the plate, respray, and try again!

12. Attach to the object of your choice

Final step: glue it on. I used Super Glue, as it's quite good at sticking mixed materials (in this case, metal to plastic) and can be removed with the appropriate solvent. I'd recommend using the masking tape as before to allow you to align the plate first, flip it up so that you can spread the glue, and then flip it back to stick it correctly into position (Super Glue dries really quickly).

Brian Tang is a student at Balliol College in Oxford, England.

contest WINNER!

Screen Printing: Cheap, Dirty, and at Home

Screen print fabric, shirts, cards, even the dog (well, maybe not the dog), with this easy home printing set-up! By Tracy Rolling

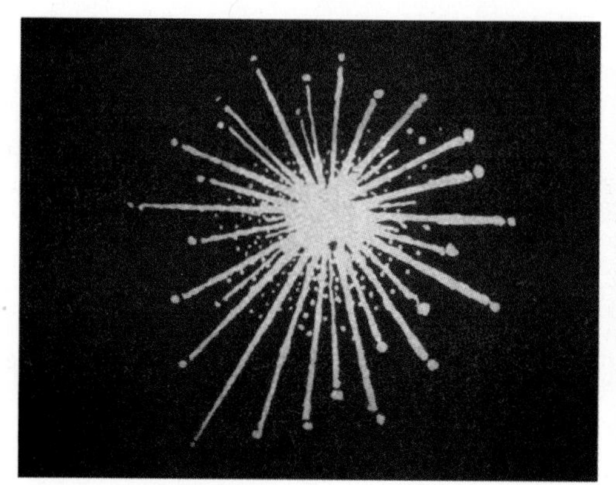

Figure A: A finished print

Figure B: Collect old frames to make your screens

You don't have to spend a ton of money on equipment or have a screen printing studio, to make some pretty good quality prints.

I taught some friends how to reuse old picture frames and curtains to make screens, burn them in the sun, and clean them with a garden hose. While we were at it we took some pictures so we could share the lesson with you.

1. Gather up your materials

- **The image you want to use:** It's best to have your image photocopied onto a transparency at maximum darkness. You can also paint or draw with white out on transparent plastic (cellophane wrap or clear packaging from toys). Another option is to make a cut-out with dark colored construction paper, or to lay some flat object (pieces of lace are nice) on the screen. Objects that aren't flat (skeleton keys, for example) can also work, but then you have to move the screen around to avoid a shadow.

  You want your image on the transparency to be super dark because the image won't transfer to the screen if light gets through. If you want subtlety and shading you can do it with dots, like a newspaper image. In this Instructable we're keeping it simple and only printing one color. Very fine lines are not recommended with this technique. Start out with something big and bold and then start experimenting.
- **Wooden picture frames that are completely flat on the front surface:** You can find these types of frames in a wide variety of sizes at the Goodwill or Salvation Army or at garage sales for about a dollar each. You will also be using the pieces of glass that come with the frames. You will need a piece of glass from a frame smaller than the frame you use to make your screen.
- **An old, gauzy curtain:** Color doesn't matter, but it does need to be in reasonably good condition. It can't have too many holes. The more tightly woven the curtain, the more fine your print can be, but you can get pretty nice results with any

For additional information, discussion, and more, please visit the Instructables project page:

Figure C: Take apart the frames

Figure D: Begin stretching the fabric over the frame

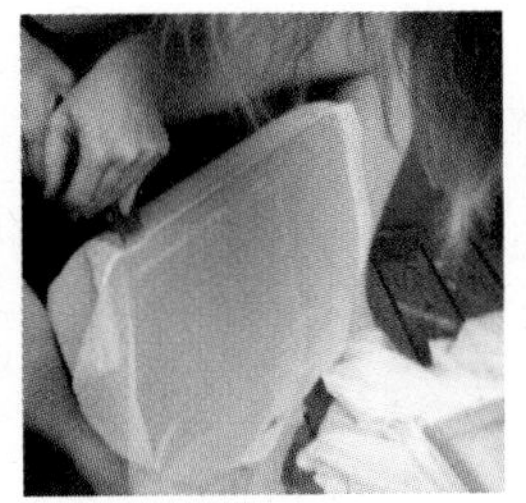

Figure E: Staple the fabric along the edge of the frame, stretching it as you go, and keeping the grain straight

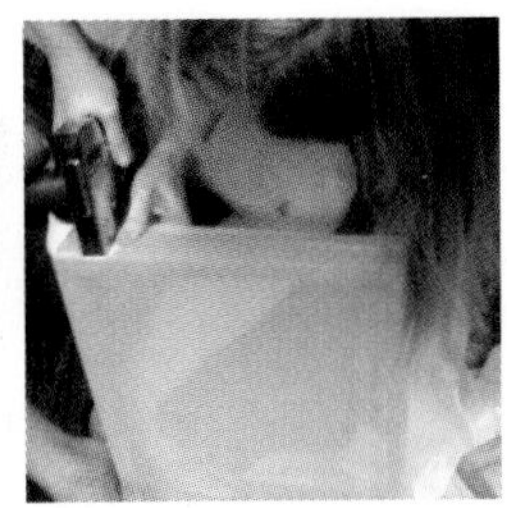

Figure F: Finish stapling the fabric on—don't staple your hair!

gauzy old thing. I keep my eye open for these at thrift stores and yard sales.

- **A piece of black or dark-colored fabric big enough to put the frame on**
- **A staple gun and staples:** Don't get staples that are too long or they'll poke out through the frame. Even that isn't such a big deal, but it's preferable to not have sharp little metal points sticking out along the inside of your screen.
- **Photosensitive goo and activator:** Speedball is the most common brand you will find for this at the art store. You need the emulsion and the activator and they come in two different bottles that you have to mix together. Don't bother with the screen cleaner unless you want to reuse the same screen for other designs.
- **Art squeegee:** I recommend buying an art squeegee especially for screen printing. You can get along without one, but it's a lot easier to print with this tool than to do it with a hunk of cardboard. But in a pinch, the hunk of cardboard will work, too. The lip of a box works best because it has a good straight edge and is rigid yet flexible.
- **Screen printing ink:** You can get this at the art supplies store. I have also printed on wood with acrylic paint and gotten good results.
- **Masking tape**
- **Old cereal boxes or similar kinds of cardboard scrap:** You'll want to have a little supply of pieces of thin cardboard around. They are super useful for all kinds of things, like scraping ink off screens and putting it back in the jar.
- **Old newspaper to protect the surfaces you're working on:** If you are printing T-shirts you'll need paper to put inside the shirt when you print so that the ink doesn't bleed through the side you're printing on all the way through to the other side. I use regular printer paper for this, but old newspaper would work fine.
- **A garden hose:** It's best to have an attachment on the hose that shoots the water out with some pressure, but you can get away with not having one. I have used the scratchy side of a kitchen sponge to help me get the emulsion off while spraying the screen with the hose. It damages the screen a little, but it works. Even rubbing with your hand helps.
- **An old rag for spills**
- **Clothes you don't care about:** You're going to mess up your clothes.

2. Build your screen

Take apart the picture frames. Remove all the little metal bits and put the glass aside. Later in the process, you will need a piece of glass from a frame smaller than the one you are using for your screen. Watch out for the edges of the glass; don't cut yourself.

Cut a piece of the curtain larger than you need to cover the front of the picture frame and wrap around the edges. You are going to stretch the curtain over the frame just like stretching a canvas for painting. It helps to keep an edge, so that you have a straight line to go by.

Stretch the fabric over the front of the frame and staple it into place. Try to keep the fabric as straight as possible. If you staple it on diagonally, with the grain of the fabric too far off from the square of the frame, you will run into trouble later. You don't have to be too anal about it, just try to put it on straight.

Staple the fabric onto the sides of the picture frame fairly close to the front edge. Pull it hard to make it as taut as possible.

I recommend putting a couple staples on one

Figure G: Mix emulsion and spread it onto the screens

Figure H: Put the screens in a fairly dark place to dry

Figure I: Burn the image into the emulsion by placing the image transparency over the screen

Figure J: Hold the image flat with a piece of glass and expose it to UV light. Watch out for shadows!

end, then a couple on another end, then on one side, and then another, and keep going round and round, pulling all the time.

Cut the excess fabric off around the sides of the frame. Don't cut too close to the staples or the fabric will fray and come loose.

That's it. Now you have a screen for printing.

3. Spread the photo emulsion onto the screen

This is where all those scraps of cardboard come in handy. The backs of old notebooks work particularly well.

Read the instructions on the emulsion and activator bottles and follow them carefully to mix them together properly. Pour a little bit of the emulsion onto the screen and spread it as evenly as possible onto the screen. You have to coat both sides. You can scoop the excess back into the pot of emulsion. Don't put it on too thick. You need a thin coat, as even as possible, and on both sides. Try to avoid drips. Again, you don't have to be a perfectionist. Just do your best.

Once you've got the emulsion on there, put the screen in a darkish place to dry. Closets work pretty well. It doesn't have to be totally dark like a photo lab or anything. I like to point a fan at the screen to help it dry faster.

Once it's dry you're going to want to go straight on to the next step so that the emulsion doesn't get exposed to the light before the image is in place. If it does, it will harden completely on your screen and not just around the image. You can touch the screen to feel if it is dry.

4. Burn your image

Once the screen is dry, take it out in the sun and lay it face down on a piece of black cloth. The back of the frame is facing up and the screen is flush against the cloth. You need UV rays to expose the emulsion. Even on a cloudy day, you can get enough light for exposure. A lamp at night will not work.

Now place your transparency (or bit of lace or leaves or whatever) on it, inside the frame and lay a piece of glass smaller than the picture frame on top of the image.

Leave it there for a while.

The sun is going to harden the emulsion. You'll notice it change from a lighter green to a sort of bluish color. Be careful of shadows. Even the piece of glass can cast a shadow. It's a good idea to carefully move the whole thing a little bit to avoid shadows. If you do end up with a shadow, you can always patch it up later by dabbing a bit of photo emulsion on the gap. Still, it's preferable not to have to do that.

Any part of the screen that doesn't get hit by the sun is going to wash clear. Be very careful not to let your image move around. The glass helps keep it in place and also makes sure no sun gets in under the edges of the image.

How long this takes depends on how much sun you've got.

5. Rinse off your screen

Once you feel pretty confident that the photosensitive emulsion is hardened in the sun, take your screen to the garden hose and start getting the emulsion off where your image was.

This is easiest to do with a lot of water pressure, but it's still possible to do without a hose attachment. Rub the screen with your hand to help the emulsion wash out. I've even gently rubbed with the scratchy side of a kitchen sponge. This can damage the screen a little bit, but it works pretty well. Don't

Figure K: Rinse the screen carefully

Figure L: Mask the edges around the image

Figure M: Spread some printing ink along the top of the screen and squeegee it over the image

Figure N: You can use several screens for multiple images!

scrub hard, or you'll screw up your screen.

I find this part to be the hardest part. I always get a little frustrated that the emulsion doesn't wash off fast enough and worry that it'll harden while I'm in the process, especially on a sunny day. If your image was opaque enough you shouldn't have any troubles, though.

6. Print!

Now that you've got a screen you are ready to print.

Mask the edges. The main weakness of these home screens is the edges. It's a good idea to put some masking tape along the edges of the screen before you start printing so you don't get any sloppy leaks off the sides.

Place your fabric on some papers to protect the surface underneath. If you're doing a T-shirt, put a piece of paper inside the shirt.

Put the screen down on your fabric, put some printing ink along the top of the screen, and pull it over the image with your art squeegee. If you can't afford a squeegee, use a piece of cardboard.

Be very careful not to let the screen move while you are printing. Hold it down firmly. You absolutely have to print on a good flat surface or you will get terrible results.

Lift the screen and admire your work. You rule!

Tracy Rolling (aka tracy_the_astonishing) is a semi-professional dilettante. She can build just about anything and she also throws a great party. Tracy currently lives in Berlin, Germany with her husband and son. She has worked as a translator, teacher, costume designer, fashion designer, waitress, photo assistant, internet community director, and detassler.

Controlled Bleaching with Discharge Paste

A quick and easy way to create designs on fabric By Ed Lewis

Figure A: A finished bleaching job

This is more of an introduction to discharge paste than a full-blown Instructable. Yes, the name is terrible and the stuff is white and gooey which makes it even worse. But it's pretty damn cool stuff, so you just accept it and move on towards greatness.

Where bleach is incredibly thin and can destroy natural fibers if you're not careful, discharge paste is the opposite. It's gooey, so it doesn't spill easily and it can be used with silkscreens. It's also nice to natural fibers, which is very good as well. It removes most fiber reactives, direct dyes, and acid dyes, and typically leaves a light golden color when it's done.

Everything you need to know about this chemical agent is printed right on the label. To use it, you just apply it to a fabric, let it dry, then iron it on the lowest steam setting for a few minutes to activate it. To make sure it works on a new fabric, test it by putting a small spot in a hidden area of your target material (assuming you're using a shirt, for instance) and see how well it works before spreading it everywhere.

Discharge Paste is available in the U.S. from Dharma Trading Co. www.dharmatrading.com/html/eng/1574-AA.shtml.

1. Apply to the fabric of your life

Put down a stencil or a silkscreen or forgo all premade plans and apply the paste directly to your fabric. If you want to let the goo seep in a little deeper, you can thin it out with some water. You can just go for it or keep on testing on more areas of the fabric.

Used shirts are cheap to play with, but for consistency and thin shirts, go to the underwear aisle and get the dyed T-shirts. They're usually less than $5 each for decent ones and they tend to be thinner than Beefy Ts that make me sweat like I still live in Southern California.

2. Make magic with your iron

At first, there's not much to see on the shirt. Even when you wait until the paste dries, there's still not much there. This all changes dramatically when you put the iron to the fabric. The color quickly disappears, and POW, there's your design!

One more thing: be sure to work in a well-ventilated area. This stuff stinks, and when your fabric is done, it too will stink until you wash it. So don't run out to a bar to show off your new threads right away 'cause you'll destroy your shirt's first impression with a resounding "P-U!"

Now clean up your mess and start planning your next project.

3. BONUS! Before and after... again

Another example of how the paste looks dried on black cotton before and after it's been ironed (Figures H-J).

Ed Lewis lives in Oakland, CA, with his wife Maria and Biscuit, the stunt cat. Being outside, cooking, making wine, and building funny objects are current favorite activities.

For additional information, discussion, and more, please visit the Instructables project page:

Figure B: A 1 qt. bottle of Discharge Paste from Dharma Trading

Figure C: Using a stencil

Figure D: Applying the Discharge Paste

Figure E: Make sure to achieve full and even coverage

Figure F: You won't see much of your handiwork until after you iron it

Figure G: Nice job!

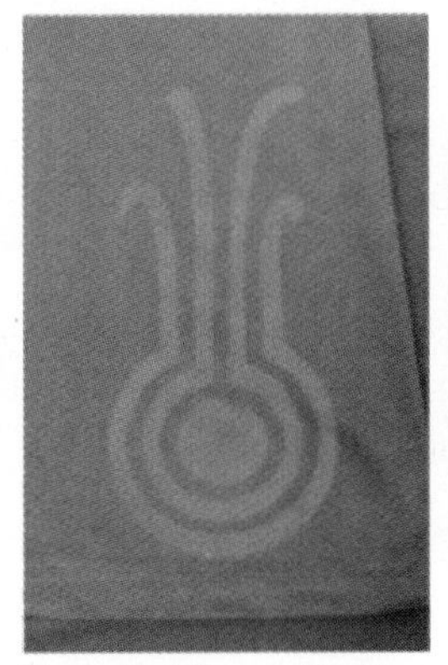

Figure H: After application, before ironing

Figure I: After ironing

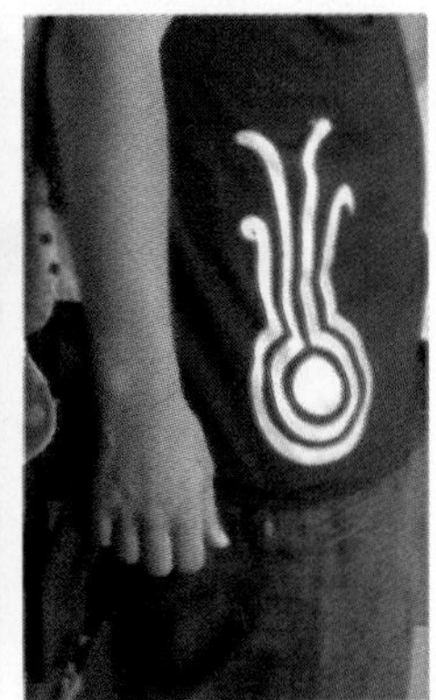

Figure J: Ready to roll

Paracord Bracelet with a Side-Release Buckle

Transform a simple piece of cord into this fantastic bracelet!

Figure A: The finished bracelet

By Stormdrane

This tutorial will show how to make a paracord bracelet with a side-release buckle. When made on a larger scale, you can make this for use as a dog or cat collar as well. I get my paracord from the Supply Captain (www.supplycaptain.com/index.cfm?fuseaction=page.Store) and the side-release buckles from Creative Designworks (cdwplus.com/singleSRB.html). More projects, links, and knot references can be seen on my blog page, Stormdrane's Blog (stormdrane.blogspot.com).

1. Materials

You'll need:

- Paracord, or equivalent 1/8" diameter cord
- Tape measure or ruler
- Scissors
- Side-release buckle
- A lighter (torch lighter works best)

The amount of cord used can vary, but for this example, we'll start with 10' of paracord. Actual amount of cord used for the bracelet is about 1' of cord for every 1" of knotted bracelet length. So if your wrist is 8", you'd use approximately 8' of cord.

2. Measure wrist

Wrap the paracord around your wrist and make a note of where the cord meets. Hold this point next to your ruler or tape measure and that's your wrist size.

3. Find the center of the cord

Hold the ends of the cord together and find the center of the loop. Take the center of the cord and pull it through one end of the buckle (either side of the buckle). Now pull the cord ends through the loop until it's tightened up and attached to the buckle.

4. Finding the bracelet length

Take the buckle apart and pull the free ends of the cord through the other part of the buckle, sliding it up towards the attached part. You're going to measure the distance between the two buckle ends for the bracelet size for your wrist. Add about 1" to your measured wrist length; this will make the finished bracelet a comfortable fit. You're measuring from the end of the female part of the buckle to the flat part of the male end of the buckle (the part with the prongs). The prongs don't count for the measurement because they fit inside the female part of the buckle when the bracelet is closed.

5. Start making the knots

The knot used for the bracelet has a few different names: cobra stitch, Solomon bar, and Portuguese sinnet. Take the cord on the left side and place it under the center strands running between the buckle ends. Now take the cord on the right side under the left side cord, over the center strands, and through the loop of the left side cord. Tighten up the cords so the half knot you just formed is next to the buckle. Now take the right side cord under the center strands. The left side cord goes under the right side cord, over the center strands, and through the loop of the right side cord. Tighten up the cords; not too tight, just until they meet the resistance of the knot. Now you have a completed knot. You will continue alternating the left and right sides as you go. If you don't alternate, you'll quickly see a twisting of the knots; just undo the last knot and alternate it to correct (Figures I-L).

6. Continue knotting

Keep tying the knots until you have filled the space between the buckle ends. The knots should be uniform from one end to the other. Tie each knot with the same tension to keep them all the same size.

7. Trim the excess cord and melt the ends

You can now use your scissors to trim off the extra cord closely to the last knot you tied. I trim one at

Figure B: Gather your materials

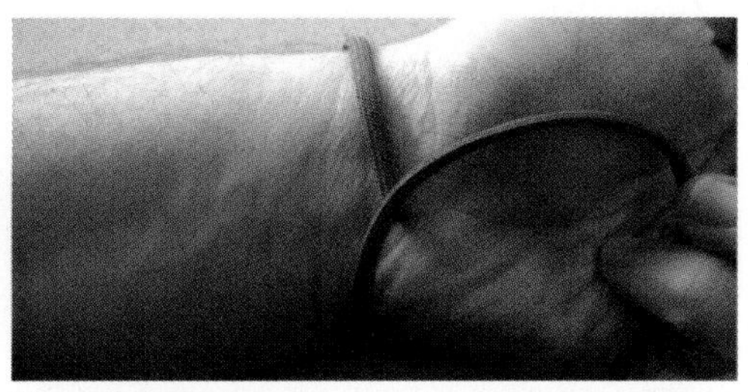

Figure C: Wrap the cord around your wrist to measure

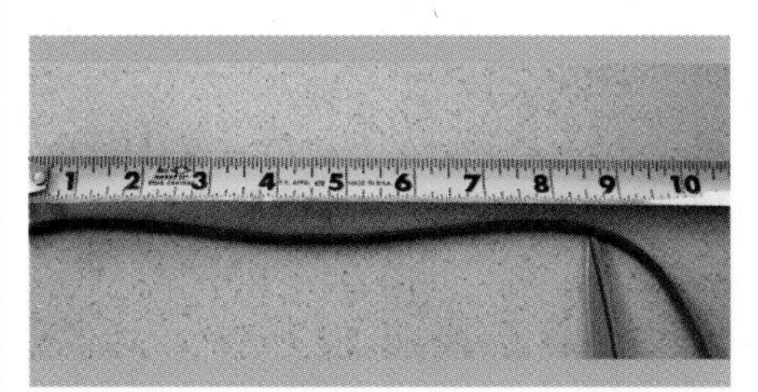

Figure D: Measure the cord to find your wrist size

Figure E: Find the center of the cord and pass through the buckle

Figure F: Pull cord ends through loop and snug up

Figure G: Thread the free ends of the cord through the other part of the buckle

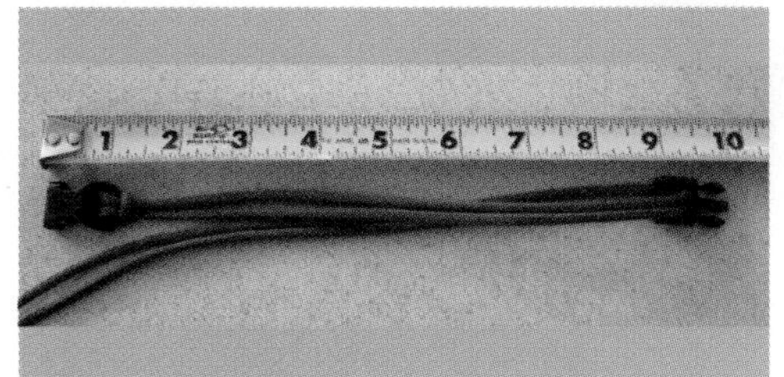

Figure H: Fold over the cord at your wrist measurement plus about an inch

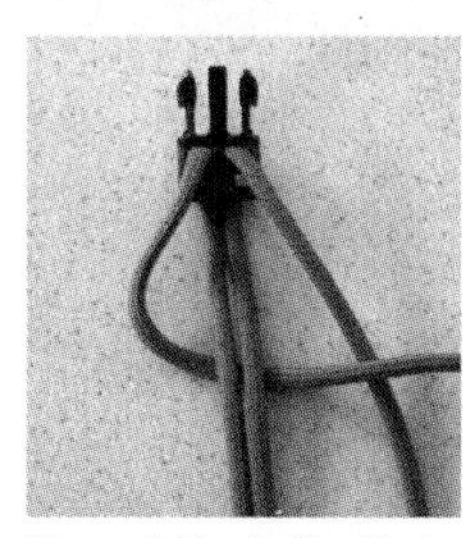

Figure I: Begin the first knot

Figure J: Thread the cord through the loop as shown and snug up to the buckle

Figure K: Begin the second half of the knot. Don't forget to alternate sides!

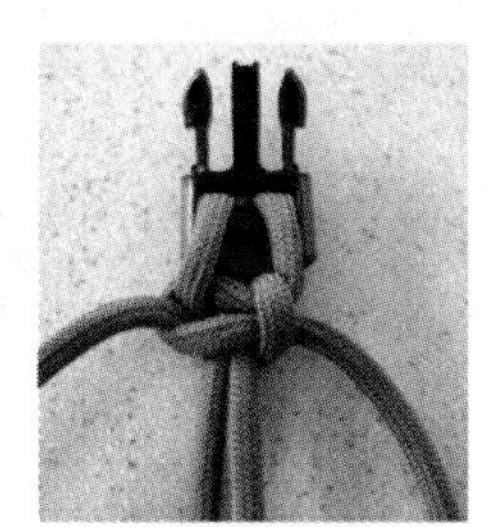

Figure L: Thread cord through loop and tighten to complete the first knot

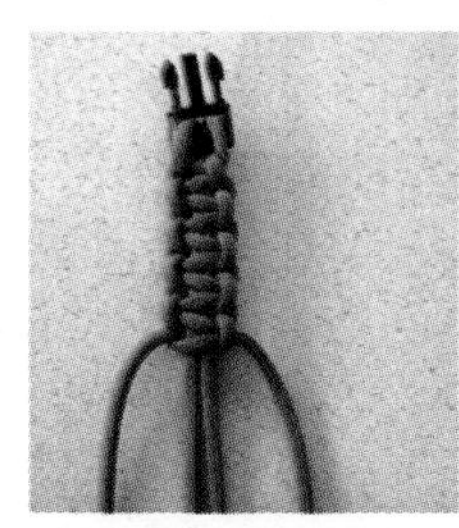

Figure M: Continue making knots to fill the space between the buckle ends

Figure N: Keep the knots uniform

Figure O: Trim the end of the cord

Figure P: Melt the end

Figure Q: Press the melted end into the surrounding cord

Figure R: Another view of the finished bracelet

Figure S: Variation with king cobra stitch

Figure T: Add interest with two colors of paracord

Figure U: Add even more interest with three colors!

Figure V: Basic bracelet and bracelet with another layer of cobra stitches

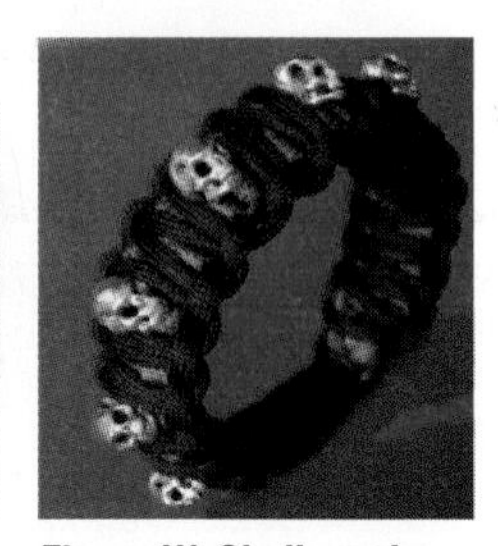

Figure W: Skulls make it scary!

a time, and use my lighter to quickly melt the end. I cut, wait a second for the melted cord to cool just a bit, and then use my thumb to press the melted end onto the surrounding cord so it hardens as it attaches. You must be careful with this step. The melted cord is extremely hot, and it's possible to get burned. Use a soldering iron for the melting step if you wish, or even something like a butter knife to flatten out the melted end of the cord to finish it.

8. You're finished

If you did everything correctly, it should look something like the finished one. Once you know what you're doing, you can vary the amount of cord used by making the knots tighter or looser. Pushing the knots closer together as you go will use more cord.

9. Other variations

Once you have the hang of the basic bracelet/collar, you can add another layer of cobra stitches overlapping the first set of knots, called a king cobra stitch. The amount of cord used for the king cobra stitch is about twice as much as for the regular stitch.

Stormdrane, aka David Hopper, is a former Boy Scout and enjoys a hobby of learning decorative and useful knot work.

A Dozen Red Origami Roses

Make this everlasting bouquet for your Valentine

By Travis Hydzik

Figure A: A lovely bouquet for your sweetie

Give a unique twist to the traditional giving of roses on Valentine's Day. Handmade origami roses are everlasting and inexpensive to make.

1. Inspiration

Looking for a unique way to convey my love, the idea of paper roses popped into my head. A quick Google search and it turned out it wasn't such an obscure idea.

The first result returned Bloom4ever (www.bloom4ever.com), a company that specializes in handmade roses, and they were even nice enough to include a "How to fold origami roses" (www.bloom4ever.com/HowToFold.html) guide. The site lists bouquets of a dozen roses for $80.

2. Equipment and materials

Only the most basic craft stationery is required:

- Scissors
- Ruler
- Craft glue
- Pens/pencils

The most important material is the paper. Unable to find any origami paper, I had to settle for colored A4 paper, and this turned out for the better. I used 80gsm A4 paper. I purchased 100 sheets of red and green for $6.50 AUS. You may be able to borrow some colored paper from your school or work. One sheet of colored paper is required for each rose and one sheet for each stem and set of leaves. You should experiment with different paper sizes and thicknesses to find what best works for yourself.

Other materials include:

- Clear transparent wrap (cellophane) to wrap the bouquet
- 1mm diameter wire—to make the rose stems; nice to have but not a requirement
- Ribbon to tie the bouquet of roses

3. Folding the rose

This rose was designed by origami theorist Toshikazu Kawasaki.

Start with a sheet of A4 paper, and cut the paper into a square. I used a 17cm square, as I found 21cm width (or A4) was too large, and 13cm (recommended by Bloom4ever) was too difficult to fold.

I will not go through each folding step as there are numerous resources already available:

- How to fold origami roses, at Bloom4ever (www.bloom4ever.com/HowToFold.html). Very good instructions, except for step 25.
- Rose, at pajarita.org (www.pajarita.org/aep/internacionales/intern1-3.pdf)
- How to Fold a Paper Rose, at wikiHow (www.wikihow.com/Fold-a-Paper-Rose)

I used the Bloom4ever instructions for the most part. Even though step 25 is broken into steps, I could not work out how to perform the "twist" maneuver.

This YouTube video (www.youtube.com/watch?v=BZnhMI85dq4) saved my life; big credit to the producer.

Figure B: Everlasting paper roses

Figure C: Red and green A4 paper

Figure D: Clear cellophane wrap

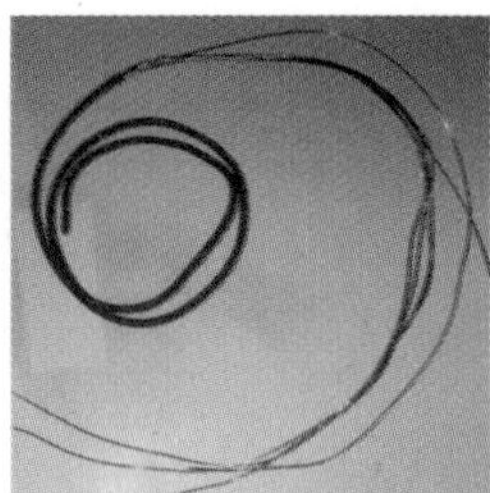

Figure E: Copper wire for stems

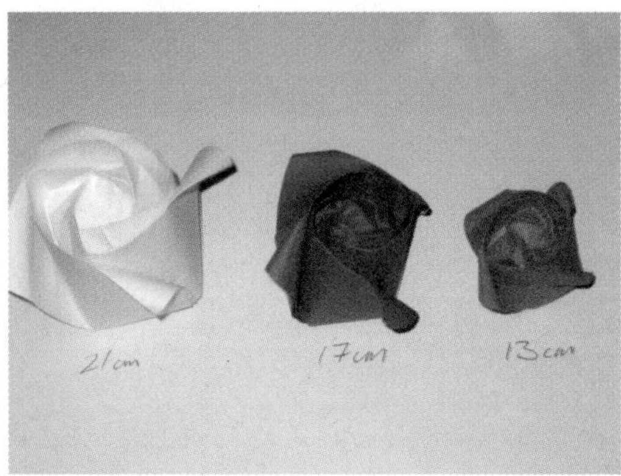

Figure F: Roses from paper squares measuring 21 cm, 17 cm, and 13 cm

Figure G: Rose heads

Figure H: Folding diagram—match the triangles; diagonal folding line; join the set of lines in the lower corner

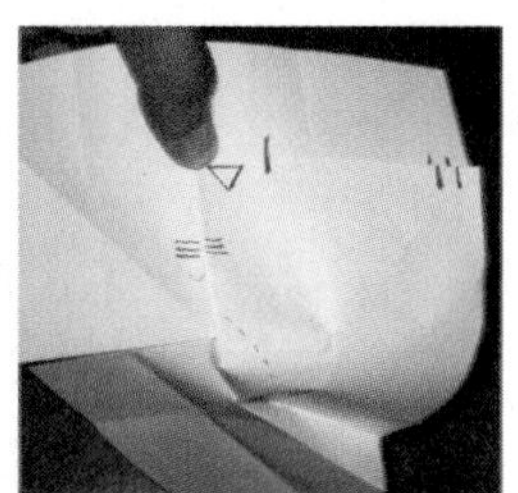

Figure I: It should look like this when folding; note the scrunched up part in the center

Figure J: Start to wrap a stem

Figure K: Glue this last bit to hold the paper in place

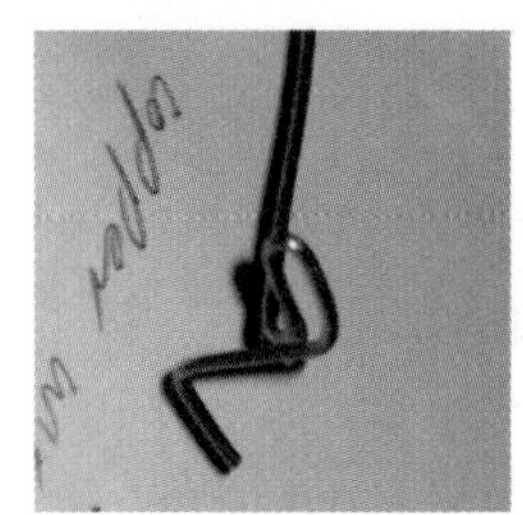

Figure L: Bend the end of the wire into this shape to fit into the rose

Figure M: Group of stems ready for their roses

Figure N: Twist the rose head onto the bent end of the wire

Figure O: Stemmed roses. On to the leaves . . .

Figure P: Draw the leaves on the green paper. Note the stipules; these will be glued onto the stems.

Figure Q: Cut lots of leaves

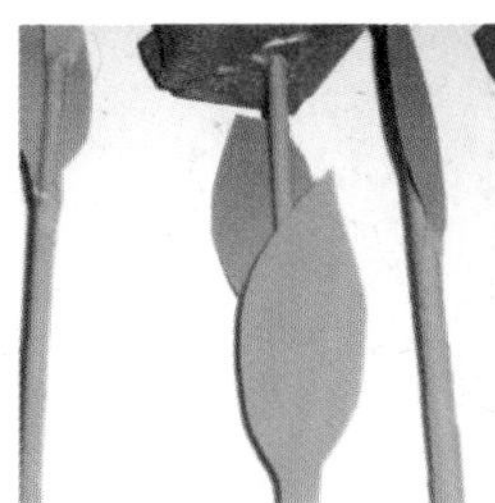

Figure R: Gluing leaves onto the stems

Figure S: Curve the leaves a bit after the glue has dried

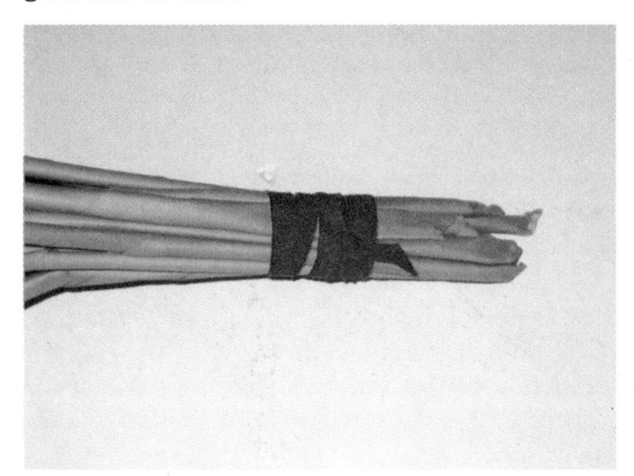

Figure T: Tie the bunch of flowers together with a ribbon

Figure U: A beautiful bouquet, ready to be wrapped in cellophane

4. Stems

I found some 1.86mm copper wire lying around that could be used for the stems. The advantage of copper wire over a paper-only stem is that the copper can be bent allowing for better arrangement of the individual roses in the bouquet.

The copper wire was first straightened by pulling on the wire with one end in a vise. Once straightened, the wire was cut into 30cm lengths. Each wire was then rolled with green paper and the end glued. It is possible to make stems without the wire by only rolling paper.

I needed a simple method of attaching the stems to the rose heads. Each wire end was bent into a shape that allowed the rose heads to simply be twisted onto the wire. Glue was then applied.

I also think it would look nice to make the rose heads from silver metallic paper, with copper or silver stems.

5. Leaves

After Googling "rose leaf" for ideas, basic leaf shapes were cut out of green paper. Each leaf had a little stipule allowing it to be glued to the main stem. There were two leaves per stem, glued in an alternate pattern.

Once the glue dried, the leaves were slightly bent through the center axis and shaped to give a more pleasing look.

6. Making the bouquet

A dozen roses were bunched and arranged together to form a bouquet. The stems were securely tied together with red ribbon. I used clear transparent cellophane to wrap the bouquet and fastened it with more ribbon. I made a simple card to accompany the flowers.

Travis Hydzik says, "These roses were to result of wanting to make something special for my girlfriend on Valentine's Day."

Blank Books from Office Paper

How to recycle paper into blank books

By Becky Stern

Figure A: The completed tome. Plenty of space for notes.

Figure B: All the materials we'll need

Figure C: Tools—a square corner ruler, scissors, large binder clip, an awl, and a cutting surface (not shown—sewing/bookbinding needle)

In this Instructable, I will demonstrate a simple bookbinding technique using a Japanese stab-binding method for making blank books from paper that's printed on one side. These books are useful for all kinds of notes, and tell an interesting story about the place they came from. I work in the computer lab at my school, where a lot of printer paper is wasted. I go through the recycle bin to find my paper.

This is a great little book for phone numbers and random notes. You can make it any size you like, and the paper never has to go to the processing plant! Using string binding instead of glue is easier on the environment, too!

Materials:

- Recycled paper (blank on one side)
- Thicker recycled material (postcards, envelopes, cardboard, etc.) for covers
- Twine, yarn, or other string

Tools:

- Awl, drill, or drill press
- Large sewing needle or bookbinding needle
- Paper cutter, scissors, or utility knife
- Cutting mat
- Ruler

1. Cut and fold your paper

Using a paper cutter, scissors, or a utility knife, cut your paper down to twice the desired size. Fold each sheet in half, and cut your cover material (one for front, one for back) down to the size of a folded sheet (Figures D-E).

Figure D: Cutting the inside page to twice the desired size

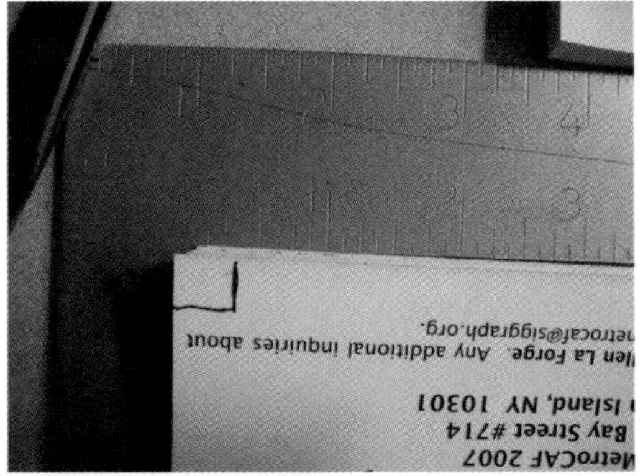

Figure G: Marking 3/8" from the corners of the bound edge

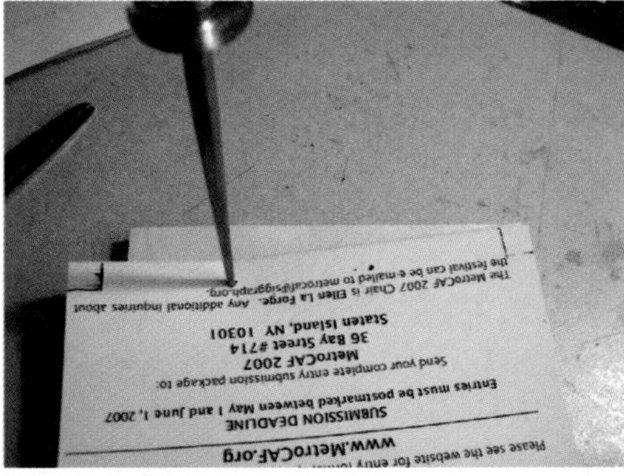

Figure H: Drill holes along the binding edge

2. Line up, clip, and make holes

Stack your cover material and pages together and line up all edges. Clip with the large binder clip to secure (Figure F).

Mark 3/8" from both to-be-bound corners (the

Figure E: Each inside page folded in half

Figure F: Lining up the edges of the inside pages and cover. Secure with clip.

User Notes

Eric Cushing says: Here's a fancy version I did a while back.

A couple tips: A nice bead of Tacky Glue on the spine works really well. Hot glue is a little stiff in my experience and yields a book that is tough to open and keep open. If you go to a paper/printing store, you can find something called Padding Compound. It's a glue specifically designed for binding the spine of notebooks.

Other cool items to consider including as pages are:

- Envelopes
- Acetate (clear sheets, very cool. You can even make a whole clear book, use a lighter to melt the spine together.)
- Large leaves
- Etc., etc....old clothing?

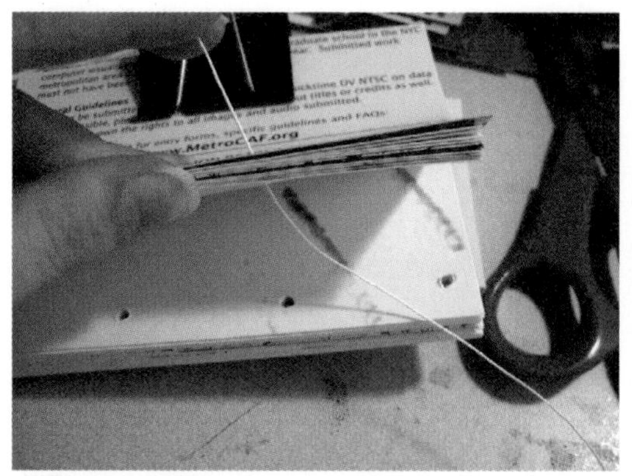

Figures I-L: Sew through the binding holes as shown in these four steps

Figure J: Keep sewing!

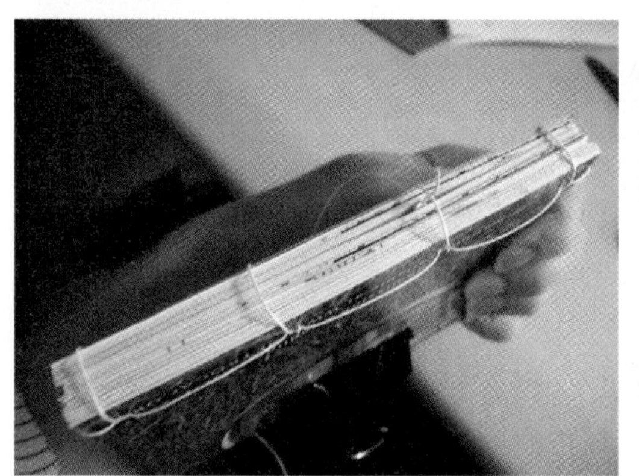

Figure K: Almost there

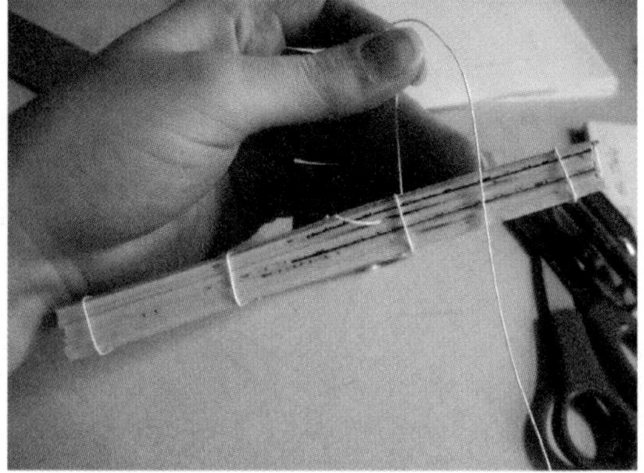

Figure L: Tie it off

Figure M: What the binding should look like when you're finished

folded side of the sheets). Divide the space along the binding between these marks into thirds, and mark those locations. These are where we make the holes for the twine binding (Figure G).

Make holes all the way through the book with an awl, drill, or drill press. If using a drill or drill press, clamp your papers down tightly to avoid any paper ruffling. Placing a piece of wood under the book helps make a clean cut in the back (Figure H).

3. Bind the book

I used this tutorial (www.sff.net/people/Brook.West/bind/bindit.html) to learn to bind books this way. It's very thorough, so I won't repeat its instructions, but basically you sew the binding in a particular way with the twine and needle. Make sure it's very tight and secure (Figures I-L). That's it!

Becky Stern (sternlab.org) is an artist and a blogger for MAKE and CRAFT. She's currently a grad student at Arizona State University, where she studies sculpture.

How to Make Your Own Plastic Vacuum Former

Make a plastic molder using parts from around the house

By Adam Harris

Figure A: A mini R/C car body copied with a peanut butter jar vacuum former

Plastic vacuum formers are an important part of the prototyping process. If you need a nice plastic robot body or custom case for a project you are doing, get your tools, 'cause this project is easy to build and fun to play with.

Vacuum formers are based on a simple concept. They use the power of a vacuum to suck heated, gooey plastic sheets very tightly around an object you place in them, making a 3D copy of pretty much whatever you want.

Plastic vacuum formers are usually big, expensive machines, however many of us don't always need to make huge pieces for our projects, so these machines would be pointless to have—or at least that's what I tell myself so I won't be tempted to buy one. Our vacuum molder will be a good size for most projects you're likely to deal with.

1. Gather the pieces

The main parts to this machine are:

- A "workspace," the place where the object to be copied is set and the magic happens. This has holes drilled in it so the suction is fairly even over the entire surface.
- A hollow cavity, like a strong, airtight box. This is also used to get the same approximate suction on all parts of the top, while adding support to the machine.
- A vacuum cleaner (shop vacs are a pretty good choice because they have a lot of suction, but a normal vacuum cleaner will work, too). This is the source of our vacuum/suction.
- Two frames to hold sheets of plastic. These can be two picture frames, or they can be made of Popsicle sticks.

The first vacuum molder I made was supposed to just be a test run but it worked so well that I now use it for any small parts I make. In this Instructable, I will first show you how to make one of these. You don't have to build it, but it will illustrate the concepts of how and why the machine works. Then I'll show you how to use it. A bigger machine is described later in this article.

Parts list for very small plastic molder:

- A plastic peanut butter jar or similar. (Don't use glass as you will need to cut it.)

Figure B: The jar lid with a matrix of holes drilled in it

Figure C: Hole cut in side of jar to accept neck from soda bottle

Figure D: Attaching the vacuum port (soda bottle neck)

Figure E: Wrapping the jar with Saran Wrap

Figure F: Sealing the jar/ Saran Wrap with electrical tape and attaching the lid

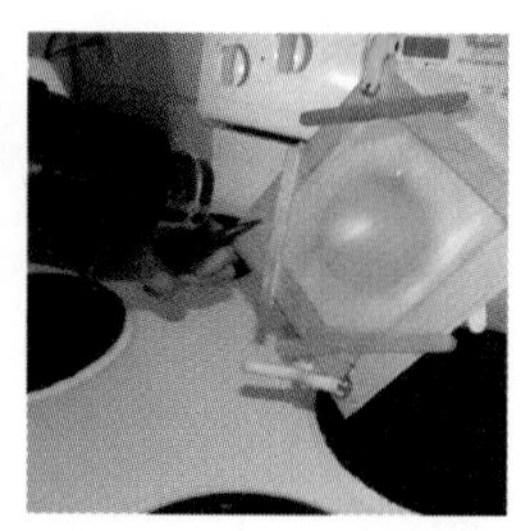

Figure G: Using a heat gun to soften the plastic stock over the frame

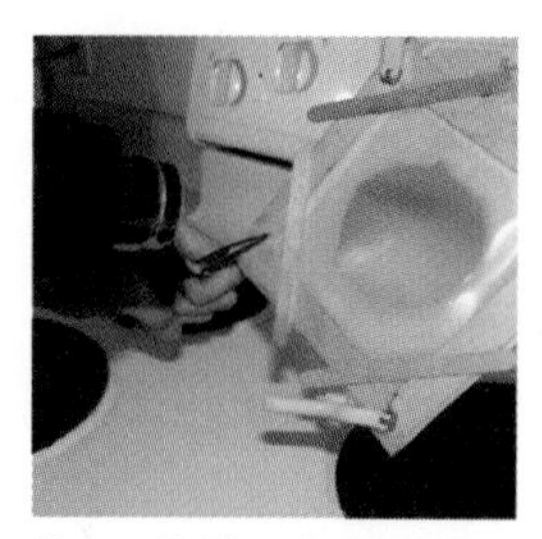

Figure H: The plastic will get gooey and flexible when it's ready

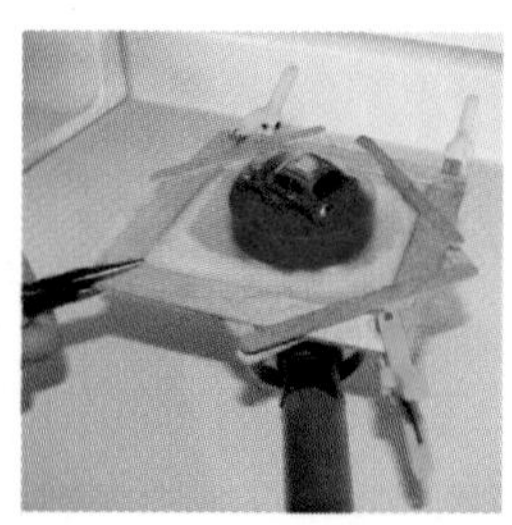

Figure I: Applying the vacuum to the vacuum former

- A 2-liter Coke bottle or similar
- A home vacuum cleaner, or similar
- A few sheets of Saran Wrap
- Some good tape, like electrical tape or duct tape

Tools needed:

- A good sharp cutting knife or razor (be careful not to cut yourself!)
- An electric drill with a small drill bit—a Dremel with the standard 1/8" drill bit works nicely

Got all that? Good. Let's start.

2. Start building

First, drill a bunch of little holes in the lid of the jar, spaced about 1/4" apart. Try to evenly space them in a grid pattern. This will become our "workspace," as shown in Figure B.

3. Prepare the jar

Cut a hole in the side of the peanut butter jar just large enough that little more than the tip of the 2-liter bottle top will fit through it. This is shown in Figure C.

4. Adding the "port"

Use a knife to cut the top off of the soda bottle. Put the top of the bottle through the hole in the jar from the inside, as shown in Figure D.

5. Make it airtight

Now Saran Wrap and tape the whole assembly. (Make sure to get Saran Wrap in the threads of the screw top of the jar.) This is shown in Figure E.

6. Put a lid on it!

Put the lid back on the peanut butter jar. The whole thing should be airtight except for the holes in the top.

7. Use it

Select whatever object you want to copy. Some tips on selecting appropriate objects:

- Make sure that the object is not tapered at the bottom. This will make it impossible to get out of the plastic shell we are making.
- Make sure that the entire object fits on the workspace, leaving plenty of holes around the edges.
- Make sure the object can stand the pressure and heat of the process, otherwise it will deform or melt.
- Make sure the object is not too tall. If it's too tall, the plastic will stretch too much and become too thin to work with.
- Make sure there isn't too much detail on the object.

For my test subject, I chose the body of a tiny R/C car. For plastic, I use the sides of 1-gallon water jugs or milk jugs. Cut off the sides of the jugs and clamp them (or hold them somehow) between the two frames. To form, place the selected object onto the workspace and put a spacer under the object so that the final product will look better. Use the vacuum cleaner's attachment hose to connect the vacuum cleaner to the 2-liter bottle top on the vacuum former. You may need to tape the hose to the vacuum port.

Heat up the plastic between the frames with a heat gun, or hold the plastic over the burner of an electric stove until the plastic starts to get gooey and sag in the middle. HDPE plastic will turn from white to clear when it's warm, this is normal. DO NOT use a gas burner; it will catch the plastic on fire

Figure J: The molded plastic copy removed from frame

Figure K: The copied object removed from the plastic shell

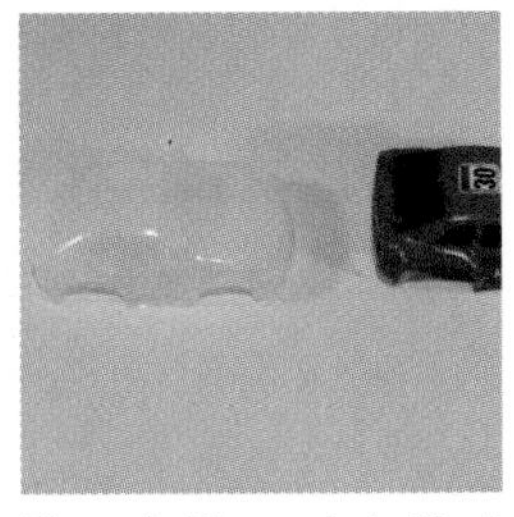

Figure L: The copied object cleaned up and ready for use

Figure M: A bigger version of the vacuum former, made from a trash can and baking pan

which is not good. Figure G shows the plastic before heating, and Figure H shows the plastic at the optimal temperature for forming.

8. Stretch it over the part

Once the plastic is good and saggy, slowly place it over the object to be copied. The plastic will stretch over it. Try to get a good seal all around the object, it should be airtight to get maximum suction. Once the airtight seal is formed, turn on the vacuum. Don't keep it on, just hit it with a good burst for about a second. This is shown in Figure I.

9. Done with molding

The plastic will be sucked tight to the object and to the workspace. When you turn off the vacuum cleaner, if the plastic is still gooey enough to come up slightly, hit it again with another burst from the vacuum cleaner. It should be done by that point. Hold it steady as the plastic totally hardens. When it's done, leave it alone for a little while so the plastic can cool off.

Once the plastic is cooled, take the frames off. It should look something like Figure J.

10. Clean up the edges

Remove the object you copied from the plastic (Figure K), cut the extra plastic off (Figure L), put it in the recycling bin, and you're done!

11. Building bigger...

Get some more practice with the smaller unit, see what you can do with it and how it all works. If you would like to make a bigger vacuum former, you'll need the following:

- One 5-gallon plastic trash can with an approximately 8" x 12" rectangular top
- One 8" x 12" metal baking pan
- One to two tubes of silicon caulk
- One 20-ounce soda bottle or similar
- Two picture frames about 8" x 10"

You do basically the same procedure as with the smaller unit, just on a larger scale. Drill a grid pattern of small holes in the baking sheet. Cut the bottom off the 20-ounce soda bottle. Cut a hole near the bottom of the trash can just large enough for the 20-ounce bottle. (Now here's where I have had a bit of a problem. You may need to reinforce or brace the inside of the trash can and/or the bottle with some wood or something before you go on. Otherwise, they might collapse under the vacuum pressure. It hasn't caused too many problems for me but it might depending on your setup.)

Put the soda bottle in the hole in the trash can and caulk the seal between them strongly, making sure you have an airtight seal. Then, turn the baking sheet upside down and caulk it to the trash can. Let it all dry thoroughly and you are done. The finished product is shown in Figure M.

12. Closing thoughts

Large plastic sheets are available online from numerous suppliers. Check out the United States Plastic Corp. for material or be creative and use things around the house.

When you're finished with the plastic mold, you can fill it with fiberglass resin, Alumilite, or other casting material to generate a nearly exact copy of your original object.

Adam Harris is a graduate student in the field of electrical engineering. He is also a freelance writer, musician, co-owner of SheekGeek LLC, and all-around hacker.

contest WINNER!

Computer Keyboard Wallet

Make a 21st century fashion statement out of techno-junk By Ryan McFarland

Figure A: The surprisingly stylish and cool-looking keyboard wallet

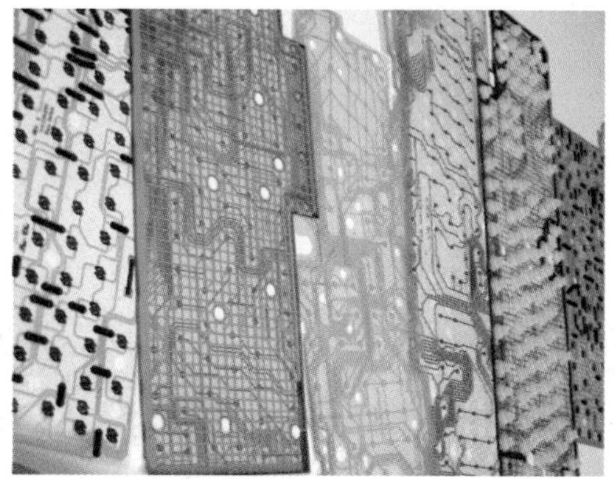

Figure B: All of the different types of circuit sheets you'll find inside of keyboards. The one on the left is from a Gateway 2000 keyboard and has a nice black and green pattern on it.

In all likelihood, there's a keyboard within a few feet of you. Inside, there's likely a circuit sheet that makes a surprisingly durable and thin material for wallet-making.

To make a keyboard wallet, you'll need:

- A desktop computer keyboard (for the circuit sheets)
- A screwdriver
- A ruler or tape measure
- A cutting board or mat
- A razor knife
- Sharp scissors
- A roll of clear packing tape

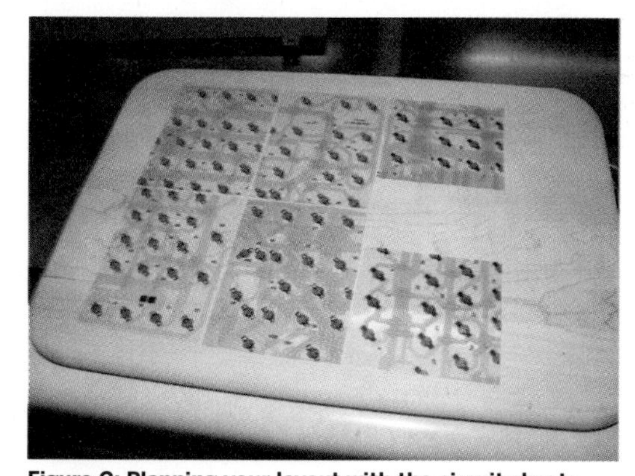

Figure C: Planning your layout with the circuit sheets you're using

1. Open the keyboard

All of the keyboards I've opened have used small Phillips screws to secure the top and bottom pieces. Remove these and open the keyboard. There's probably a sheet of metal secured with more screws. Remove this and set the metal aside to use as a straight-edge during cutting. Next is the circuit sheet, which should be quite easy to remove. Some keyboards have two.

Note: Keyboards with the numeric keypad extension have a larger circuit sheet. Using a 3"x 4" billfold and two 3" x 3" card/ID pockets requires more than one sheet, even of this size though. Hope for two circuit sheets inside the keyboard you crack open.

2. Plan your wallet

At a minimum, you'll need four 3" x 4" rectangles of the circuit sheet to form the billfold area. I only

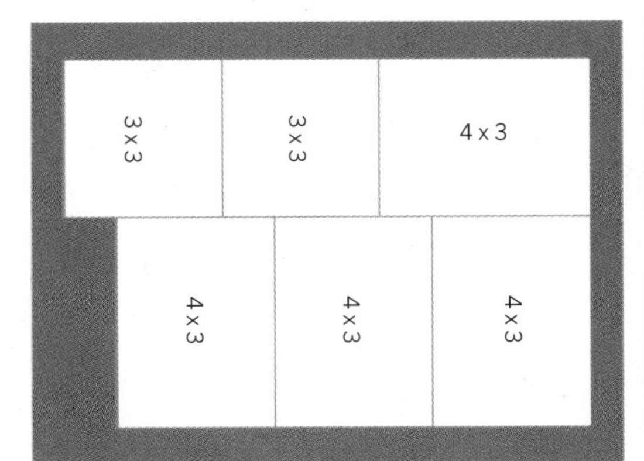

Figure D: A template to help you plan. Make a paper version first to work everything out.

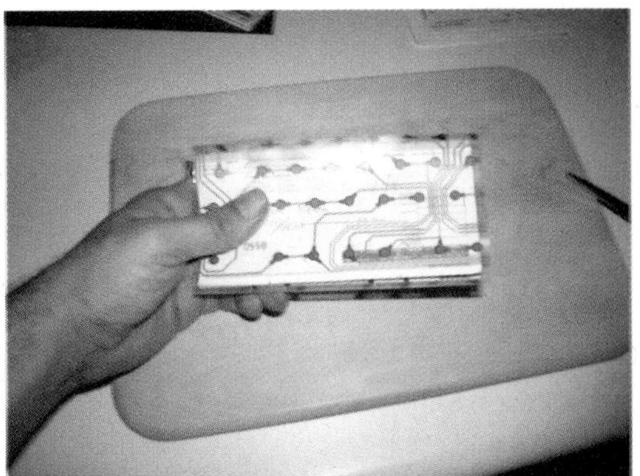

Figure E: A circuit board checkbook cover

carry a few credit cards and my ID, so I added two 3" square pockets. It might help if you make a template on a sheet of paper and tape the pieces together to better visualize your end product. The circuit sheets are very forgiving and tape can be pulled off without damage.

Use thick packing tape for the best durability. I've been using my wallet daily for six months now and have only had to add one small piece of tape. Where the wallet folds, leave a few millimeters between the circuit sheets to act as a hinge. The more space, the more cash (or receipts) the wallet can carry without being forced open by the thick bills. Remember to try and build your wallet so that the tape is folded over any seams. Tearing the packing tape is much more difficult if it's folded onto itself.

3. Taping instructions

For a detailed sequence of taping instructions, check out the project page for this Instructable.

4. Suggestions/tips

- Look over your wallet to see if any spots need reinforcing. Another piece of tape will not add much to the thickness.
- If you have a lot of shopping club cards, check out Just One Club Card (www.justoneclubcard.com) to consolidate them.
- Use the same materials to create a matching circuit sheet checkbook cover (Figure E).
- Other things to do with the keyboard parts:
  Keys—make fridge magnets, bracelets, clocks, or 3D notes to loved ones (I Ctrl U! :-))
  Plastic casing—If your area accepts #6 or #7 plastics for recycling, do that
  Metal—Recycle as scrap or use as a surface for projects that use glue, solder, or other messy materials.

Ryan "Zieak" McFarland lives in Alaska and spends his days as the parks and recreation director for a small town. Nights and weekends he does computer stuff, tinkers with projects, and spends quality time with his friends.

User Notes

Looking for ideas on what to do with the other parts of the keyboard (or boards) you cannibalized to make your wallet? How about making the keys into fridge magnets:
www.instructables.com/id/Keyboard-Refrigerator-Magnets---New-Method

Or try Instructable member **TangMu**'s snazzy keyboard bracelet to go with it? Instructions can be found here:
www.instructables.com/id/Wintereenmus-ideas-Keyboard-Bracelet

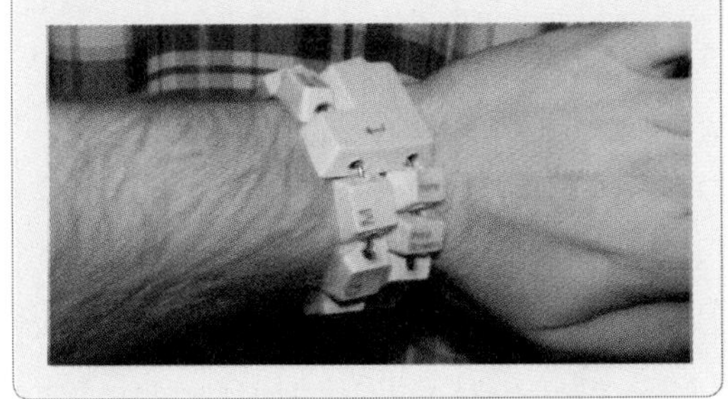

Duct Tape Rose

Duct tape is not just for repairing ducts/cars/books/broken bones; it's also for making roses

By Josiah Bradbury

This Instructable will teach you how to make duct tape roses; they're great for Valentine's Day and lots of other events, like fundraisers. The original idea came from the Duct Tape Club (www.ducktapeclub.com), but after making these roses for a year or so, I've developed my own style.

Figure A: An assortment of roses

1. Get your stuff together

You will need:

- Duct tape in several colors; I use Duck brand Duct Tape
- Floral wire, found almost in any floral department

It takes about 10 minutes to make each rose.

2. Rip tape

Start by ripping a 2" x 2" strip of tape. It doesn't have to be exact; in fact, varying the size of the petals makes the rose seem more real.

3. Making the petals

Fold the top right corner over on itself making sure you leave some adhesive on the left side. Fold the top left corner over on itself making a triangle with some adhesive left on the bottom (Figures D-E).

4. Start making the rose

Take your newly made petal and wrap it around the top of the wire stem (Figures F-G).

5. Building off the base

Repeat steps 2 and 3 to add petals to the rose. I have found that it works well to add them at an angle and then wrap them (Figures H-J).

6. Finish the petals

You can make smaller roses by using just a few petals or make a larger rose by adding more; it's up to you. You should end up with something that looks like the rose in Figure K.

7. Stem

Now you have a pretty good rose, but the wire stem is pokey and hurts when you get stabbed with it. Let's make it a little safer.

Take your stem color and rip a strip that's half the width of the roll and longer than your stem (Figure L). Start at the top by the rose and wrap tightly down the stem. If you have some leftover, keep it for the next step.

8. Finishing up

Take the tape you have left after wrapping the stem, or rip some more little squares of tape, and apply them to the underside of the rose (Figure O). This helps keep the rose on the stem.

And there you go! Go crazy; use any and all the colors you want.

Josiah Bradbury lives in Iron Mountain, MI, with his wife Amy and their cat Koneko. Geocaching, drawing, gaming, and duct tape arts are just a few of his interests, and Instructables is just one of his many outlets.

For additional information, discussion, and more, please visit the Instructables project page:

Figure B: Duct tape and floral wire

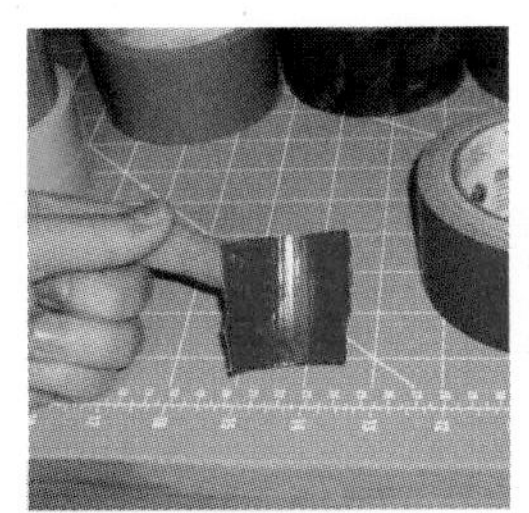

Figure C: Rip tape into squares

Figure D: Fold over, leaving some glued surface along both sides

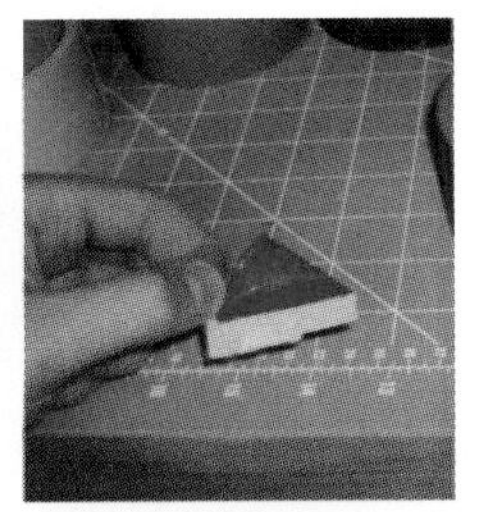

Figure E: Fold over the other side, leaving a strip of the glued surface at the bottom

Figure F: Start wrapping the petal around the stem wire

Figure G: This first petal should be tightly wrapped

Figure H: Continue wrapping petals around the stem wire—note the angle

Figure I: Wrapping petals at an angle gives the flower a more realistic look

Figure J: Continue wrapping petals—your flower should start to look like this

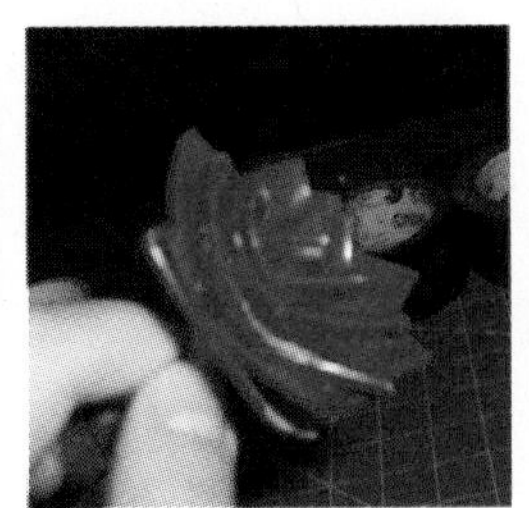

Figure K: A decent rose

Figure L: Rip a long strip of tape for wrapping the stem

Figure M: Start next to the rose and wrap down the stem at an angle

Figure N: Continue wrapping the stem tightly

Figure O: Use small squares of tape to help attach the rose to the stem

Figure P: The underside of the rose

Figure Q: A finished rose

The Best of Instructables
Entertainment

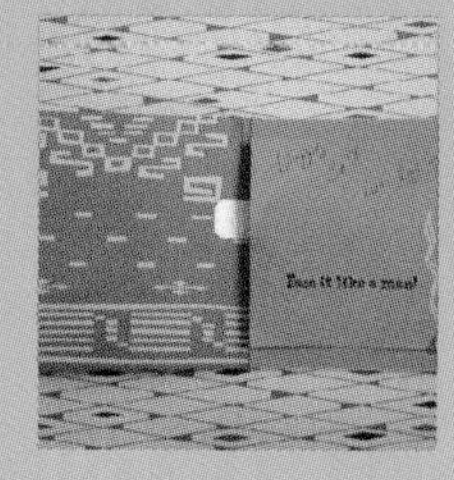

I've got a bin full of headphones that came with various devices, some still deliver great sound, others are barely working. Until I read the Instructables in this section, I never gave much thought to customizing my listening experience, let alone the look of my gear. But why wouldn't you want to personalize the experience a bit?

Imagine taking your old speakers out of their drab enclosures and sticking them into the heads of vinyl dolls. Or cross-breeding your headphones with a couple of stuffed animals. The impact of these projects might be unusual, to say the least, but nobody will accuse you of conformity in either case.

And if personalizing your media is too internalized for you, how about scaling it up so that everyone can enjoy it? With the Guerilla Drive-In Instructable, you'll be able to blast audio and video from your vehicle for the rest of world to experience—whether they want to or not.

Several other Instructables here approach audio and video a bit differently than usual. Hopefully, these projects will give you new ways to enjoy the media that you daily pump into your brain.

—Brian Jepson (bjepson)

Stuffed-Animal Headphones

Warm ears, cool music By WurlitzerGirl

Figure A: Adorable headphones

This project will show you how to make a quick (10-minute), cuddly audio accessory out of a pair of cheap headphones and two small stuffed animals. Not much in the way of skills is necessary. Your friends and acquaintances will coo with delight!

1. Procure the supplies

What you will need:

- One pair cheap headphones (mine were of the five dollar Amtrak variety, I believe), preferably small-ish with removable foam ear discs, although this is not a necessity.
- Two stuffed animals. You want these to be medium-small and (preferably) cheap. I don't want you cutting into any Steiff bears, now. You'll save yourself loads of time if the animal has a seam up its back. You can find them in places like drugstores and five-and-dimes.
- Seam ripper
- Needle and thread
- Fabric glue (Optional. I didn't use any, but it's up to you.)
- Quality music

2. Un-suture the poor bastards

Use the seam ripper to tear through that handy back-seam, pulling out cut threads as you go. If you didn't get an animal with a back-seam, well, shame on you. In this case, I would suggest you make a very neat, straight cut along where the back-seam would be if it had one (down the middle of the toy's back). You don't necessarily need to undo the whole seam, just enough to fit the earpiece of the headphone in. Remember to take care and make it neat, because you're going to have to sew it all back up again in a moment. If your stuffed animal seems ill-at-ease, reassure it with a comforting pat on the head and tell it, in soothing tones, that everything will be all right. Repeat for second animal.

3. Insert the earpiece

First of all, check to see a) how thick each animal's "fur" is, and b) what kind of earpiece you are dealing with. For example, my earphones had a kind of

Figure B: Representative stuffed Koala

Figure C: Oh, the humanity! Un-suturing.

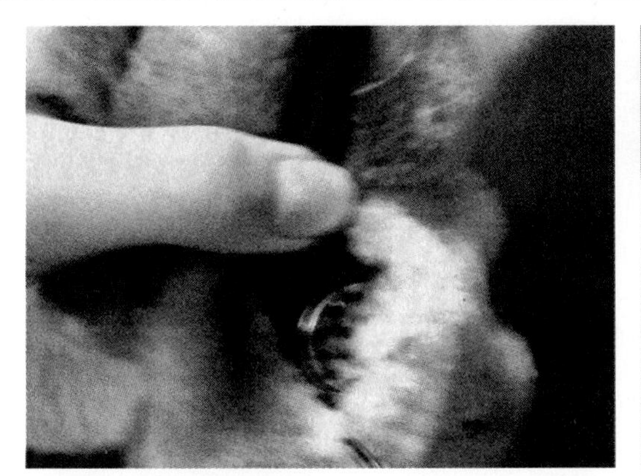

Figure D: Inserting earpieces

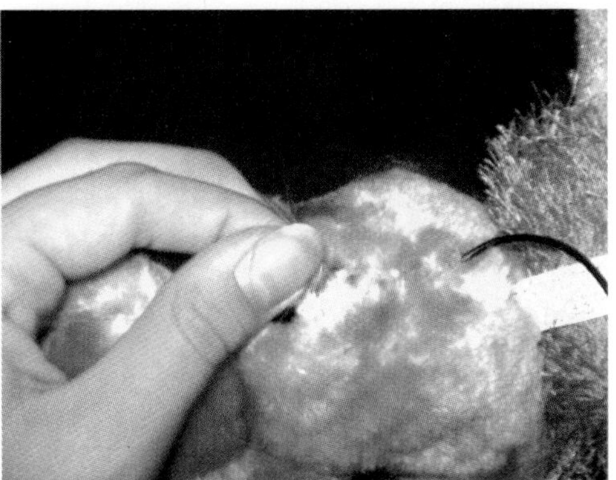

Figure E: Closing the wound

Figure F: The finished headphones!

User Notes

Paulathing made Wild Things headphones!

Wild Things!

hard, thin, plastic cup around the speaker itself. Not something I would want jabbing into my head. The tan-colored bear had thin-ish fur, so when I inserted the earpiece I left the foam disk on. The koala, however, had very thick fur, so to avoid muffling the sound I took the foam disk off. Due to the thickness of the fur I, thankfully, could not feel the speaker inside. But anyway, insert the speaker, noisy-side up (meaning facing the back of the animal). This is where some glue could come in handy, but again, I didn't need any.

4. Stitch it back up

Use your needle and thread to sew up the back-seam (I used a simple whipstitch), adjusting the earphone as necessary as you go. Leave the wire hanging out of the bottom of the seam and the connector for the earphones (the headband part) out of the top. Again, glue? Do it neatly, and don't try to make the scar too big. That won't feel very pretty on tender ears. Repeat for both earphones.

5. Plug in your freshly minted plush headphones and jam

May I suggest some nice theater organ music? http://tinyurl.com/5pchw6. And have fun! I would welcome any feedback, and if you do decide to do this project, I would love to know how it turns out!

Kat Brightwell, aka WurlitzerGirl, lives in Seattle. When not busy crafting, practicing the theater organ, volunteering at the Seattle Architectural Foundation, shopping for vintage clothing, playing in the marching band, hanging out in old movie palaces, and urbexing, she moonlights as a high school student.

Earbud Cord Wrapper

Make an earbud headphone cord wrapper in less than five minutes By Jeremy Franklin-Ross

Love your shiny new iPhone, but sick of tangling up that darn cord on your earbuds? Grab an old credit card and a pair of scissors. You are about to solve one of life's least important problems.

1. Snip the card in half

Cut the credit card in half, lengthwise. Want bonus points? Round them corners. Want *more* bonus points? Use a defunct Baby Bell phone card!

Figure A: The earbud cord wrapper in action

2. Notch away!

At either end, cut two notches on opposing sides. The notches should be angled roughly as shown in the pictures. Take care to make the opening of your notches only slightly wider than the cord is thick (about 1/16th of an inch).

Make the interior of the notch wider than the opening, but narrower than your plug (about 3/16th of inch).

Note: Instructable member Spandox suggests using a hole punch. Great idea!

3. Wrap cord

Insert your plug end into the notch and wrap your cord up neatly. When you get to the buds, tuck them into their notch. Voilà!

4. Usage

You can unwrap the cord on either end to a length that suits your need. I usually keep my iPhone in my shirt pocket, as you can imagine, without this cord wrapper my cord is always in the way.

Jeremy Franklin-Ross (throbbing.org/thejer) is a co-founder of HazardFactory, Seattle's dangerous arts collective. He is a lifelong artist, hacker, technologist, and entrepreneur. Jeremy writes software for CultureMob.com, an events discovery site that he co-founded. He takes pleasure in making fun of other drivers.

Figure B: Card cut in half lengthwise, with corners rounded.

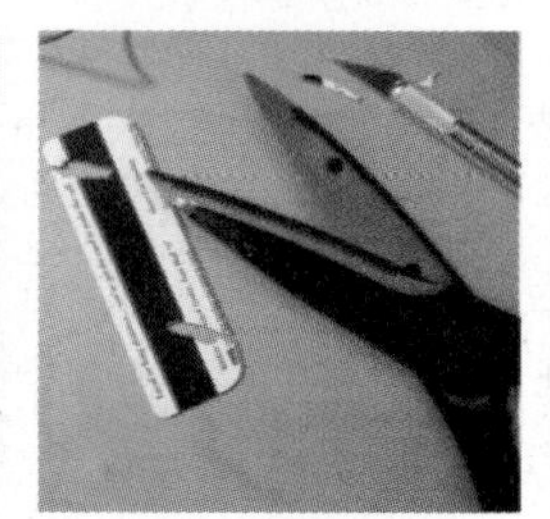

Figure C: Cut two angled notches as shown here.

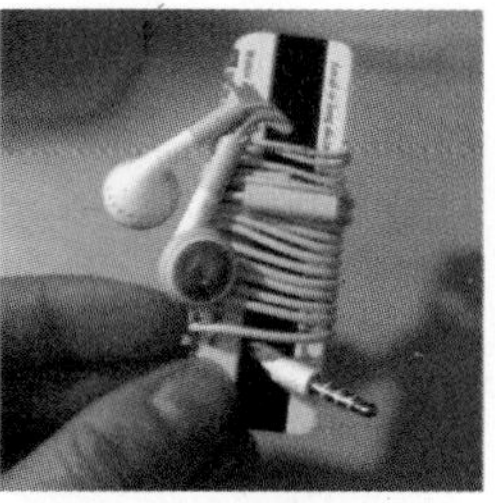

Figure D: The finished product, wrapped and ready to rock.

Figure E: Reel out only as much as you need.

Instructables project page: www.instructables.com/id/Earbud-cord-wrapper-in-5-minutes-or-less!

Make an iPod Speaker from a Hallmark Music Card

Don't throw away that old music card—it's chock full of hackable gadgetry! By Justin Seiter

Figure A: The completed project

Figure B: 1. Electrical tape 2. Utility knife 3. Glue gun 4. Cereal box from Kellogg's Variety Pak 5. Hallmark Musical Card 6. Old headphones (you know, the ones that make that "ting!" sound every time the bass hits)

Ever get one of those cards for your birthday that plays music when you open it? Don't throw it away! With a little help from Tony the Tiger, you can use it as a speaker for your iPod.

1. Required materials

- Hallmark music card
- Old headphones
- One empty cereal box from a Kellogg's Cereal Variety Pack—I used Frosted Flakes, but you can choose your favorite cereal. It will not impact the outcome of this project. :-)
- Glue gun
- Electrical tape
- Utility knife
- Not required, but helpful: soldering iron

2. Remove the speaker from the card

Use a utility knife to cut along the top and bottom edge of the card to expose the speaker.

Cut the speaker wire at the base of the circuit board *first* and then gently remove the speaker from the card. If you don't cut the speaker wire first you could end up ripping the wire from the base of the speaker and then you'd really need a soldering iron.

The speaker is fixed to the card with a small circular piece of double-sided tape, so it shouldn't be too difficult to remove. Avoid the urge to use your utility knife to "cut away" the speaker—you could accidentally cut right into the speaker itself. Finally, strip about a quarter inch of insulation from the wires.

Note: In case you're wondering, the Hallmark card is powered by a CR2032 3V Lithium Battery. This is the key ingredient for making the LED Throwie Instructable (see "LED Throwies," page 108) and is the powerhouse for a bunch of other Instructables. The battery isn't necessary for this Instructable, but holding on to it could prove to be handy down the road.

Figure C: Opening the card

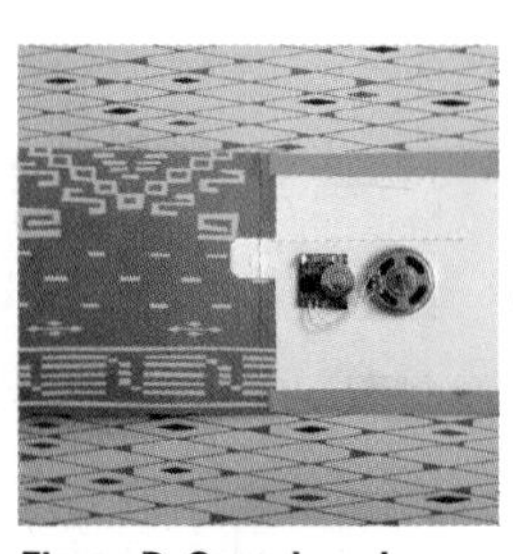

Figure D: Sound card, battery, and speaker

Figure E: 1. Cut here first, then remove speaker

Figure F: Strip the insulation from the wires

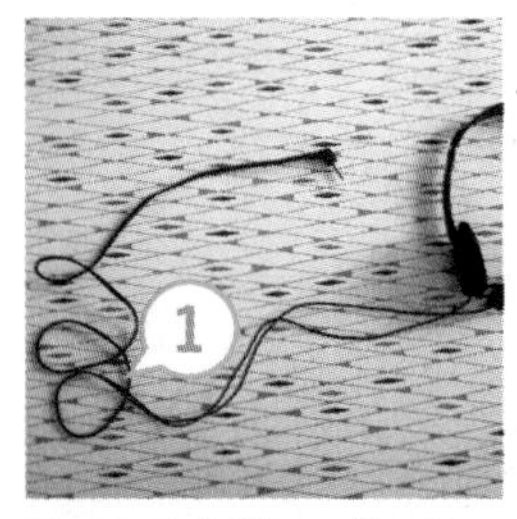

Figure G: 1. Where the line splits—you only need one of the lines past this point

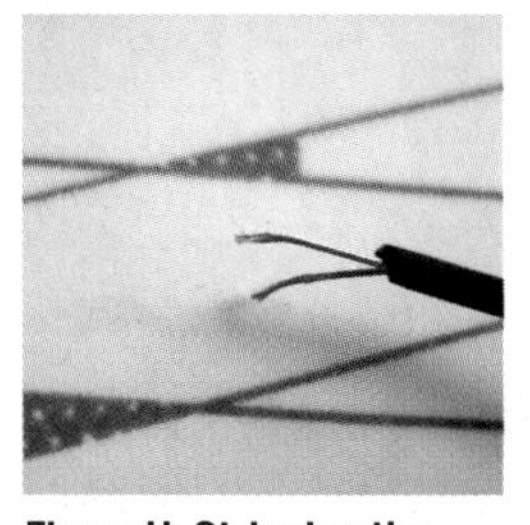

Figure H: Stripping the wire

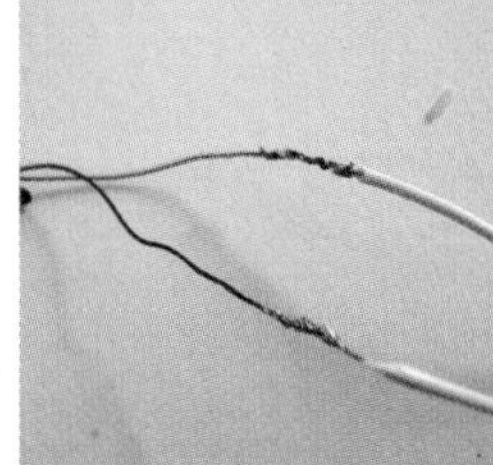

Figure I: Connecting the headphone jack to the speaker

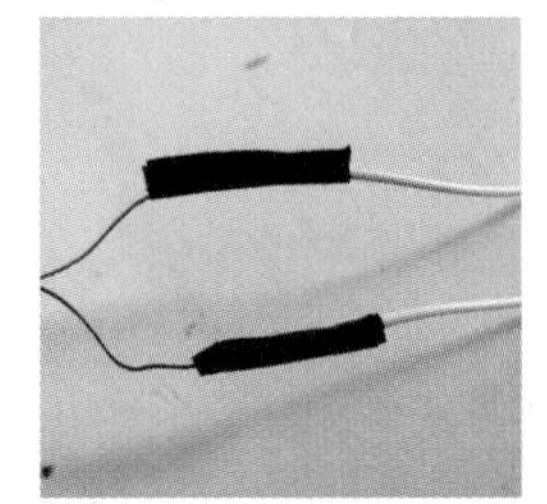

Figure J: Insulate the wires

Figure K: Close-up of the insulated connection

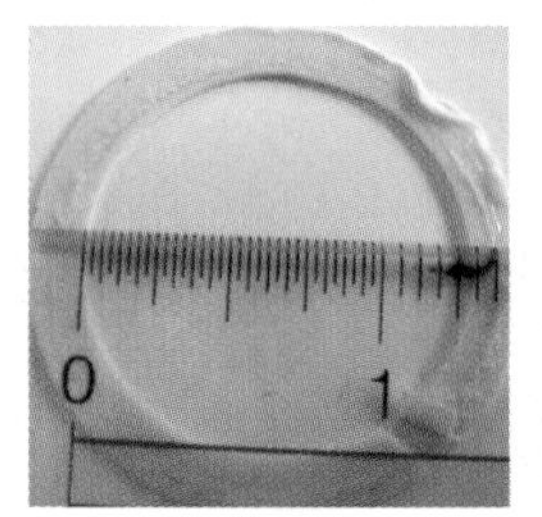

Figure L: Measuring the speaker diameter

Figure M: Use the scored line in the box to help center your hole

Figure N: Running the wire

3. Prep the headphones

Cut the headphone wires at the base of each headphone (left and right). You'll only need one of the two lines, so pick the one you want to use and cut the other where they converge mid-way down the length of the chord.

Next, strip the wires. You may notice some fiber-like material interlaced with the leads themselves. Strip this stuff out with your utility knife or you *will not* be able to get a reliable connection from your headphones to the speaker (thanks to Richard at my local RadioShack for that tip). Alternatively, you can use a soldering iron to burn this stuff away. I found the soldering iron method much *easier* and a bit more effective.

4. Connect

Connect the headphone wires to speaker wires. I'm not sure if it matters which ends you connect, but make sure you're getting sound before you seal the deal, so to speak.

Once you've connected your leads (again, soldering here is a good idea, but not necessary) insulate and tidy up the exposed connections with some electrical tape.

5. Prep the cereal box

You'll want to cut a hole slightly smaller than the full diameter of the speaker. Remember the nice beveled edge that Hallmark utilized to fix it to the card with double sided tape? You need to do the same thing here. A 1.25" diameter hole should be

Figure O: Mounting the speaker (1. shows where I got a little heavy handed with the glue)

perfect. There's a dashed line that runs through the middle of the box that will help center you up before you cut.

You'll also want to cut a small hole towards the bottom rear of the box to feed the headphone jack through. Feeding the wire through the back/bottom of the box not only makes it look nice, but also weights the box so that it doesn't fall forward on its face when you stand it up on end.

6. Secure speaker, seal box, and rock!

Run a small amount of glue along the beveled edge of the speaker and *quickly* attach it inside the box. Hold it to give the glue a chance to dry. Secure it by adding some spots of glue along the top and bottom edges of the speaker where it meets the box. Try not to glob it on—this will avoid adding more weight. I ended up using way more than I should have, but it didn't make the box unstable.

Finally, glue back together the top and bottom of the box as it was originally sealed, connect an iPod, and kick out the jams.

Justin Seiter loves sushi, hates snow, and wishes AC/DC's "Thunderstruck" could be played every time he entered a room.

Figure P: All done!

User Notes

drchucklesme says: The only problem with this speaker is it will only play one channel—this will often leave the vocals or other instruments missing.

If you want both channels to play through the speaker, don't remove one of the ends of the headphone cable. Instead, splice the corresponding wires from each end of the cable (that is, the wires from the cable that used to connect to the left headphone and the wire that used to connect to the right headphone) together.

For instance, if the two ends of the headphone cable each contain a white and a red wire, you should solder/tape the two white cables together, then do the same with the red cables. THEN use the spliced wires as you would have used the single red and white cable from the one end of the headphone as explained in the Instructable.

Munny Speakers

Give your speakers some extra personality by making them out of a vinyl doll By Ed Lewis

Kid Robot (kidrobot.com) makes the easily hackable Munny doll and I've been meaning to cut one up. The combined need for some new speakers created a happy union of doll and speaker.

Now, I'm not the first person to think of this. I've seen at least three different versions of a Munny with a speaker in its head. Still, the only one I've seen in person was just one doll for mono sound and the driver was pretty worthless. I wanted mine to sound at least decent and I'd say I surpassed that.

1. Get a pair of Munny dolls and the drivers of your choice

I went to the Kid Robot store in San Francisco to offset the $25 price tag by not having to pay for shipping. At the time I couldn't find any of the black Munnys online and kept calling till they were back in stock.

Each Munny comes with a few random accessories: a big pencil, a cape, goggles, all sorts of stuff. The drivers I got are Tang Band 3" Bamboo Cone Drivers (PartExpress.com) and have an extended range of 105-20,000 Hz.

2. Trace the circle

I bought drivers that are reasonably small, at 3". But this was still pushing it pretty close to what would fit on the Munny head. To be sure everything was where it should be, I measured the diameter of the back of the driver and cut that out of card stock. I then put this card template onto the face of the Munny and traced out the circle (Figure C).

3. Heat 'em up!

The vinyl that the Munny is made of is pretty firm when it's at room temperature, but heat it up with a hair dryer or a heat gun, and it softens up nicely. After that, an X-Acto cuts through it like butter.

Cutting is all about careful subtraction. I cut within my safe area quickly, but when I was getting close to my line, I started whittling down the space with constant checks against the drivers, eyeballing it from the side to see where I needed to remove more vinyl next (Figure D).

4. Cut the faces off

Here's how the Munny heads look without their faces. Cute, huh? (Figure E)

5. Wire it and fill it

Thread the speaker wire through the neck and stuff the head with some polyfill (Figure F).

6. Seal it up

I was considering using bolts to keep the head in place, but the look was a little too S&M for my taste so I decided to seal the connection with some Lexel. Run a bead around the whole driver, tie it up, and let it sit over the weekend (Figure G).

7. Hook it up and enjoy the music!

The Munny speakers are all set and ready to go. Hook them up to an amp (I'm using a Sonic Impact T-Amp), run in an input from an MP3 player or your computer, and you're good to go!

Ed Lewis lives in Oakland, CA, with his wife Maria and Biscuit, the stunt cat. Being outside, cooking, making wine, and building funny objects are current favorite activities.

For additional information, discussion, and more, please visit the Instructables project page:

Figure A: Give your speakers some extra personality by making them out of a vinyl doll

Figure B: The Munny doll ready to be hacked for sound

Figure C: Using a template to position the speaker hole

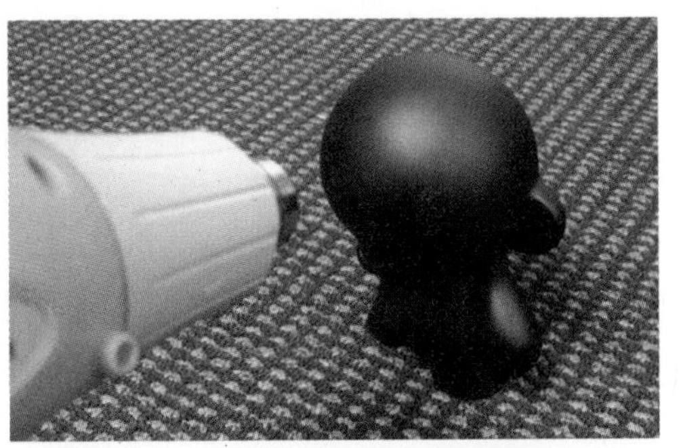

Figure D: Softening the vinyl with a heat gun

Figure E: The Munny heads with holes for the drivers cut

Figure F: Wiring the speakers and adding polyfill

Figure G: Using Lexel sealant to affix the driver

Figure H: The finished Munny speakers, ready to rock

Start a Guerrilla Drive-in (aka MobMov)

View-and-run with your own mobile theater

By Bryan Kennedy

Figure A: Let's go to the movies!

Figure B: Equipment setup

Have you ever wanted to run an outdoor theater à la MobMov.org or the Santa Cruz Guerrilla Drive-in? This Instructable will tell you what equipment you'll need and how to set it up. Cyberpunk urban theater, here we come!

1. What this guide is for

Guerrilla drive-ins come in two flavors: a true **drive-in**, or a **walk-in**. Walk-ins were the first guerrilla movie method, and involve a big grassy area, some blankets, and a movie projected on a screen. The projectionist powers the projector using a generator, and big speakers are used to project the sound. The Santa Cruz Guerrilla Drive-In and Dolores Park Movie Night are walk-ins. A true guerrilla "drive-in" is just that—a bunch of cars and a movie. The projectionist usually powers the projector with his or her car or a small generator and an FM transmitter transmits the soundtrack to the other cars. The Mobile Movie (mobmov.org) was the first guerrilla drive-in. I'm rather partial to drive-ins over walk-ins because of their high-tech nostalgic feel, and the fact that your car can become your own private theater! It sure beats the pants off that cineplex with its screaming preteens. This guide will tell you how to set up your own guerrilla drive-in using your car. If you'd prefer to set up a walk-in, portions of this guide may be helpful, but much of it will be inapplicable. There are guides online more suited for this purpose however, see: http://tinyurl.com/6gfv2r. Parts of this tutorial were taken from the tutorial (www.mobmov.org/manifesto) that I wrote for my MobMov members. Hey kids, let's go make a drive-in!

2. Gathering the equipment

To run a guerrilla drive-in, you'll need:

- Bright projector (1200+ lumens)
- FM transmitter
- Inverter or power generator
- DVD player or laptop
- Car

I'll detail these components below, so you'll know what to look for.

Projector

The cornerstone of any mobile movie, the projector makes the magic happen. As such, you should plan on spending the most on this part of the kit.

That said, you can find the best deals by searching Craigslist for used projectors. With how rapidly projector technology is improving, you can get a $1400 projector from last year for less than half that price today. However, keep your eye on the bulb price. If the bulb is anywhere near death, it's going to cost you $300 or more to replace.

When choosing a projector, select one with the highest lumens you can afford. 2000+ is optimal, but 1200 or more will suffice if you're on a budget. Other useful features include lens shift (which lets you reposition the video without moving the projector itself) and a zoom lens. Both combine to make it much easier to just drive up and start projecting films. A zoom lens in particular is vital if you want the most flexibility in where you show your films.

I've had lots of luck with Epsons and Panasonics, and some luck too with InFocus. If you'll just be using the projector for movies or games at a drive-in, you don't really need a true movie/HDTV projector, as they generally have less lumens. You'll be projecting onto suboptimal walls anyway, so all those extra pixels won't get noticed. Oh, and LCD projectors often have a brightness/color advantage over DLP. Expect to spend anywhere from $500 to $5000 or more on a good projector.

FM transmitter

You'll need some way to get the soundtrack to your audience. You *could* set up a bunch of speakers on hooks like days of old, but why do that when you live in the future! Now, don't run out and grab yourself one of the $30 iPod transmitters. It won't work. You need a transmitter that can transmit sound 150ft+. The only one that I am familiar with is the Ramsey line—I use the FM25b model, but I hear the FM30 has more features. These units come in parts, meaning you'd need a soldering iron (and an EE degree!) to put them together! Luckily, they're readily available pre-assembled on eBay for about $150.

DVD player or laptop

I personally use my old and crusty laptop to show movies, because it offers the best flexibility in film formats. I often open up a show with the mobmov logo and cartoon short, and then do a 10-minute intermission in the middle.

If you don't have a laptop, any old DVD player should work as well. Another viable option is to use an old Xbox or other gaming system—you can often use it to play DVDs (especially if you get a "chipped" version), and you have the added bonus of being able to play games during intermission! Super Mario, here we come!

Power inverter or power generator

To power all this fancy equipment, you'll need to bring your volts with you. It's very unlikely that you'll find a suitable outlet anywhere nearby your urban cinema. Luckily, your car has a built-in powerplant that can be conveniently harnessed to power all this equipment!

All you need for this is a basic power inverter. Always opt for more power than you need, so in this case, choose an 800w+ (continuous, not "peak") unit. I'm using a Coleman 800w and it does the trick. Whatever inverter you buy, be sure it's a "modified sine" or "pure sine" inverter. A good quality inverter will set you back about $50-$100. Square sine will cause interference and may damage your equipment.

You'll have to connect the inverter to your car battery, which in your car is probably under the hood. If it is, you'll need to run a cable yourself or head over to BestBuy or someplace and have them run it for you. For my car, this only cost $20. If you do power your equipment in this manner, you'll need to keep your car running, as car batteries don't like to be drained all the way down.

A far more efficient and environmentally friendly (albeit more complicated) method is to use a power generator with built-in inverter. These relatively small beasts are made by Yamaha and Honda, are pretty quiet, and start at around 1000w. I'd choose the smallest-wattage unit your set-up requires, as more power typically means a louder generator.

I personally chose to power my equipment with my idling car, just because of the convenience of doing so—but this is probably the smartest option if you plan on doing many shows. Car engines are not designed to idle for long periods of time, and you'll be putting undue stress on it if you idle it for too long. Generators, on the other hand, are meant to idle.

Your car

If you already own a Toyota Prius, you're in luck! You've got the perfect car for powering your drive-in! Basically a battery on wheels, the Prius has the battery in the trunk, so there's no drilling required to plug in your inverter. I personally use a small SUV, so pretty much any car is capable.

Figure C: Connect the parts

Figure D: Dusk

If you don't have a car, consider renting one from Zipcar or another hourly "car sharing" shop. Especially if you rent a Prius or use a generator, you don't need to make any "modifications," so it's perfect!

That's about it for the equipment. Now, put on those gloves, cause this is where the real work begins.

3. Now for the hard part...

Just kidding! Setting up the equipment is ridiculously easy. In fact, if you've ever plugged in a new DVD player or TV, this step will probably seem a little obvious. There's really no magic to hooking up your new equipment—video out to video in, audio out to audio in, and power out to power in.

The most finicky part is the FM transmitter. Make sure it is positioned outside your car, and high enough so the top of the antenna is above all the other car antennas. FM antennas transmit sound in an umbrella shape—any antennas positioned above the transmitter will get much less reception. Also, electronic interference is a common problem—be sure to wrap the ends of your audio cables in Ferrite (magnetic) filters, which you can get from RadioShack or somesuch geek playground, and choose thicker cabling when you can.

If static proves to be abundant, try isolating the cables and equipment from the metal frame of your car (if the interference is coming from the car's power) or grounding the transmitter.

The diagram is taken from my tutorial on mobmov.org and shows how you'd connect it up.

4. Running your mobile movie

I don't profess to have arrived at the "best way" to run a mobmov. But after two dozen or so shows, this is how I do it now and it seems to work well.

Getting the word out

I announce showings online through the online mailing list software I created. It allows people to sign up for areas where they want to see movies, and then notifies them when a showing is in their area. If you're interested in using this software, sign up to be an official mobmov chapter. It's free of course.

You can try new inventive forms of advertising that I have yet to attempt. Why not post some appropriately campy posters around town? Believe it or not, some movie licenses actively prohibit some forms of non-theatrical publicity, so be mindful of this.

Getting set up

I generally try to arrive a few minutes ahead of time, mostly because I know if I do, I'll actually arrive on time! My particular set up is so easy to connect (most of it stays connected) that it takes me all of about 3-5 minutes to get going. This is actually part of what I think makes the mobmov such a success—if it took me 30 minutes to haul out a projector and batteries, I'd do it a lot less often.

While I'm setting up, I have a mobmov welcome title showing from my laptop, so people know that they've reached the right spot and not some "other" guerrilla drive-in. :-) It also tells them what radio station to tune to. Free title files can be downloaded from our website as well.

I usually park in the front-center, and cars generally line up on both sides first, and then to the back

Figure E: Getting dark

Figure F: The projection room

when the front row fills up. My car is rather tall (I have a mini SUV), so that limits the number of people that can park directly behind me.

The show

I'll generally wait 5-10 minutes after the announced showing time. During this time I show an intro title with information on the movement, and play some good music in the background to let everyone figure out the radio reception and talk to me if they need help. I always like to take the time to meet any newcomers and make sure everyone is good to go.

Then I start up the openers. These usually consist of one or two old-school B&W commercials and a cartoon or other short film. Sometimes this is a newsreel. All of these can be found on archive.org in the public domain. The movie ends up starting about 20 minutes after the announced showtime, which ensures that any stragglers won't miss the film. The old-school openers really set up the air of nostalgia and I've gotten a lot of positive comments about them.

Intermission

In my opinion, the intermission is really the most vital component to an enjoyable mobmov experience. About mid-way through the movie (usually at a cliff-hanger if I can find one), we break for a 10-minute intermission. A title pops up with some good music, and people are encouraged on-screen to get up and meet people, say hello to the driver, buy some snacks, and donate if they can. Fueled by such an intoxicating atmosphere, I find that people are anxious to do all of this!

Donations

I usually tell people on-screen during the intermission that there is a donation mug on the lead car, and ask them to contribute a few bucks to help support costs like gas and projector bulb replacement. People have been quite generous in the past. Be wary of charging mandatory admission—not only will this water-down the "guerrilla" atmosphere of the whole thing, but may get you in trouble depending on your licensing agreements for the movie you're showing.

Snacks

Selling snacks is a good way to recoup some costs without directly asking for money. I usually head down to Costco the night before and grab a big bag of pretzels or assorted chips, along with some candy bars and other movietime snacks. Recently I've started giving out the chips for free (they cost me hardly anything) and sell the candy bars and soda for a buck apiece. This gets people down to the "snack bar" to buy the other stuff, and I think it makes them more generous with the donations. Plus, it's just a good thing to do.

Now, on to the important stuff—like where to show and what to project.

5. Legal issues

Running a mobmov does entail its share of legal issues, and I'm not going to interpret the law for you. However, I will relate to you my personal understanding of the issues at hand. Of course, don't take my word for it, and always check with the appropriate authorities before actually running a mobmov of your own.

Figure G: Don't run down your batteries with your parking lights!

Figure H: Intermission

Copyrights

If you're running a mobmov, it's very important that you respect copyrights. A mobmov can attract a lot of attention, and as the movement grows, the powers that be may take a glaring notice of not just you, but mobmovs in general. If we do this thing right though, the movie studios will have more reason to cooperate with us and might even appreciate what we're trying to do.

The bottom line is that you must respect the copyrights of the movies you show—just because you aren't charging admission doesn't mean that you can show the movie without paying for its use. I personally have had the best luck contacting the studios directly. They are usually so excited by the idea, that they approve it for a limited audience without charging any fees. Your mileage may vary; mine has. As the novelty of the mobmov wears off, it may also become more difficult to get free showing rights. A lot of movie studios will just direct you to an independent licensing house, which is usually a dead end. Most movie licensing houses do not permit outdoor showings of films, even if you offer them lots of money. Without attempting to preach, it's a very silly and outdated system.

Swank Pictures is the *only* movie distributor that I know of that will license non-theatrical outdoor showings (non-theatrical means that you don't charge admission). They have a bazillion rules, including a stipulation that you must show in a predetermined location. Licensing fees, though, are quite affordable, ranging from $100-$300 depending on the movie. Read the rules that govern outdoor movies. You should contact them to set up an account if you wish to take this route.

I have secured a limited number of independent movie licenses for mobmov use that you can project for free to cheap if you're a mobmov driver. That said, there are numerous sources of freely available creative commons movies and media available online. Check out archive.org for listings of such movies.

FCC rules

It is my layman's understanding that the FCC permits unlicensed broadcasts as long as they are under a certain power and do not interfere with existing radio stations. See www.fcc.gov/mb/audio/lowpwr.html for details.

Police

From my experience, the police are not too concerned about mobmov activities. It is very clear at a glance what we are doing, and that it is safe and legal. But I always bring along any licensing documents to prove myself just in case of inquiry.

In the course of running my own mobmov, there have been two police drive-bys. Both times, the friendly cop slowed down for a harrowing few seconds, took shape of the scene, and then drove on their merry way. Once a policeman approached me after a show and asked me what station we transmit on, so he could tune in next week!

I suppose that at 10 p.m. in a dark area of town, we really are the least of their worries. If anything, our presence makes it safer. However, there are three main points to keep in mind to minimize the possibility of attracting (negative) police attention:

Noise pollution: One of the important features of a mobmov is there is usually a very low level of noise generated. Walk-ins, on the other hand, must blare their audio over loudspeakers. This sort of distraction is the kind of thing that will attract the attention of the police, but as a mobmov owner you can be less wary. Even still, a mobmov showing can be very noisy depending on the environment and the patrons. Keep an ear on it and you should be okay.

Light pollution: Always be very keen about where you show your movie. I highly recommend against showing in a residential area or worse—on a residential building. This can generate some very strong complaints, and might get you fined for disturbing the peace.

Trespassing: If you will be inviting cars into a parking lot or other space owned by someone other than you, make sure you get permission to show there first. We've had reasonable success projecting onto dimly lit walls from the street, so that no one is parked on private property. As far as I'm aware, there is no law yet prohibiting trespassing with light. Please be aware that any public property, such as schools or parks, are usually heavily guarded against after-hours trespassing. You will very likely be dispersed and questioned if caught showing a movie on public grounds (this happened to the Santa Cruz GDI). When in doubt, check with the city or your local police.

If you'll be showing where there are neighbors, invite them to attend. Not only may you grow your audience, but if people know about it and feel included, they'll be less likely to cause a fuss.

6. That's it!

I hope you enjoyed this tutorial. If you have any questions or suggestions, mail me here on instructables or leave a comment and I'll be sure to get back to you! Also, check out mobmov.org, where budding urban projectionists like yourself can join our growing movement. Now, go light up the night!

Bryan Kennedy started the Mobile Movie, a worldwide collection of guerrilla drive-ins, because he was tired of the high ticket prices and bland experience of the cineplex. The MobMov has since grown to over 200 chapters and 8,000 members across the globe, with monthly shows in San Francisco drawing 100-150 participants.

User Notes

Nick Whitworth says to add headphones for walk-ins: Rent cheap personal radios with headphones to walk-ins (clean the headphones between movies). Also, patch a small mixer between the laptop and the transmitter, plug in a microphone, and you'll be able to make announcements over the radio. The Behringer Xenyx 502 Mixer looks like it would be perfect. Just patch the laptop to the Tape In, and the FM transmitter to the Tape Out.

Jaime Guerrero on how to go from laptop to projector: If the video source is a laptop, use its VGA or DVI connector to the projector. You'll get a better quality picture than using S-Video, because the laptop's CPU will be used to extrapolate to the additional pixels available on the laptop/projector screen.

Most DVD viewing software for laptops does an excellent job of expanding the DVD video image into a very high-quality image for the laptop screen. Take advantage of that software!

Remember that NTSC video—which is what S-Video cables transmit—was invented in the '40s with 200 (noninterlaced) / 400 (interlaced) vertical lines with fairly limited analog horizontal resolution—a far cry from the usual 1024 x 768 or higher resolution of most laptop screens and projectors.

But make sure that the projector's native resolution matches the laptop's native resolution or you'll defeat this trick (the projector will have to expand or contract the image in its CPU, adding distortion and noise).

If your laptop has a video or S-Video output, its video card has already "compressed" the on-screen VGA image to video resolution for the signal it outputs on that connector, which the projector will then "expand" back to the native resolution of the LCD/DLP. A lot of quality loss in those two conversions.

Home & Garden | Food | Photography | Science | Computers | Electronics

The Best of Instructables Fun & Games

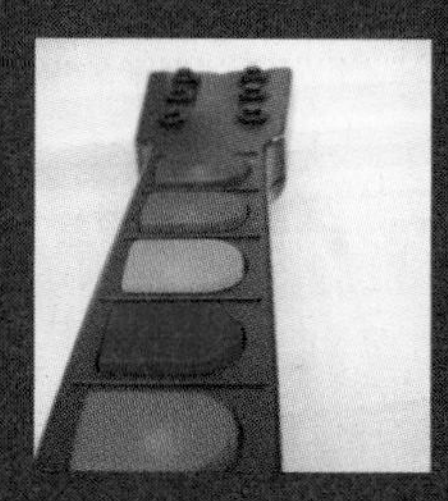

Many DIY projects have noble goals behind them. You can improve the world, learn a vanishing artisanal craft, or get the exact look you want without someone else messing it up in translation. Then there are the ridiculous projects that won't save the planet, improve your mileage, make your hair shinier, or do anything useful at all, but that's exactly the point! Who wants to be so practical all the time? That can wear you out and make you just a bit grumpy every morning.

Sure, you can recycle that cardboard, but why not turn it into a costume instead? You can help your compost with your newspaper, but wouldn't you rather wield a magic wand with it? And since a Halloween is always on the horizon, how about taking your time this year to build a full-on custom werewolf costume to scare all the children away for good. They have it coming and you know it.

Taking the time to seriously build something so fantastically unique and often "useless" is a primal need that we all should indulge in once in a while. If the idea of the finished item makes you smile, even just on the inside, go for it! An appealing idea is good, but making it real can be absolutely amazing.

When all is said and done, what you make will likely inspire yourself and others to make something else even more bizarre and wonderful. Sometimes your own creations can show that good stuff can come from anywhere and almost anything. And in that way, you *did* improve the world after all.

Now go and make that desktop trebuchet or decorate your house with Blinkybugs. Take a holiday from worrying about the function and just explore for a little while. You'll enjoy it. Trust me on this.

—Ed Lewis (fungus amungus)

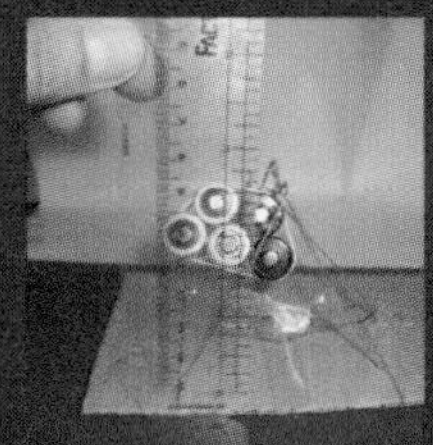

contest WINNER!

The Marshmallow Shooter

How to make a marshmallow gun By Eric J. Wilhelm

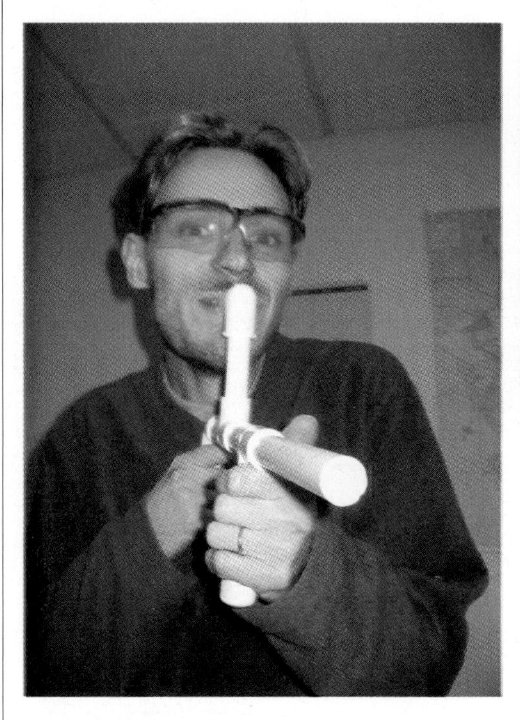

Figure A: Testing out the marshmallow shooter. The face shield was made from a plastic 2-liter bottle.

This marshmallow gun (marshmallow shooter) will completely surprise you with its accuracy, range, and ease of construction. Plus, it's tons of fun and a lot better than any store-bought toy because it encourages modifications.

1. Get the plans

You can find the plans for the marshmallow shooter on the Instructables page for this project (see URL below).

Materials:

- 22" of 1/2" PVC pipe (1/2" is a the nominal diameter, its actual outer diameter is closer to 7/8")
- PVC end caps (2)
- PVC three-way junctions (2)
- PVC elbows (2)

2. Cut the PVC and assemble

Cut one length of 7" and five lengths of 3". Hacksawing is a good choice for this step. A How to Hacksaw tutorial can be found here: www.instructables.com/id/Hacksaw.

Lay out your pieces and assemble them. Friction should hold them all together.

3. Test your shooter

Put on your safety glasses. Point the gun in a safe direction. Load a mini-marshmallow into the mouthpiece, seal your lips to the mouthpiece, and give the gun a quick burst of air. Yes, the marshmallow does go around all those curves. Pretty cool, huh?! Keep your ammunition sealed. Dry marshmallows don't work very well.

Clean up your marshmallows when you're done. Especially, don't leave any around roads: they will attract animals that could be hit by cars.

Eric J. Wilhelm is the founder and CEO of Instructables. He has a Ph.D. in mechanical engineering from MIT and is a co-founder of Squid Labs.

For additional information, discussion, and more, please visit the Instructables project page:

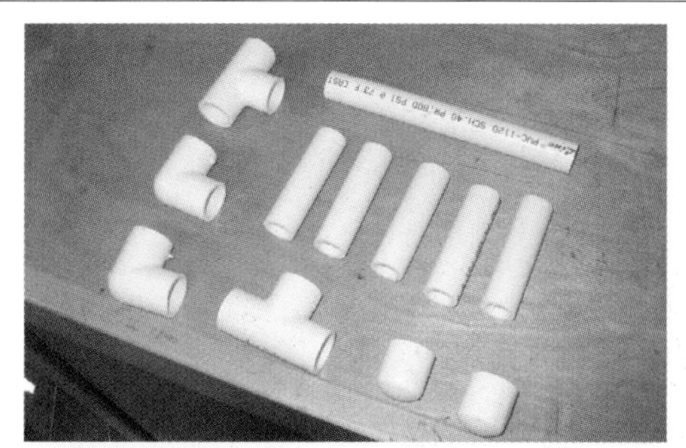

Figure B: All of the parts cut out

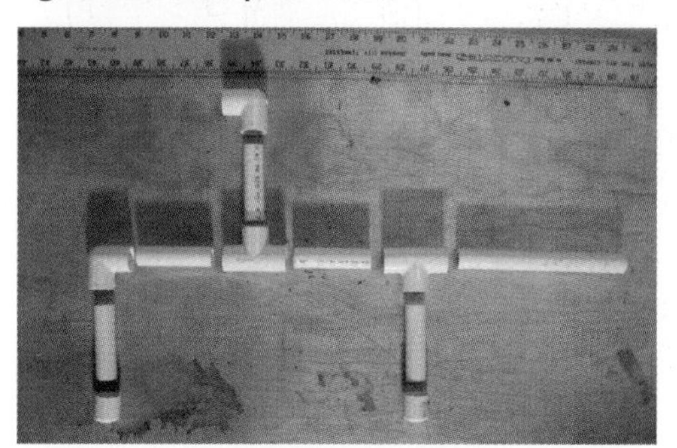

Figure C: The pieces laid out and ready for assembly

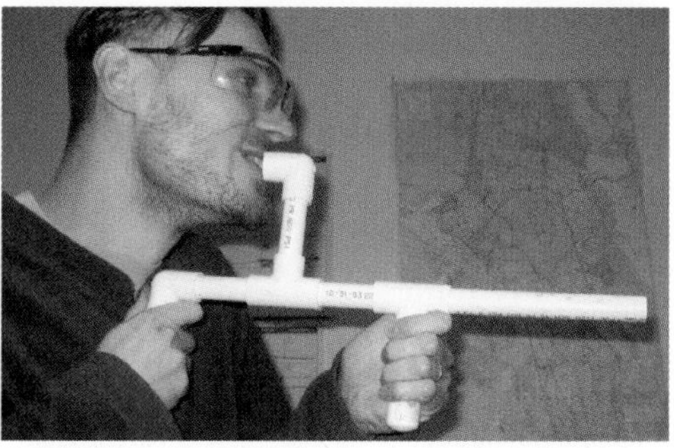

Figure D: How to hold and fire the marshmallow shooter

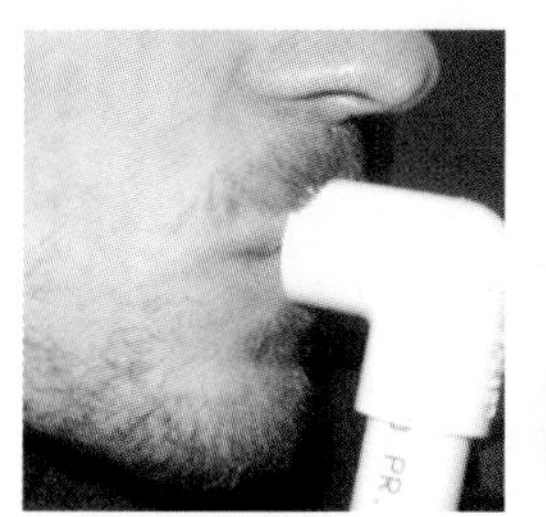

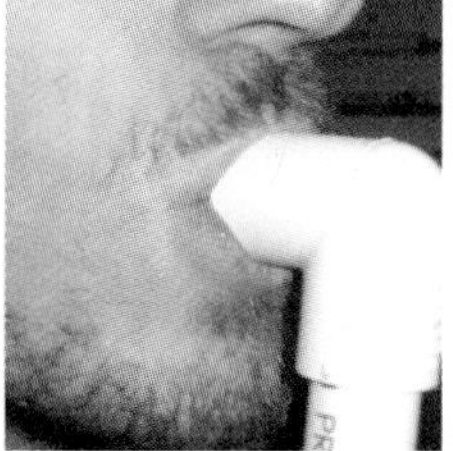

Figure E: Seal the mouthpiece. You can do this either by putting your lips inside of it or sealing your lips around it.

User Notes

Marshmallow Shooter Hack Attack!
By Zot O'Connor

About a week before taking my family of five kids to Las Vegas for the BlackHat and DefCon computer security conferences, one night, my son Aidan went off on a tangent about this S-shaped thing he'd seen online that fired mini-marshmallows. He asked if we could build one. I was skeptical, but as home schoolers, I said we could sure give it a try.

I researched online and found the "Marshmallow Shooter" Instructable. That night, we built one and I was shocked at how well it worked. We built another. I used this project to show Aidan how to use a hack saw, mitre box, rasps, and other tools. The next evening, the whole family built more and a gunfight broke out. Mayhem (and fun) ensued. "We should make a bunch and bring 'em to DefCon!" Aidan shouted. And with that, a plan was hatched. Before we knew it, we'd packed ourselves and our candy arsenal into the van and were heading for Vegas.

At DefCon, my first order of business was handing a gun to conference organizer Dark Tangent. Aidan showed DT how to work it. We heard of him using it for days afterward. By Saturday, I was having way too much fun shooting people in the hallways. Soon, we'd given out all our guns. We still had a long night of parties...so instead of dinner, my wife and I drove to Home Depot and bought materials. We made about eight more shooters and took them to the parties. I'd love to say that eventually we grew tired of blasting people with mini-marshmallows, that we acted our age, but I can't claim anything but that we had tremendous fun.

Realistic Werewolf Costume

Freak out your cat, scare small children, and make grown men weep with this terrifying costume!

By Missmonster

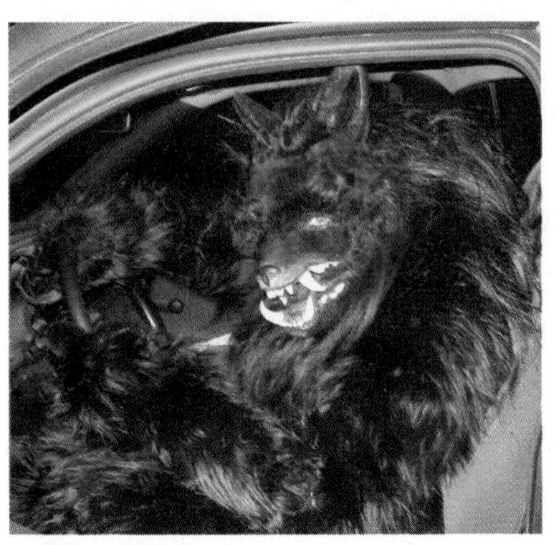

Figure A: Cruisin'!

Create your own werewolf costume with a moving jaw, padded body, and lots of fur! This is my second full, crazy costume, the first being an armored demon. I do costuming as a hobby, but I'm a full-time artist. If you have any questions, feel free to ask via Comments on this Instructable's project page and I'll try to answer them as best I can.

Thank you for checking out my Instructable. I hope it inspires you to make a big mean creature of your own. Have a great Halloween!

1. Start the head

You will need:

- Upholstery foam
- Plaster gauze strips
- Lotion
- Elastic
- Hot glue
- Needle and thread

Using plaster strips, make a little half mask. Don't forget to lube your face with some lotion!

When the half mask is done, start adding some foam to protect your noggin. Nothing too thick, you don't want the wolf to have a huge balloon cranium (or maybe you do, whatever). With the foam added, now make a little elastic bit to hold it on. Sew it into a ring and add a little hot glue to the sewn parts to make sure it stays. Add some strips over the foam to make it more solid.

2. Add the snout and jaw

You will need:

- Celluclay
- Aluminum armature wire
- Pourable plastic resin (optional)
- Resin epoxy glue
- Hot glue

Basically I formed a little wire armature for the snout, you can see it running along the sides there, nice and ugly. I then glued it down with the epoxy and glued some foam over the wire armature with hot glue. Snip the foam snout down to the shape you want. Make it a little skinnier than needed since we are going to add to it.

I then coated the entire thing with resin to make it super strong!

Mix your celluclay and create a basic snout. Let this dry for awhile.

3. Add the jaws and teeth

You will need:

- Sculpey
- Dremel tool or sand paper (Dremel tool will save a ton of time)
- Aluminum wire armature
- Key ring (minus keys)
- Elastic
- Epoxy glue

For additional information, discussion, and more, please visit the Instructables project page:

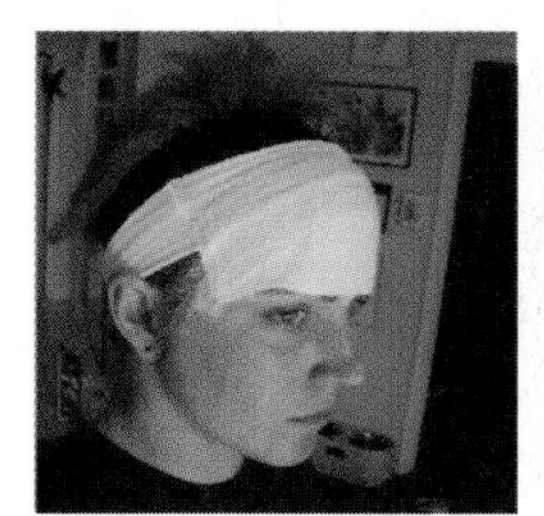

Figure B: Starting the mask with plaster strips

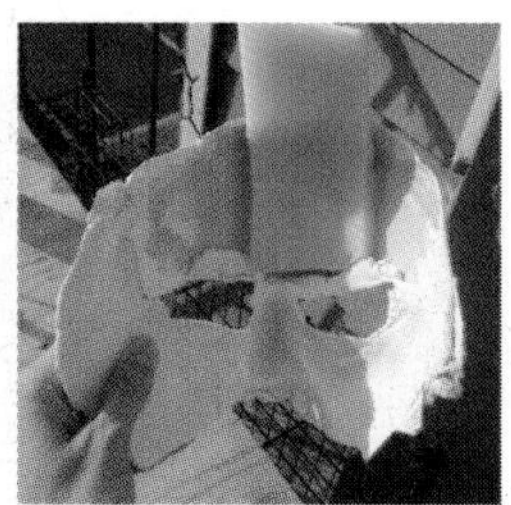

Figure C: Placing foam inside the mask for cushioning

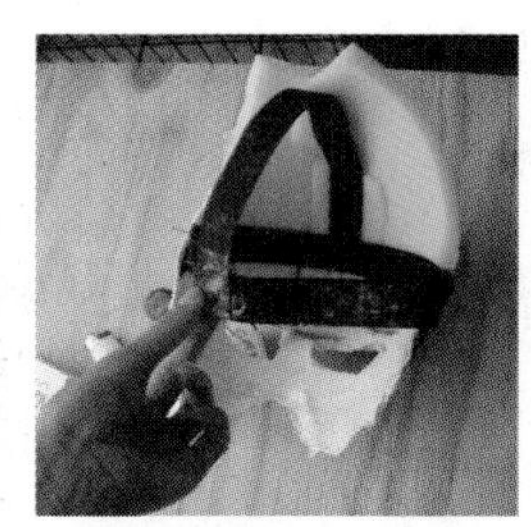

Figure D: Adding elastic to keep it on your head

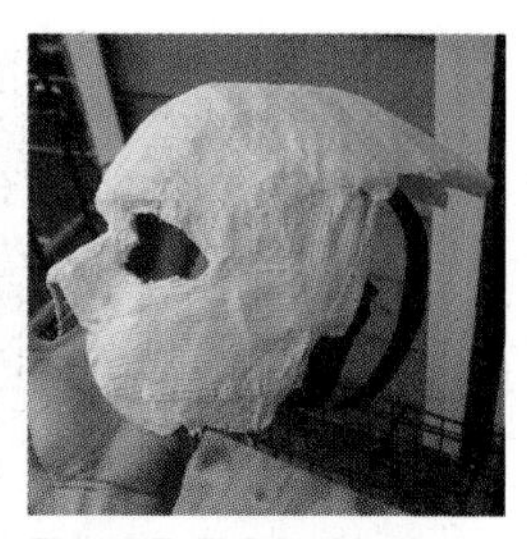

Figure E: Finished half-mask

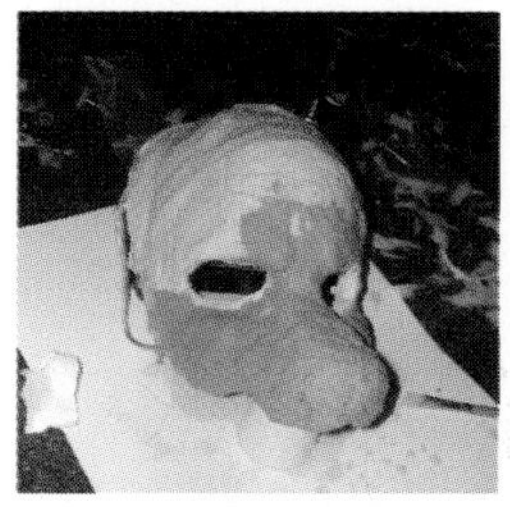

Figure F: Building the snout over an armature

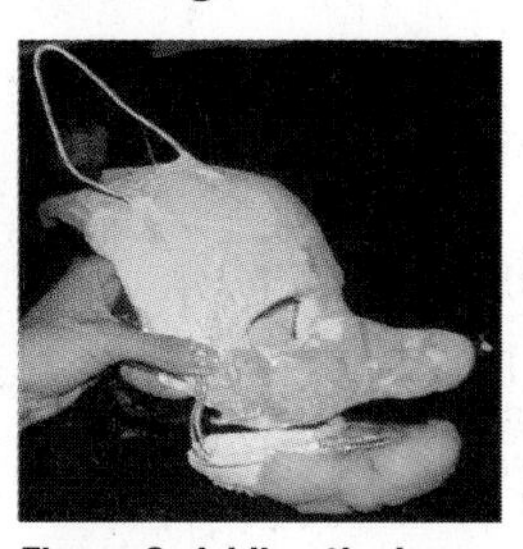

Figure G: Adding the lower jaw

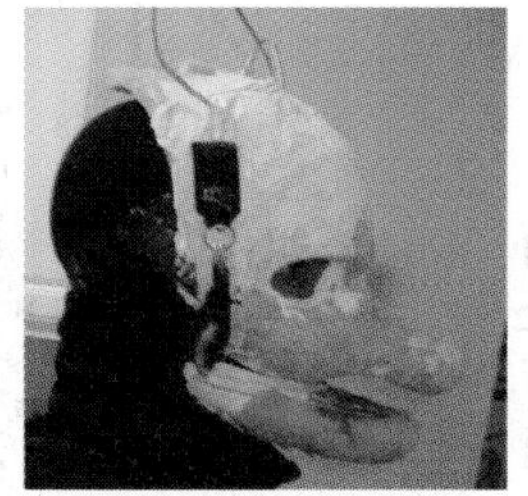

Figure H: Try it on for fit

Figure I: Snarly teeth!

Once the clay dried I went and Dremeled down a lot of the texture. I didn't think I'd be covering the whole thing with fur, my plan was to hand glue hair and then leave most of it smooth...so I really got into smoothing the face which was sort of a waste. Oh well, that's what happens when you make it up as you go along! I also added wire ears and the jaw. The jaw was also made with plaster strips molded onto my actual chin. A wire serves as the support for the foam mandible, just like with the upper head.

The jaw works by making a little ring that attaches to the upper head—the lower jaw hooks in and is removable. I made the hook with some wire formed into a hook shape, attached to an elastic loop.

4. Detailing the head and making the teeth

You will need:

- Sculpey
- Oven to bake Sculpey
- Sharpie marker
- Dremel tool
- Spray primer
- Epoxy glue

I added plaster strips to the ears as well as paper clay. Lots of shaving, sanding, and swearing later, and we kind of have a werewolf on the way. Sculpting with paper clay sucks. I just did a little, let it dry, sanded, added more—repeat until happy. It's frustrating but it's worth it to me. This paper-mâché clay is REALLY light and really strong. It takes patience. I'd work on it between other projects so I didn't get too annoyed. Oohhhh man, this is where it got fun. The teeth are Sculpey, molded to the jaw, then baked and glued in. I also put a little snarl over them in paper clay later on. I also used a ton of primer on this thing while I was obsessing over the smoothness, that's why the whole thing is suddenly grey. Notice how I marked the teeth so I could glue them in easier later on. I didn't detail the teeth as much as I could have but it works anyway.

5. Finish up the head

You will need:

- Acrylic paint
- Brushes
- Clear epoxy
- Hot glue
- Fake fur
- Scissors
- X-Acto knife

Time to paint him! Again, I had thought most of the front face would be naked so I cared way too much about painting it all nice. It's not a total flat black, there's actually a little shading and highlight-

Figure J: Paint the face black so it doesn't show through the fur. Wear a ninja mask under it!

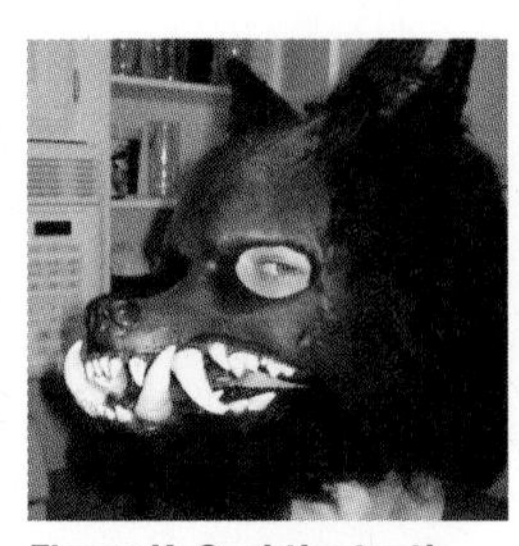

Figure K: Seal the teeth and start adding fur to the face

Figure L: Looking hairy!

Figure M: Don't forget the back of the head—use hot glue to stick it all on.

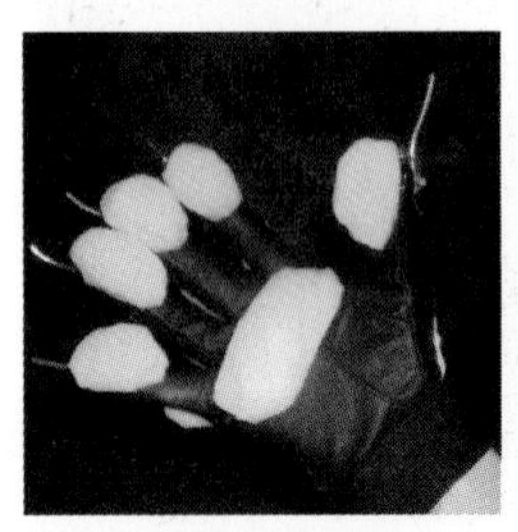

Figure N: Glue foam pads onto a baseball batter's glove and add the claws

Figure O: Furring the paws

Figure P: Paint the pads and claws black

Figure Q: Furring the arms

Figure R: Furred arm—look at the paws on that thing! Note the opera glove on other arm

ing. That red stuff in the jaw is tool dip, I figured I'd seal the inner face since it can get pretty nasty in there with all the sweating I'd be doing. Easier to wipe off and makes it a little smoother.

I also have some black eye makeup to blend things a little better...and a balaclava to hide my mouth (ninja mask type hood). The teeth have been sealed with a glossy clear epoxy. Do a test on your epoxy first, because the brand I used for the bottom teeth yellowed within a month or two and they no longer matched. The teeth are not a pure white, either. I did stain them a little but it doesn't show up too well in the flash. The hair was weird—at first I added mohair that I dyed and glued in piece by piece. Like I have mentioned, I was going to have a somewhat naked face and was going to blend the hair into the skin. Dumb waste of time. I ended up hot gluing on fake fur and trimming it to become short on the face and snout, and blended that into longer hair towards the top of the head. I also bought some long high-quality fur from National Fiber Technology for the back of the head. I used hot glue to attach all of the fur onto the face and head. Cut fur through the backing with an X-Acto, it will help keep the fur intact; if you use scissors to cut through the fur, it will make it difficult to blend pieces.

6. Make some paws

You will need:

- Baseball batters' gloves
- Epoxy
- Dremel tool
- Plastic craft claws
- Foam
- Scissors
- Needle n' thread
- Fake fur

Figure S: Glue together a block of foam for the feet

Figure T: Carving the toes

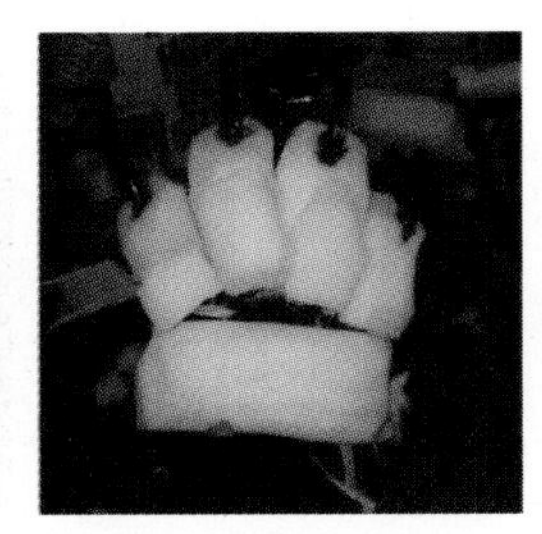

Figure U: Adding claws

Figure V: Trying them on for size!

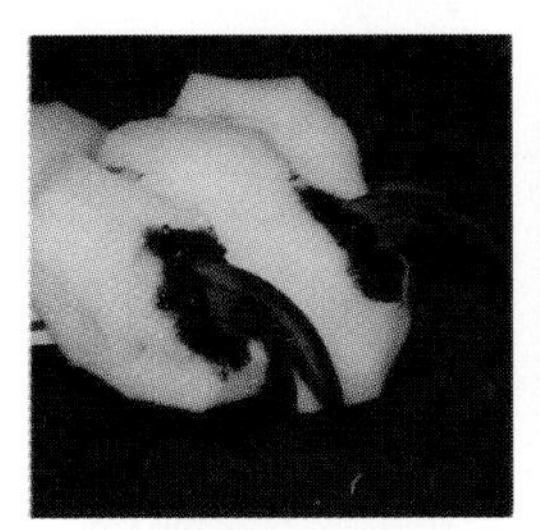

Figure W: Claw detailing

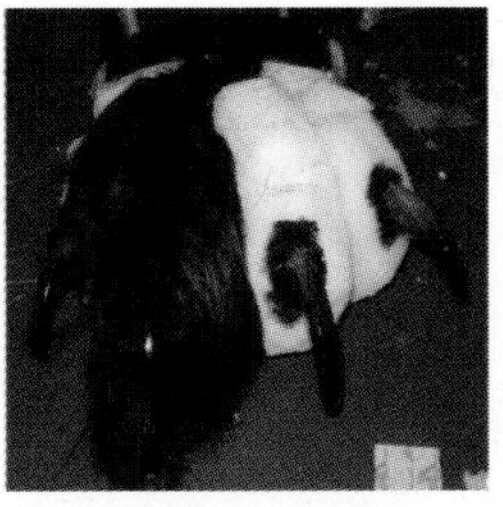

Figure X: Don't forget to fur the feet

Figure Y: Feet strapped onto shoes with fur toes trimmed

Figure Z: Shoes with Velcro straps to keep the feet on

- Armature wire
- Opera gloves
- Acrylic paint
- Clear polyurethane

Put the foam over a baseball batter's glove. Hot glue it on. The wires make the claws sturdy on top of the fingers (the claws will fit over the wire; create a hole in the claw with the Dremel tool). The foam makes the hand look like it has different stubbier joints and hopefully a little less human (Figure N).

Glue the claws onto the finger wire with epoxy. Paint the claws and seal with the clear epoxy. I painted the paw pads first with the latex to reduce the pores of the foam, then went over that with black tool tip (there are better ways to make paw pads but this is what I did). I shaved down the finger tip fur and let it fade into longer fur on the back of the hand (Figures O-P).

Cut the fingers off of the opera gloves, put one on and insert it into the glove. Using hot glue—yes, hot glue—glue on strips of fur up to your elbows. The trick is to use a low heat glue gun (NOT the large high heat!). Apply your glue onto the strip of fur, let it cool for a minute and then apply it to the opera glove. It will be warm but won't burn you. You just need it to stick enough to hold in place. Once all the fur is attached, remove the opera glove with the paw still attached and sew all of the fur seams together. Lower arm done! (Figures Q-R)

7. Make the werewolf feet

You will need:

- Shoes
- Foam
- Hot glue
- X-Acto knife
- Plastic claws
- Epoxy
- Acrylic paint
- Dremel tool
- Clear polyurethane

Add Velcro around the top of your shoe, then some around the sides for the end of the fur legs to Velcro on to. I also made a strap for this brand of shoe, which also has Velcro on top of it for the legs to attach to.

I riveted AND glued the Velcro on—don't want it going anywhere! Plus rivets are fun to install. Toes! Basic block of foam, carve with scissors to round them out a little. Don't bother too much with making it super smooth, it's going to be covered with fur, anyway. :-) I glued my claws in by cutting a slit

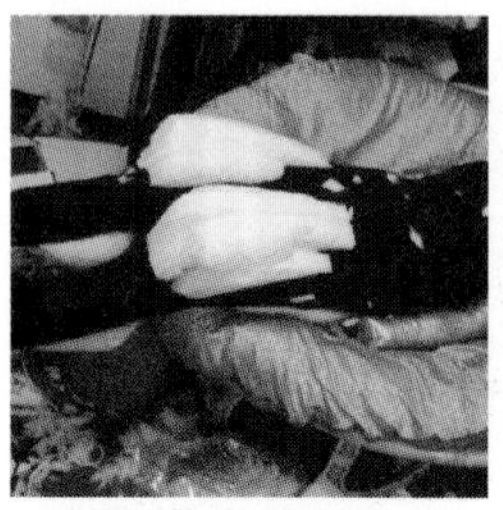

Figure ZA: Adding foam to fronts of the legs to look like dog legs

Figure ZB: Foam on the backs of the legs helps get the digitigrade look

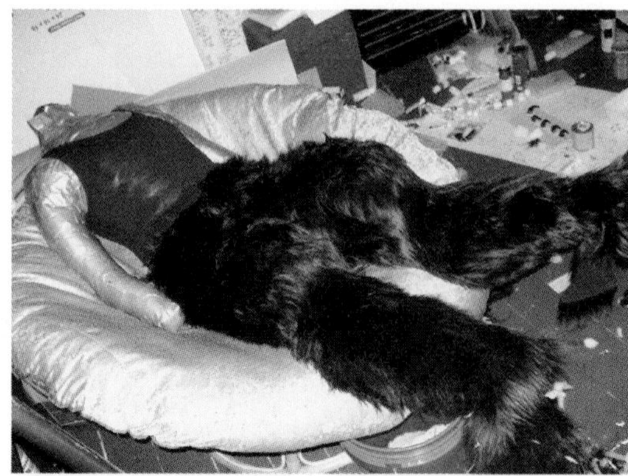

Figure ZC: Furring the body with more hot glue

Figure ZD: Body parts

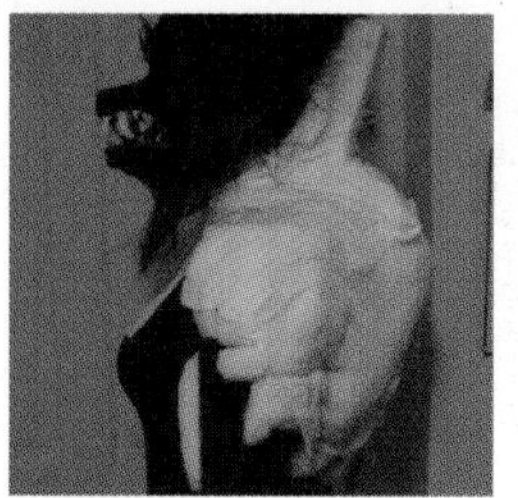

Figure ZE: Building up the shoulders and upper back—werewolves are beefy!

Figure ZF: Trying it on

into the foam, filling it with hot glue, and inserting the claw. Paint your claws and varnish them with polyurethane. Add strips of fur, hot glue it all down, trim the toe fur, and you are done! (Figures S-Z)

8. Make the body

You will need:

- Fake fur
- Hot glue
- Leotard
- Under armor-type shirt
- Foam

I used a duct tape dummy to help out with sizing. I put a leotard with an under armor shirt over it. Then I hot glued foam shapes over the shoulders, chest and back to make a beefy upper chest. Then to create a digitigrade leg (dog leg appearance—not a backwards knee like everyone thinks, they just have shorter lower legs, longer feet, and stand on their toes). I added foam to the front of my top leg and the back of my calf to make a shape like I'm bending my knees, even when the leg is straight.

Try the suit on every so often with the head on to make sure the proportions are okay. Trim accordingly—get the foam right because you don't want to have to go back a step to fix things. Remember, you will be adding fur on top of the foam. This will add more bulk, so slight shapes are best. The shirt and leotard connect with Velcro. I sewed and then glued the corresponding sides to the inside of the shirt and outside waistline of the leotard. Also add some Velcro to the elbows of the shirt; this is where the lower arms will connect.

Cut shapes of fur with your X-Acto (cut the back as to not mess up the fur) and hot glue it all on. Place the pieces close together, fill gaps with smaller pieces. After you are satisfied, sew all of the glued on pieces together so when you flex and bend, the fur does not show seams and gaps. I also included a photo of all the parts we have covered so far.

9. Try it on

You will need:

- Colored contacts
- Kryolan aqua makeup (buy at sillyfarm.com)

Figure ZG: Scary contacts add attitude

Figure ZI: Strike a pose!

You are pretty much done. Get some colored contacts to look less human and really freak people out! I blended my face into the black of the mask with some Kryolan makeup. It mixes with water and can be applied with a finger or brush and does not rub off too easily.

Add a tail if you feel like it! I didn't document that process but just sew some fake fur, stuff it then sew it to your wolf's butt.

Try on your new suit! You may need to make some changes and adjustments, but congrats!

Figure ZH: Come closer...

You are now a werewolf. Stay hydrated and be very careful, this suit gets really hot. You will lose a lot of water and moving will take a little more out of you. Take frequent head breaks (take off the head and hood) in front of a fan and have someone around to help you get in and out of the suit, aka a wrangler. A wrangler can also help you spot trouble where you can't see it (people tugging on the suit from behind or kids getting a little too aggressive), or do simple things like open doors for you when your claws can't. Most people will be really cool and want to pose for photos or pet you, but be aware of people just wanting to make trouble with an easy target. Especially drunk people! Having some back-up can make the difference between having a great time out or getting your costume damaged and you possibly hurt. I have had some bad experiences being out alone with another costume and I'd hate for anyone else to have the same trouble. But don't let all that scare you. It's going to be awesome and you will have a great time. Go scare some kids and have a great Halloween!

Missmonster, aka Melita Curphy, is a freelance artist living near Chicago. She paints, sculpts, makes lots of monsters, and loves to scare kids as the werewolf. Check out her art at www.missmonster.com.

Guitar Hero Hack: Key Molding

Add transparency to your virtual shredding with these lightable fret buttons

By Matthew Borgatti

Figure A: The finished neck, ready for lights

I recently molded a Guitar Hero controller at the behest of fungus amungus. I'm going to show you how to create duplicates of your GH Controller keys in frosted clear plastic.

I'll review two types of molds using the same materials and principles. The first will be a simple block mold for a single key. The other will be a single mold for multiple keys; this is often called a gang mold.

1. You're going to need

- RTV silicone: I will be using Smooth-On OOMOO two part silicone molding compound. This is fairly easy and inexpensive to pick up. You can order it online or at some craft and plastics stores. Any RTV (or Room Temperature Vulcanizing) silicone will do. For this process, the polyester casting resin I will be using has more cure inhibition (meaning that it will remain sticky) than tin cure silicone (your standard inexpensive RTV). This means that to get good parts out of the mold, you will have to coat the mold in sealant before casting your parts, bake the parts after they're cured, or apply a clear sealant to the parts after they are molded.
- Latex gloves
- Foamcore
- Stirring sticks
- A digital scale or graduated measuring cups
- Bondo car body filler: You can pick this up at most hardware stores and auto shops.
- Clear polyester resin: find it at a plastics store, or order it online. Sometimes craft stores will sell this as a way to make cutesy paperweights with flowers trapped in them, etc.
- MEKP catalyst: This will often come with your resin. You can also find it as fiberglass catalyst in most hardware stores.
- Polyester resin dye
- Respirator, ventilation system, or open well-ventilated space to work in: be mindful of your neighbors, as the polyester resin smells something awful, and will cause some people with sensitivities to feel ill.
- Hot glue
- A mold board: any piece of smooth wood or plastic will do. Remember to seal the wood before casting on it, or else the mold will be difficult to remove.

Figure B: Opening the neck

Figure C: Take out the circuit board

Figure D: Take the rubber bumpers off the keys; save all the parts!

Figure E: The five original keys

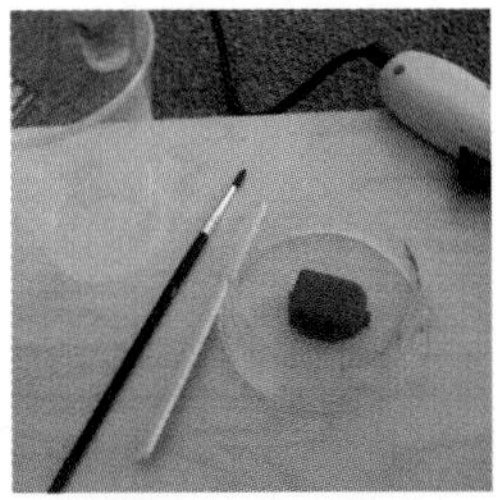

Figure F: Setting up for a single key mold; glue the piece of cup to the mold board

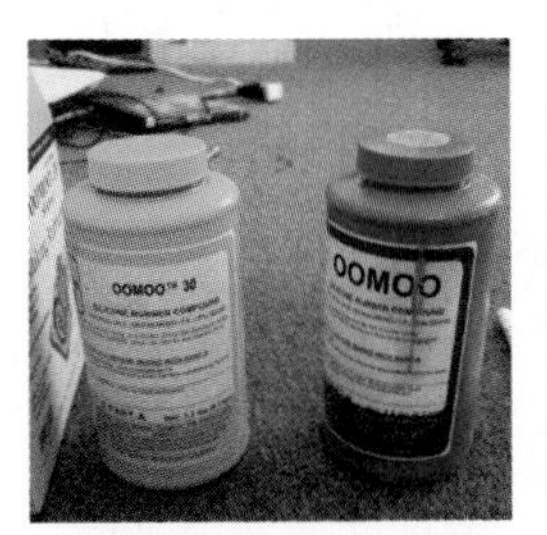

Figure G: Magical OOMOO silicone rubber compound

Figure H: Pour in equal parts of compounds A and B

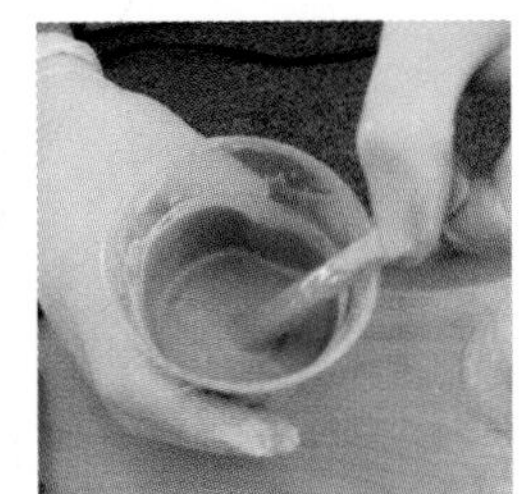

Figure I: Stir for at least a minute; try not to stir in bubbles

- Disposable paintbrushes
- Plenty of paper towels
- The touch
- The power

2. Dismantle your controller

It's time to suck it up and take the screwdriver to this fancy bit of extruded dinosaurs. You must remove the main body panel before the neck will unfasten. Once you've opened the neck, take a small electronics screwdriver and carefully take out the circuit board (you *are* saving all the parts, right?). Remove the keys, and take their rubber bumpers off. The rubber bumpers will fit perfectly into the new keys once everything is done.

3. Single key mold

This is pretty simple and straightforward. You lay a single key down on your mold board. Then you cut the bottom off of a disposable cup, and hot glue the top portion to the mold board around the key. That's it. You're ready to mix up and pour.

4. A high pour

Check over your mold and part to make sure there isn't any dust, grit, or fingerprints that will screw up your casting. Remember that whatever you see here will show up in your final parts, and there's no way to correct a screwed up mold. A few extra seconds of inspection will save you the massive headache of recasting the whole part.

Mix your silicone according to the manufacturer's specifications. The OOMOO instructions called for equal parts of compounds A and B. Stir your mixture for at least a minute, making sure to scrape down the sides and bottom. Avoid whipping bubbles into the mix by moving the stick in and out of the mixture.

One of the crucial factors in making a successful mold is making sure there are no voids or bubbles in your casting. This means you pour your mixture from high above the mold in a very thin stream. This stream will pop the large bubbles that would have been included in the casting. Start in one corner of the mold, and let it fill evenly to the other side. When you've poured enough that you can only see the top of your key, go in with the paintbrush, and make sure there's silicone filling up the small divots where the rubber bumpers were. Now you can fill the mold until your silicone is about 1/4" above the level of your key.

Figure J: Pour in a thin stream

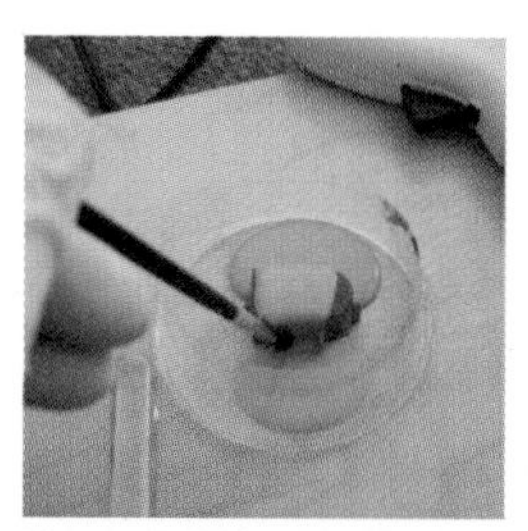

Figure K: Use a paintbrush to make sure the silicone is getting into all the small spaces

Figure L: Continue pouring in the silicone

Figure M: Make sure the silicone covers the key by about 1/4"

Figure N: After the silicone sets, remove the mold

Figure O: Turn the mold over

Figure P: Remove the key carefully

Figure Q: Preparing a gang mold; building a mold box from foamcore

Figure R: The finished mold box with the keys glued down

Figure S: Starting to pour the silicone

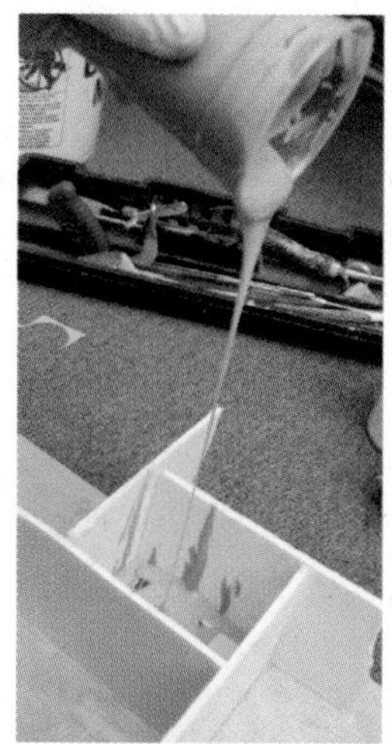

Figure T: Pour in a thin stream to cover keys completely

Figure U: Remove the box from the finished mold

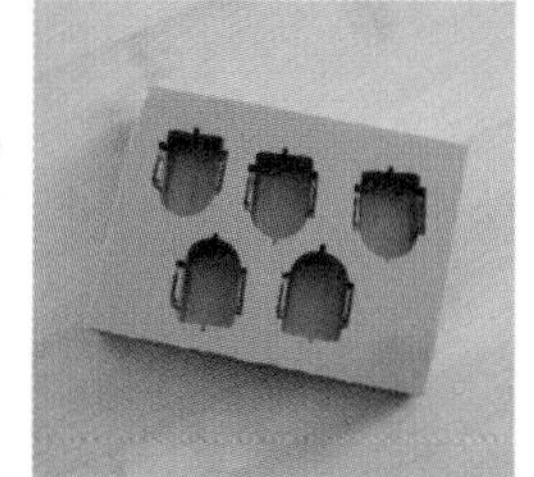

Figure V: The finished gang mold with keys removed

5. Gang mold

This mold is simply a slightly more complex version of your block mold. Start with several rectangles of foamcore. Adhere them to the mold board and seal the foamcore joins with hot glue. Run a bead of hot glue around the base of the mold, where the foam-core meets the mold board. Spray with a sealant or shellac to seal. Then glue down your parts with a daub of hot glue; this makes things a little simpler to arrange and cast. Fill this mold with silicone exactly as described in the previous step.

6. Casting the plastic keys

The Polyester resin I used was clear. For the translucent frosted look I was going for, it had to be modified a little before casting. For a frosted look, you can apply a crystal clear spray into the mold in a thick coat before molding. When the parts come out of the mold you can wipe the spray off of them

Figure W: Materials for casting the keys

Figure X: Mix the resin, adding dyes and a bit of Bondo to add opacity

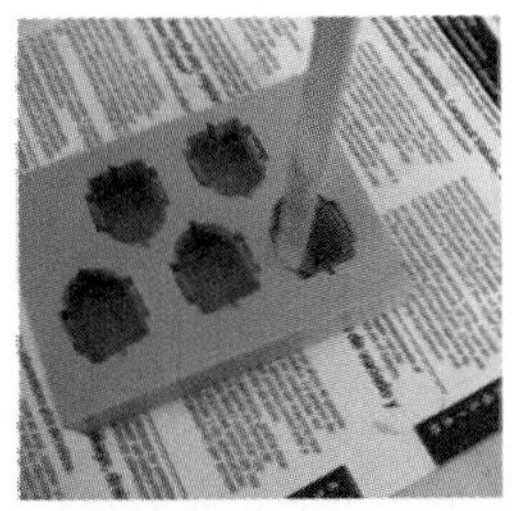

Figure Y: Pour resin into molds, removing bubbles with a stirring stick

Figure Z: Resin-filled molds

Figure ZA: Rock on!

User Notes

isuark made one with clear keys (he later used Bondo to frost them)

with acetone. For a super smooth look, you can spray PVA mold sealant into the mold using a Preval Sprayer (http://tinyurl.com/5gfwr2). It will form a super shiny water soluble layer in the mold. When your parts come out, rinse them in water to reveal a remarkable shine.

To begin, I measured about an ounce and a half of resin on a digital scale. I marked how much resin was in the cup with a marker and measured its height. Then I transcribed that line onto another cup and cut it in half vertically. Now I could put this half cup over each cup I'd use to mix and just draw the line on it, meaning that I didn't have to weigh out my resin each time.

To color the resin, I used red, yellow, and blue dyes made especially for it. I also made the resin slightly more opaque by adding a pea sized daub of Bondo to it before mixing. Make sure you mix your resin thoroughly before adding your catalyst. Each of my 1.5oz servings of resin would need six drops of MEKP. Be careful when using MEKP, as it burns when it contacts the skin, and is very bad for your health. Make sure to work in a well ventilated area or use a NIOSH rated respirator.

After you've mixed your resin and catalyst, you can pour your keys using the same high pour method described before. Make sure to go slowly, and to take out all the bubbles you can with a stirring stick or other small instrument.

7. De-mold and admire

Since the silicone mold is very flexible, you can simply bend it until your parts fall out. Make sure to insert the little rubber bumpers in your parts before putting the guitar back together. Look here (www.instructables.com/id/Guitar-Hero-LED-Mod) for the details on lighting up your guitar for night shredding!

Matthew Borgatti says: "I started tinkering in the interest of bringing ideas I drew as a child to life. That, and it's a perfect outlet for my plethora of absurd and disparate knowledge. I also do Dinobot impressions in the car."

The Office Supplies Trebuchet

Who knew how entertaining (and possibly annoying) paper clips and tape could be?

By Alexander Palfreman-Brown, aka Scissorman

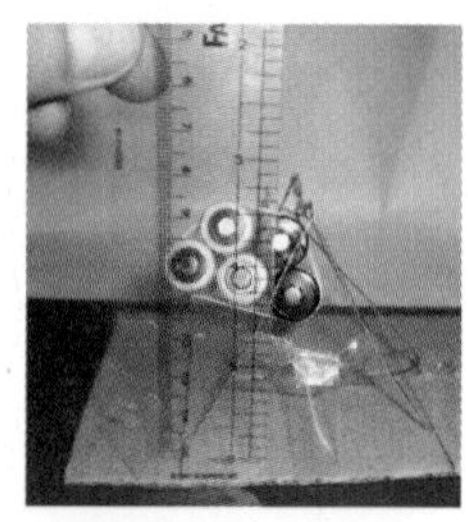

Figure A: The finished trebuchet

Bored at work? Build your own 3" trebuchet out of paper clips and throw balls of Blu-Tack up to an amazing 4 feet. WOW!

1. Tools and materials

- Pliers (needle nose would be best)
- Scissors
- 8 paper clips
- Cellotape
- Thin string
- Blu-Tack (poster tack)
- Corrugated cardboard, approx. 6" x 6"
- Ballast (I used a bunch of batteries from our recycling box)
- Rubber band

2. Straighten and shape the paper clips

Use the pliers to make the paper clips as straight as possible. At the very end of four of the paper clips, make an eyelet big enough to insert another paper clip through (but not much bigger). At the other end, make a dog leg. Try to make it on the same plane as the eyelet!?! If you lay it on a flat surface and it lays flat, it is ok.

3. The axle

Next, shape the axle. The distance between the hooked ends is about 2". It should be wider than your intended ballast. The U shape in the middle must be centered.

4. The arm

Make a loop in another paper clip about 1/2" from the end and then make another eyelet in the same end. Make a little hook at the other end (Figure G).

5. The trigger

The trigger is a little hard to describe; Figure H shows the two shapes. The bump on the left is to pull. The loop on the right is there to catch the little hook on the arm.

6. Beginning the assembly

Take the axle and insert it through the loop in the arm (not the eyelet) so that it sits in the U (Figure J). I wrapped cellotape around the axle on either side of the U just to give it some thickness, so it doesn't slide all over the place.

Figure B: Tools and materials

Figure C: Straighten the paper clips with pliers

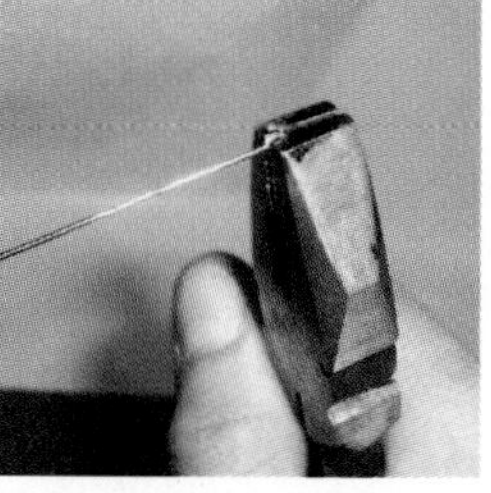

Figure D: Make eyelets in the ends of four paper clips

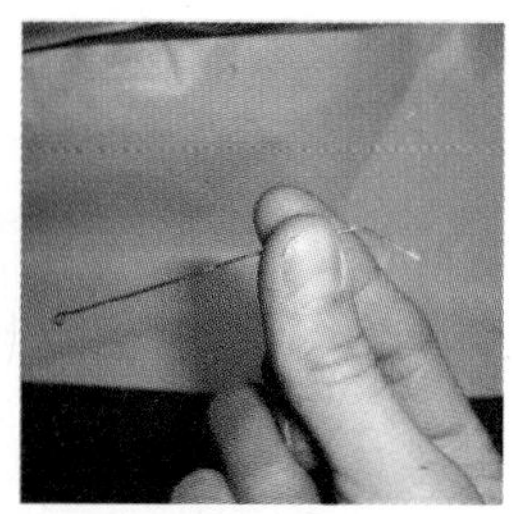

Figure E: Then make a bend like a dog leg at the end

For additional information, discussion, and more, please visit the Instructables project page:

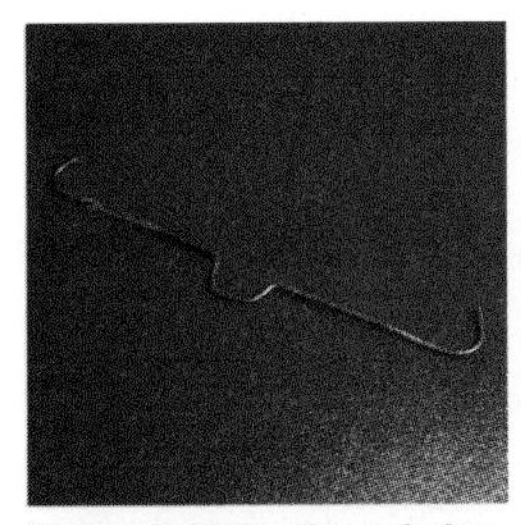

Figure F: Make the axle by hooking over the ends and making a "U" shape in the middle

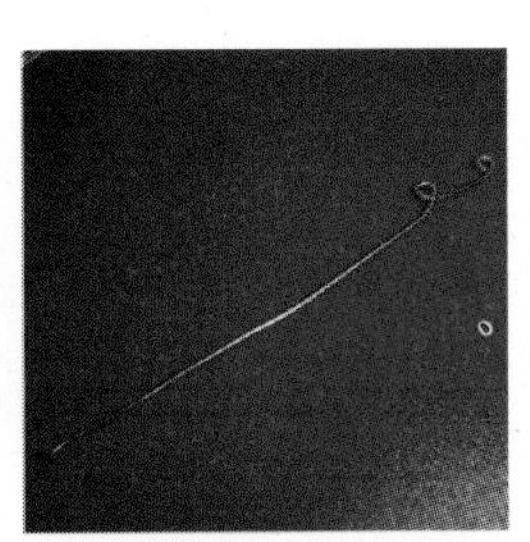

Figure G: Make the arm; it needs a loop near one end, and an eyelet

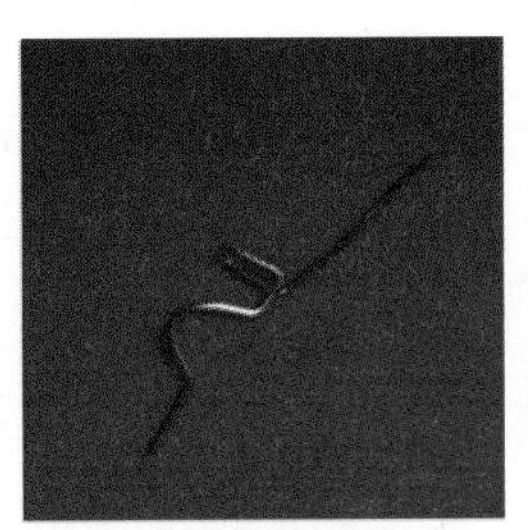

Figure H: The trigger—the most difficult shape

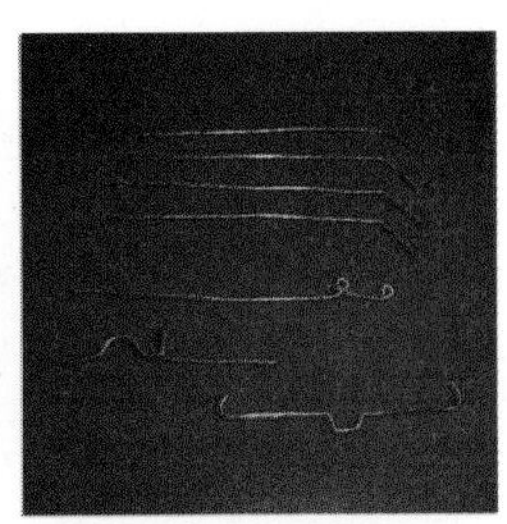

Figure I: The components, ready for assembly

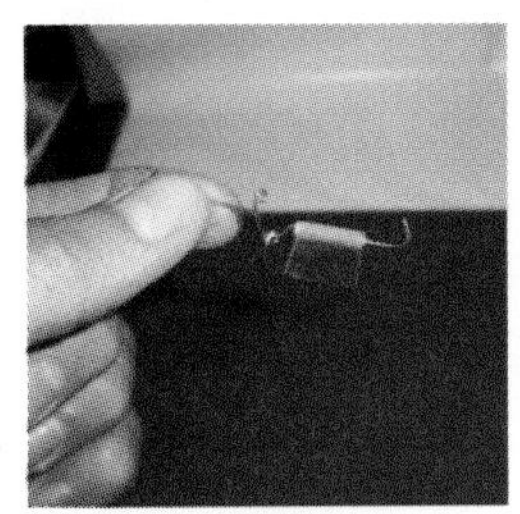

Figure J: Insert the axle through the loop in the arm

Figure K: Mark the base

Figure L: Setting up the legs on the cardboard base

Figure M: Mount the axle on the legs

7. The base

Mark the cardboard base with a center line. The two lines are about 3" apart and the little Xs on those lines are also 3" apart (Figure K). Poke the dog leg ends of the supports into the four Xs so that the ends are between the layers of the corrugated card (Figure L). If you made the dog legs at the right angle (not *a* right angle, though), they should sit nicely.

8. Mounting the axle

This is a little bit fiddly, but poke the hook ends of the axle through the eyelets in the supports, then bend the hooks down with the pliers (Figure M). Tape between the legs to make an A-frame, then run some tape from the crosspiece of the "A" shape to the sides of the cardboard base to prevent side to side motion (Figure N).

9. Mounting the trigger

Push the end of the trigger into the layers of card where the center line intersects the support line (the side that the arm hook is on). The other end of the trigger needs to run in a slit so that it can slide back and forth. Don't make the slit too long or else you risk the trigger being pulled out (Figure O).

10. The ballast

Make a little hook with paper clip number 8, and tape to the battery. Use the rubber band to attach a whole bunch of batteries together. When they're all assembled and hooked on the arm make sure the ballast is not touching the base.

11. The ammo

Take the string, put a loop in it and push the free ends into a pea-sized lump of Blu-Tack. Wrap it around a couple of times and then smoosh the Blu-Tack into a ball. The extra string inside will allow you to experiment with the best length for maximum distance when firing. A good length is about 2/3 the distance from the hook to the axle along the arm.

12. Using the trebuchet

Hook the ammo over the arm, pull the arm down, and trap the hook with the trigger. Position the ammo on the center line and at full extension of the string. Aim and fire by pulling the loop on the trigger towards you.

Warning: The arm moves fast and has a sharp little hook on it. Because this weapon has limited accuracy, you can't predict the path of the Blu-Tack. When arming or firing, protective eye gear needs to be worn by the user and anyone within its range.

Figure N: Stabilize the legs with some tape between them and over the edge of the cardboard

Figure O: Mount the trigger on the cardboard base

Figure P: Make a hook and tape to the ballast

Figure Q: Attach the rest of the ballast with the rubber band

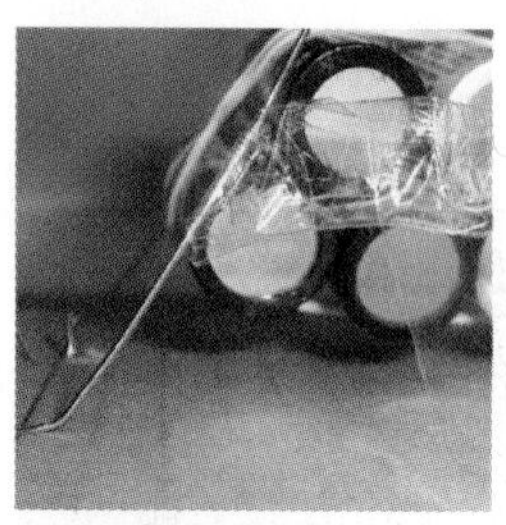

Figure R: Hook the ballast on the arm

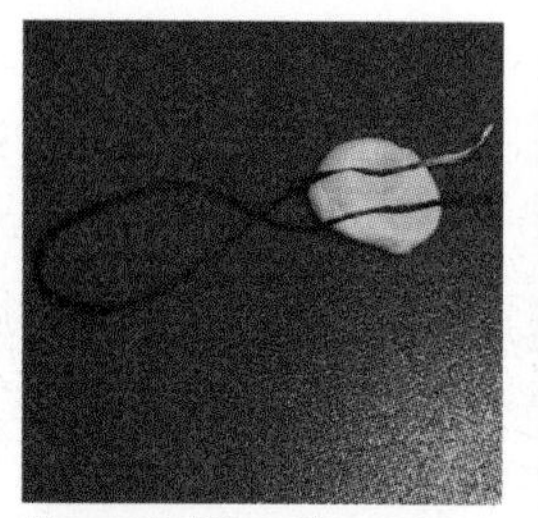

Figure S: Place the ends of the string in a blob of Blu-Tack

Figure T: Smoosh the Blu-Tack into a ball

Figure U: Hook the ammo over the arm and trap the hook with the trigger

Figure V: After the launch; note the wrap of tape that prevents string from sliding up the arm

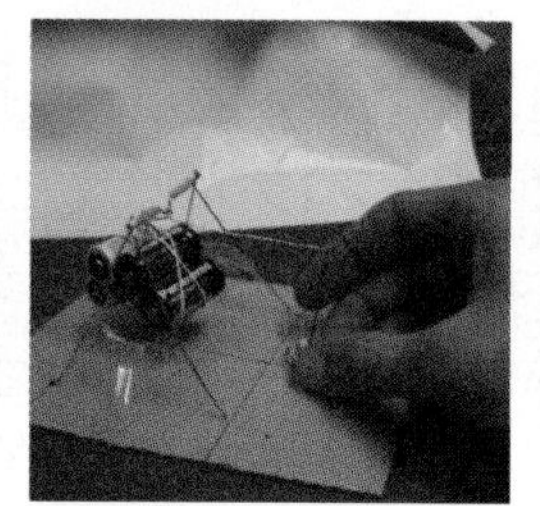

Figure W: Fine tune for accuracy

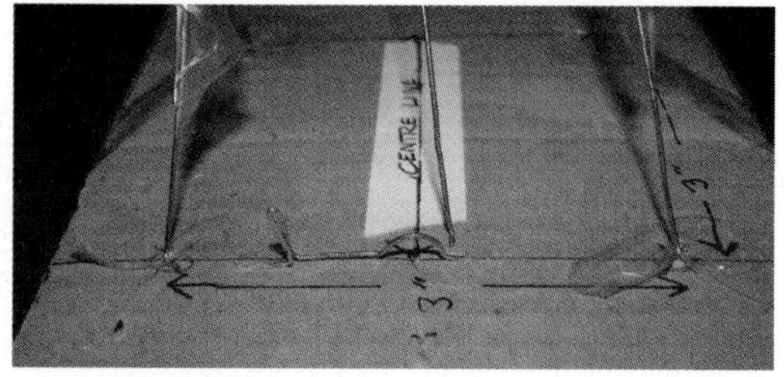

Figure X: Release pin modification; rotated 90 degrees

To prevent the string from sliding up the arm (which will interfere with a successful Blu-Tack launch), wrap tape around the arm below the hook.

13. Extras

I've already mentioned changing the string length to change the way the ammo flies. Changing the weight of Blu-Tack, changing the weight of ballast, and altering the hook angle will all affect the flight path of your ammo. Adjusting the angle of the hook will change the point at which the ammo is released. Release too early, and the ammo will fly straight up (if not backwards); release too late, and it thumps down right in front of the trebuchet.

14. Release pin modification

When I was building a new trebuchet, I found a more reliable way of launching the ammo. I rotated the release pin 90 degrees so that now instead of drawing back like a bolt, it's hinged and rolls back off the hook of the arm. You launch by pulling back on the lever. It's a very simple modification, and you can use the original pin, though it may need some reshaping. Just make sure that it's placed far enough toward to release the hook, and that it doesn't snag the ammo on its journey past.

Alex Palfreman-Brown lives in Hertfordshire, England, with his wife and daughter. His mantra is "What would MacGyver do?"

Fix the Red Ring of Death! (without towels!)

Fix your Xbox 360's RRoD by yourself, without towels or soldering irons!

By Dr. Professor Jake "Biggs" Turner

Dead Xbox 360? If you no longer have a warranty on your Xbox, and you've gotten the dreaded Red Ring of Death (RRoD, shown in Figure A), there's still hope. The RRoD is displayed by the 360's "Ring of Light" when there's a general hardware failure. While this can be caused by any part (or lack thereof), the general cause is excessive heat, which stresses the solder joints on the CPU and GPU. Over time, these joints can become separated, causing the Red Ring to show up, usually when you have a new game to play and you've taken the day off...

What this fix involves is nothing but mere screws and washers. Seriously, no overheating your console, adding new fans (although you can add them if you like), or towels. Ever since I installed it I haven't had the console freeze up once.

Figure A: What the hell are you so green for?!

1. Before we begin...

Before you even think to unplug your 360, let's work in a bit of intelligent forethought.

For starters, why would you need to repair the console yourself? If you have a working warranty, I highly suggest you send the console to Microsoft for repairs, as a DIY repair will void your warranty. While Microsoft dropped the ball on product quality in efforts to get the 360 out before the other consoles, they picked up the proverbial dirt-covered ball by extending all warranties to three years.

If you plan on actually reviving the 360, don't use the towel trick that's all over the Internet. What that does is overheat the inside of the console by blocking fan intakes, a temporary fix to make the 360 run for a brief period of hours to days. While it may be the quick and easy approach, it also overheats every other component in the 360, which is never good. It can cause chips to fail, capacitors to dry out, release the magic smoke, and possibly spark a fire. So unless you hate your Xbox something fierce, don't do this.

And of course, if you're on your fourth Red Ringed Xbox 360, and are considering buying a PS3... give this Instructable a try. I would hate to see someone give up the pinnacle of gaming due to a simple flaw, and trade it in for a Blu-Ray player with gaming functionality.

So, if your warranty is void, or you're about to join the dark side of gaming, read on!

2. The suspected cause

So, knowing what I've written, you ask yourself, "Self, if the 360 is baking its processors to the point that they'd separate, why aren't the heatsinks holding them tightly to the main board?" That's a very bright observation!

The problem is pictured in Figure B. Those two X-shaped pieces of metal are what try to hold the heatsinks onto the CPU and GPU. Problem is, they aren't springy enough to do the job. The heatsinks have a tiny amount of wiggle room, and the motherboard is free to warp from heat. That

Figure B: These X-clamps provide some tension to keep the heatsinks pressed to the processors, but not enough, and not evenly

Figure C: Yeah, none of these parts are the actual ones I used. I forgot to take photos of them in their little plastic baggies. Silly me.

prohibits proper heat dissipation, and allows the processors to break away from their connections. I blame the bad design on Microsoft's need to get the 360 out before the PlayStation 3. But now isn't the time for blame, I bet you're about ready to buy one of those Blu-Ray players that come with gaming functionality.

So now we've determined that those "X-Clamps" need to go. The next step lists what you need to get started.

3. Parts and tools needed

To secure your heatsinks *tightly* to the motherboard (and the metal case in the process), you will need:

- 5 x 20mm panhead machine screws—keep length between 20 and 40mm (4)
- 5 x 15mm panhead machine screw—must be no longer than 15mm! (4)
- #10 washers (44)

And for your English measurements (thanks to Instructables member ajmontag for providing these):

- 3/16" x 1" panhead machine screws (4)
- 3/16" x 1/2" panhead machine screws (4)
- #10 washers (44)

These 5 x 20mm (3/16" x 1") screws will be used to secure the CPU heatsink. The shorter screws will be used for the GPU, and 15mm is the most you can get into it. The 1/2" fits fine with room to breathe.

The 44 #10 washers will be used to keep the motherboard firmly in one position. Nylon washers aren't necessary, since there are no traces or components to be touched around the screw holes. But if you feel you must have them, I will not stop you

Figure D: This screwdriver is way too big. But it gives you the idea.

from purchasing them.

And while you are working with the heatsinks, you may want to apply new and better thermal paste to them. I suggest picking up a tube of Arctic Silver 5.

Toolwise, you'll need:

- A drill or drill press to widen screwholes in the metal case (with a 3/16" drill bit)
- A TORX 9 and TORX 6 screwdriver to remove screws from the 360's case
- A 1/4" wrench or socket to remove the X-Clamp posts from the heatsinks
- A Phillips-head screwdriver for those machine screws (or flathead if that's what you got)
- And a tiny flathead screwdriver to pry off the X-Clamps
- Anything else like tape and steel scrubbers you have kicking around in your house

4. Gettin' to business: remove the heatsinks

Right, first thing we gotta do is open this sucka up. I'll point you to the tutorial I read, because opening the 360 is an Instructable in itself: www.makezine.com/go/OpenTheXbox360.

You need to remove *everything* from the metal case (if you can't get the fans out, that's fine). Set the motherboard on a clean surface, like a newspaper. Put the DVD drive, screws, plastic case and other parts someplace where they can't get harmed.

Now you must remove the heatsinks from the processors. This is very nerve-wracking. I used a small flathead screwdriver, and pried two or three of the X-Clamp's legs from the posts attached to the heatsinks (Figure D). But if the blade of said pointy object were to slip, it could take out a bunch of tiny parts as it scrapes across your motherboard. The solution?

Put a piece of corrugated cardboard over the motherboard, where the screwdriver blade would otherwise rip through. So if it were to slip now, it'll just hurt a severed piece of a box.

You'll want to pop off three of the bracket legs, because the fourth one won't have anything holding it on. Once you get the clamps off, the heatsinks

Figure E: 1. Yeah, I know there's still thermal paste on here. But this is the good stuff that I put on from when I patched my DVD drive's firmware. So I'll just smooth it back out. 2. These are what you're removing, stash 'em somewhere just in case.

will come off on top side of the motherboard (you may need to tug on them a bit if the thermal paste is sticky). The next step deals with the heatsinks themselves.

5. Stripping the heatsinks

Once the 10-dollar heatsinks are freed from the 200+ dollar motherboard, you need to remove the 1/4" thingamabobs (X-clamp posts). Use a 1/4" nut driver, wrench, or adjustable wrench to get them off. Store them and the X-clamps in a bag/parts drawer, since you won't be needing them again. Take a metal scrubber and remove the thermal gunk from the heatsinks. And while we're at it, get a toothpick and carefully scrape the gunk off the CPU and GPU (Figure E). Nothing better than powering on a reborn 360 with clean heatsinks, no?

6. Drill bigger holes

To use those 5mm machine screws, you'll need to widen the eight screw holes in the metal case that previously secured the X-clamp posts. They are highlighted in Figure G.

To widen them, use a 3/16" drill bit with a drill press (Figure F) or hand drill.

If you're using a hand drill, set a wooden block under each hole as you drill it, to avoid warping the case. If you're using a drill press as I did, there should be a steel pedestal with a hole for the drill bit to pass through, saving the case from serious FUBAR-age.

Figure F: 1. Gotta love that Harbor Freight. 100 bucks. 2. What's that, you say? A countermeasure against heating.

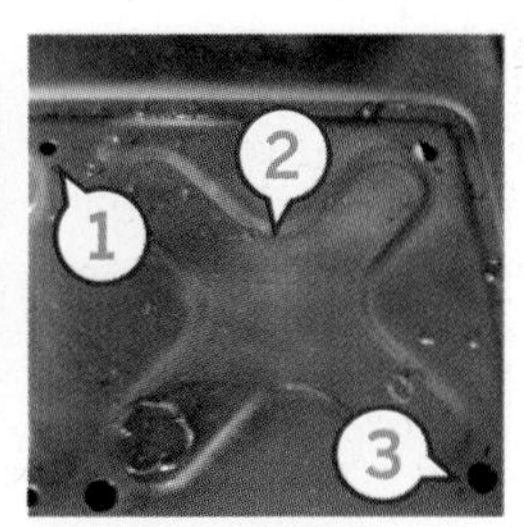

Figure G: 1. Original hole 2. Make sure you get ALL the metal shavings out 3. Brand-spankin' new hole

Have the bit spinning before you push into the center of the hole. Repeat this for all the holes. Make sure there are no burrs left on the holes or shards kicking around in the case afterwards. Bang it around a few times to get them out: the last thing you'd want is to have your 360 die from little pieces of metal shorting it out.

7. Screws, washers, and more screwyness!

So far, we've prepped the heatsinks and the motherboard case for these 5mm screws. Now we get to make sumtin of it!

With the case sitting flat, fan hole in the back on the right, take note of where you need to stick these screws. The 4 holes on the left X are for your GPU, the four holes to the right will be for the CPU screws.

Now put the 5 x 15mm screws in the GPU screwholes, screwheads on the outside, so that they come into the case (Figure H). Put tape over the heads to keep them from falling back out, as shown in Figure I. Now put the 5 x 20mm screws into the CPU screwholes, and tape them just like the GPU screws. This tape is very important, so don't take it off until I say so! Ha Ha.

Lay the case flat again, and all the screws should be poking up at you, as they are in the main photo. Place three or four 5mm washers on each of them, the idea being to get them level with the motherboard standoffs. Use a straightedge to check. If they come up too short or too high, the motherboard will get warped, and you won't be fixing anything.

Now the fun begins. You must get the motherboard back in the case, with the goal of getting the screws into those orange holes that the X-Clamp posts went through. All while you try to keep those washers on the screws! If they fall off at any point, you'll have to take the motherboard out, put them

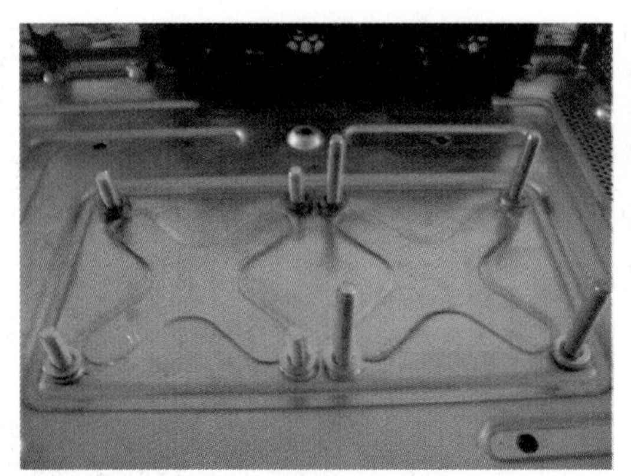

Figure H: There are three washers on each of these screws, but I ended up needing four later

Figure I: Put tape on the screws to keep them from falling out until you get the heatsinks installed.

Figure J: In a bit, you won't be seeing these...

back on the screw(s), and repeat. So try not to do it.

Tilt the case onto its right side (where the hard drive connector would be), so that you can reach the screws underneath. Angle the back of the motherboard into the case, and push it all the way to the rear of the case. Lower the board until it gets held up by the rear CPU screws you installed (they should be the CPU screws if the screws were taller than the GPU screws). Now, untape one of those rear CPU screws, and point it into its hole. Give it a few twists to hold it in place, and thread the other one in. Now, while carefully holding the motherboard and those untaped screws, lower the motherboard until more screws hold you up, and get them into their holes. Don't let those washers fall out!

When it's all said and done, you'll have screws poking through those holes. Tape the heads back to the case so they don't fall out. Now press on the motherboard around the screws, it should be solid at each screw. If not, you'll need to get back to those washers and pop another one on.

But if it's all nice and immobile, throw two washers onto each GPU screw, one onto each CPU screw, and head to the next step!

8. Re-installing the heatsinks

Now we get to mess with those screws one more time, before the tape gets taken off for good. Before you put the heatsinks back on, put thermal paste onto the CPU and GPU dies (the silver shiny things). A paper-thin coat will be good. If you're using a silver-based paste, make sure you don't put on a ton, because if it gets onto anything, the silver's conductivity will mess with the other components.

Pick whichever heatsink you wanna put on first. Make sure the CPU's copper heatpipe is facing away from the GPU heatsink when you put it on. The GPU heatsink won't let the CPU heatsink on if you put it on wrong. The easiest way to get this right is to look at the smudge marks (Figure J) on the heatsinks, and match 'em with their processors. When you thread the screws on, start each one off lightly, then get them up tightly. Try to get them evenly tightened to avoid having too much pressure on one side of the processor, and too little on another. When it's all done right, the heatsinks should not move *at all*. Now we get to test it out!

9. Testing, testing, three, six, tee...

Hook your 360's AV and power cable in, and plug the RF board (the circuit board that has the Ring of Light LEDs) back into the 360, otherwise you won't be turning it on! (Figure K)

Turn it on without the fans plugged in, and check to see how fast the heatsinks heat up (carefully, as they can get quite hot). Getting hot in less than a minute is excellent. Now plug in the fans and put on that plastic fan shroud. Turn it on again, and it should boot up normally, assuming you did the heat test (if you didn't plug in the DVD drive, the center LED will blink green).

If it goes RRoD instead, unplug the power cable, re-insert it, and try again. If you cannot get the console to boot, press down on both the heatsinks with even force, and power it on again. If it boots, power it on with only one heatsink pressed upon. If it boots with that particular heatsink pressed on, power the 360 off, and keep re-booting until you find the problem corner(s) on the heatsink that need tightening.

The idea is to tighten any heatsink corners that

Figure K: 1. It'll be handy to have this plugged in, otherwise you won't be booting the console

aren't tight enough, in order to allow a good connection for that processor.

So if it boots up normally, and you can play a game for at least an hour without locking up, you've worked your magic, and the 360 can be re-assembled! Have a beer, rent a movie, eat some gummy bears, do whatever it is you do to treat yourself, because YOU DID IT!

The next step goes over some measures you can take to prevent overheating.

10. Afterthoughts and shout-outs

Now that you've resurrected your console, what can you do to prevent such a travesty from happening again? As you see in Figures L and M, I attached a small fan from a PCI graphics card onto the CPU heatsink, and wired it into the 360's fan power supply. With it pushing air through the heatsink (or pulling it in if that's your view), the air coming out of the exhaust is actually cool!

Note: There is rumor on the Interwebz that Microsoft will ban you from Xbox Live for installing new fans. You have been warned.

Since there's no room to easily add a fan to the GPU heatsink, the next best thing you can do is optimize airflow. While the 360 has dual exhaust fans, most of the airflow goes to the CPU heatsink. By adding cardboard to the fan shroud and covering the top of the GPU heatsink as depicted in Figures N and O, you can improve airflow even more, as you've just devoted a whole fan to it.

And if you really don't like the stock fans in the 360, you can always buy third party replacements. There's Talismoon's Whisper brand of fans, with LED accents. I can't say anything on quality, as I haven't used one. But they seem to be quite popular. They can be found at Divineo (www.divineo.com).

And now onto the shoutouts.

Thanks to Google first and foremost. LOLz. Xbox-Scene.com and its members, for their endeavors in resolving this issue. RBJTech (http://rbjtech.net/xbox) for the idea of adding cardboard to the fan shroud.

Thanks to both those sites for all the info that I mixed and matched to create this tutorial. Oh yeah, and a special thanks to Cheerios for providing better airflow than Microsoft could.

Instructables FTW!

"Dr. Professor 'Jake' Biggs" Turner is a 17-year-old college freshman from a small town in North Carolina. He's tinkered with electronics since the age of 10, and is currently studying for his associates degree in science.

Figure L: This fan was taken off an old Matrox PCI video card. Yes, apparently video cards from '96 did need heatsinks and fans. That 33MHz was a heat factory!

Figure M: I powered the fan by jamming the stripped ends of the wires into this connector. (Match the red to red, unless the fan spins the wrong way.)

Figure N: 1. The cardboard in the next photo is taped onto here. 2. This cardboard separates airflow between the heatsinks. Trim it to clear some components on the motherboard.

Figure O: Honey Nut Cheerios. Lower your cholesterol, and help your Xbox 360!

Make Your Own Star Wars–Style Sword

Build an elegant weapon for $33 in 33 minutes By Craig Janson

Use this easy step-by-step instruction on putting the parts together to build your own project. It requires no soldering, gluing, or anything other than the following parts. Everything is in the plumbing area of a hardware store such as Home Depot (Figure A):

- One pop-up basin drain
- One drain extension tube with threads and slip nut
- One slip nut and washer
- 1-1/4" disposal gaskets (6). These are for the grip. There are a few different kinds. What you select depends on what you'd like it to look like.
- 1-1/4" slip joint washers (6). These come in packages of three.
- One faucet hole cover (If your hardware store has a separate sink section, you may need to mosey over there.)

Note: I used a plastic chromed pop-up basin drain. There are some that are made of all brass or all metal that looked great but they were around $24 and we were looking to save money. The rods and brackets shown next to it are part of the package, but are not used in this project.

1. Head—blade length adjustment ring

Put aside the drain cap, rods, and brackets that came with the pop-up basin drain. You can use them for other projects, but you don't need them here. Slide the slip nut onto the pop-up basin drain, and then slide its washer onto the pop-up basin drain until the slip nut is against the ball valve. Figures B, C, and D illustrate these steps.

2. Butt and handle

Unscrew the slip nut from the extension tube,

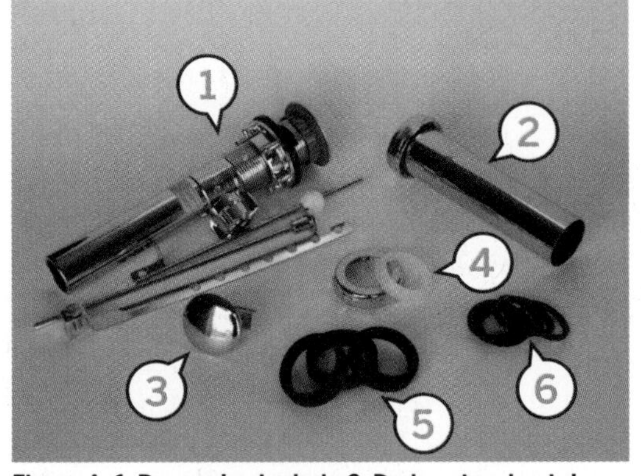

Figure A: 1. Pop-up basin drain 2. Drain extension tube with threads and slip nut 3. Faucet hole cover 4. Slip nut and washer 5. Slip joint washers 6. Disposal gasket

remove and put aside the wing nut on the faucet hole cover, and insert the faucet hole cover into the threaded end of the extension tube. Replace and tighten the slip nut. You may have to put the faucet hole cover into the slip nut first and then fit it onto the extension tube to make it easier (Figures E-F).

3. Preparing head for connection to handle

Slide two slip joint washers over the pop-up basin drain on the long end; put them about 1-2 inches apart as shown in Figure G.

4. Adding the grip to the handle

Slide a disposal gasket over the non-threaded end of the extension tube, then slide a slip joint washer. Repeat this five more times. You can vary this depending on how much of a grip you want relative to the grooves in the grip. You may need more (more of a black fully covered grip) or less (more chrome showing through if you space them apart). Figure H shows the handle with three grips.

For additional information, discussion, and more, please visit the Instructables project page:

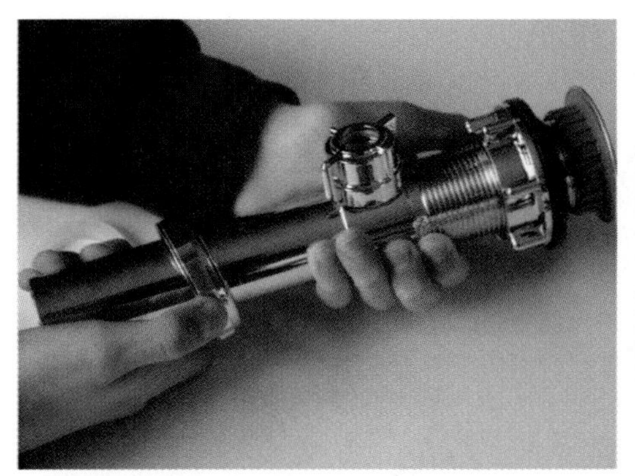

Figure B: Attaching the slip nut

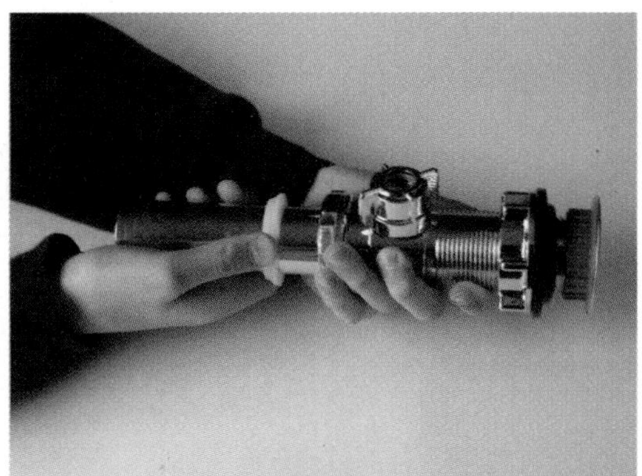

Figure C: Attaching the washer

Figure D: Slip nut and washer in place

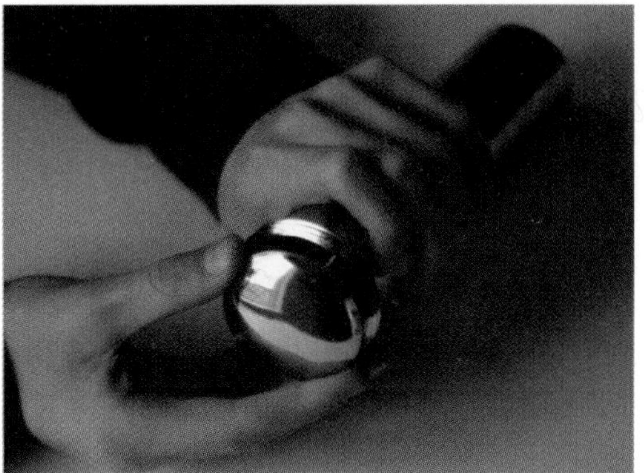

Figure E: Attaching the butt

Figure F: The butt in place

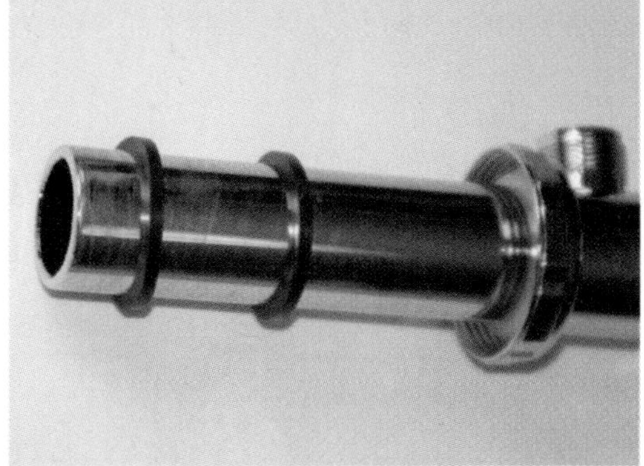

Figure G: Two slip joint washers in place

5. Connect handle and head

This is the hardest part. Slide the pop-up basin drain into the extension tube. You'll need to feed the slip joint washers into the tube because the fit will be tight. After you get the first one in, press down until the next one is lined up and get that fed in. Be careful to work it so the two tubes aren't crooked with respect to each other.

Then get your weight over it and push: be careful before you push, and make sure all the rubber is inside the tube or the hilt will be crooked. I found that when I worked the last little bit of the second

Figure H: Attaching the grips (disposal gaskets and slip joint washers)

Figure I: Connecting the handle to the head

Figure J: This spindle and wire wrapping are from an old 3 gig hard drive. A CD-ROM drive spindle will also work and fits nicely into the hole. We used hot glue to add this feature.

Figure K: The finished project (1. This was just a small plastic cylindrical bead from some craft materials we had.)

Figure L: The battle

gasket, if I applied pressure on the opposite side of where I worked the last part of the first gasket, it gave me a straight line-up of the two tubes.

If you are using the plastic pop-up drain, be careful to not break it. I was able to put a fair amount of pressure on it though. Figure K shows the assembled project.

6. Optional items

Instead of the disposal gaskets, you could wrap the hilt with leather or cord to give it a multimedium kind of feel. Or just leave it plain chrome.

As shown in Figure J, if you have a spare hard drive or CD-ROM drive lying around, you can take out the spindle hub and use it as the emitter on the business end.

You can make a button by inserting something into the ball valve on the pop-up basin drain, replacing the nut (or you could just use the nut as the on/off button or even remove it altogether), as shown in Figure K.

7. The battle

Unknown to me, my kids had some Adegan/Ilum crystals and inserted them into the blade matrix when I wasn't looking. I was able to snap this image shown in Figure L before breaking it up. It's all fun and games until someone loses an arm. (Please note: the blades in this image are Photoshopped. Technology hasn't come *that* far.)

Craig Janson is owner and chief electron wrangler of a technology consulting firm in Castle Rock, Colorado. His interests include learning, reading, music, photography, and the creation of projects at the whim of his wife, Diane, and their two boys, Andrew and Wesley.

User Notes

nstru, Instructables member, says: Thanks for the great design. I used foam weather stripping for the hand grip, and had some issues with the washers (either too big or too small!), so I ended up putting three strips inside the handle and the handle came together easily and stayed perfectly.

Also, I went to RadioShack and got a push-button switch, and screwed it into the button spot and now there's a "working" button.

One other small change: I put in the flat drain cover under the washer at the end, one with holes above the washer under the slip nut—it makes a cool vent-like effect, and I might be able to thread a D-ring in there too.

Blinkybug, Maker Faire Version

Create fuzzy little bugs that blink all day long

By Ken Murphy

Blinkybugs (www.blinkybug.com) are small, electro-mechanical insects that respond to stimulus such as movement, vibration, and air currents by blinking their LED eyes. They're incredibly simple, yet have a certain lifelike quality.

I'd been making variations of these for a while now, and showing others how to make them at museums, fairs, workshops, etc. It wasn't rocket science, but there was some tricky soldering involved, and they usually took a person at least an hour to put together for the first time. I wanted to come up with a solder-free version for the workshop I was organizing for the Maker Faire (www.makerfaire.com). So after a bit of experimenting, I came up with this simpler, solder-free design shown in Figure A.

Note: Blinkybug Kits, which include all the parts to make four bugs, are now available through MAKE magazine's Maker Shed at www.makershed.com/ProductDetails.asp?ProductCode=MKKM1

Figure A: The Blinkybug

1. Tools and parts

Tools you will need: (Figure B)

- Glue gun + glue sticks
- Rotary tool w/metal cutting blade (a hacksaw or similar may work)
- Safety goggles
- Metal file
- Measuring tape or yardstick
- Wire cutters
- Needle-nose pliers (2 pairs would be nice)
- Permanent marker
- Scotch tape
- Scissors

Parts: (Figure C)

- .009" guitar string
- Coin-cell battery (3V CR2032)
- 5mm LEDs (2 per bug)
- Pipe cleaners (aka "chenille sticks")... assorted colors
- Thin copper tubing: 1/16 x .014

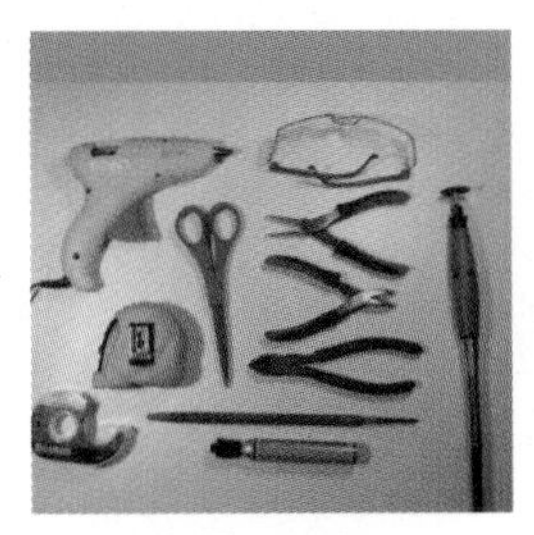

Figure B: The tools

Figure C: The parts

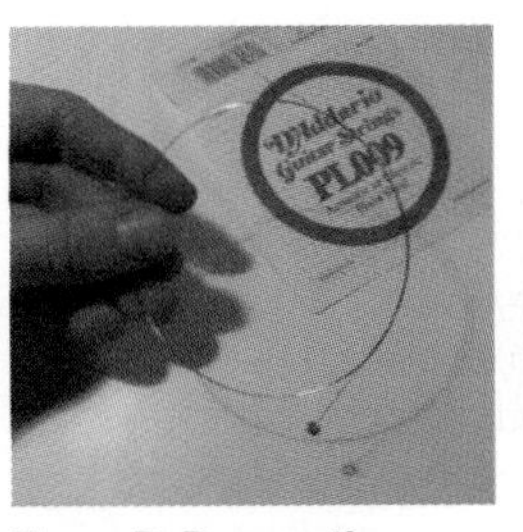

Figure D: Prepare the guitar string

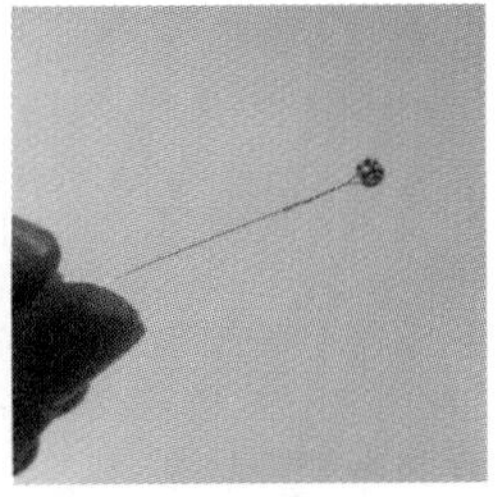

Figure E: Trim this end of the string

I obtained the copper tubing at a hardware store, and it should also be available in hobby shops. 1/16 is the outer diameter in inches, and .014 is the wall thickness. Its inner diameter happens to be about .035", which is important because the guitar string (which is .009") needs to fit through.

LEDs can be found anywhere on the Webernet, and are actually fairly cheap at RadioShack if (and only if) you buy the variety pack. You can obtain the coin-cell batteries at your local drug store, and they are also available widely online, such as at DigiKey. You should be able to get "singles" of the guitar strings at any music shop. The pipe cleaners can be found at tobacco shops and arts & crafts suppliers.

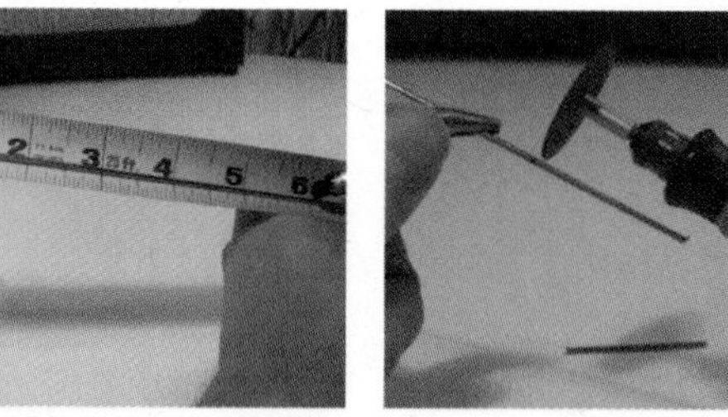

Figure F: Measure the tube

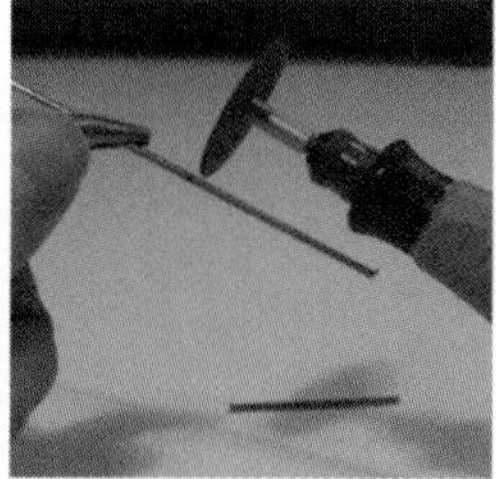

Figure G: Cut the tube

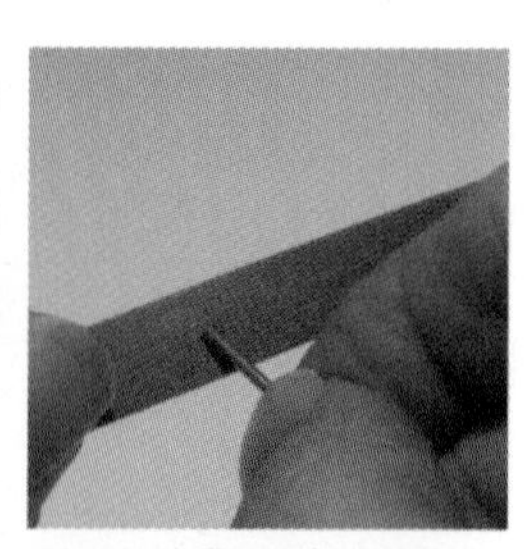

Figure H: File the ends

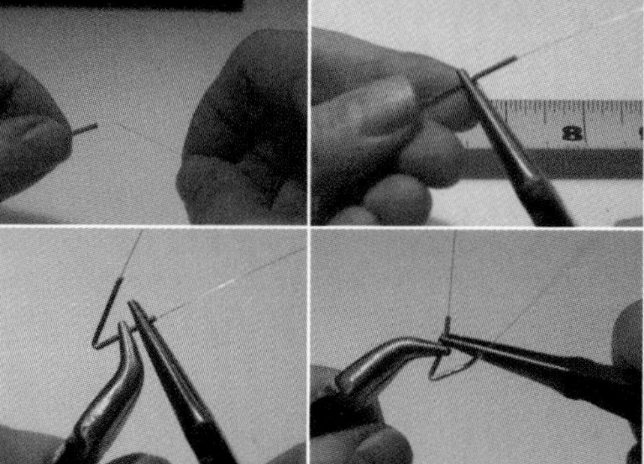

Figure I: Make the antenna guide

2. Preparing the guitar string

The guitar strings I use (D'Addario, shown in Figure D) are 39" long, so you can cut them into three 13" pieces (one bug requires a single 13" length). First, trim the end with the ball and the little extra twisted bit (Figure E), then measure and cut. Put these aside for now.

3. Prepare the copper tubing

The copper tubing is cut into 1-1/2" lengths (1 piece per bug). This will act as the guide for the guitar string/antenna, so the guitar wire must fit through the tube *after* the pieces are cut. This turns out to be tricky as it's very easy to block the tube opening when cutting. I find that my Dremel tool does OK most of the time, but on many pieces I end up bunging the thing up. Wire clippers definitely do NOT work, as they unfailingly crimp the ends closed.

Measure and mark the copper tubing at 1-1/2" inch intervals as shown in Figure F. Now carefully cut the copper tubing with your rotary tool (Figure G)... of course while wearing eye protection as little copper bits will be flying every which way! Set it at a high speed, and let the tool do the work... don't force the tubing against the cutter.

I tried this with a hacksaw with little luck... maybe you'll do better! For the Faire and kits, I'm getting them pre-cut at the factory to bypass this trouble.

Finally, clean up the cut ends with a metal file (Figure H).

4. Antenna guide

This piece will serve as a stable base that holds the guitar string antennae in their proper position.

Grab a single length of guitar string and a single piece of the cut tubing. Thread the string through the tubing, and pull the tube to the middle-position of the cut guitar string (centered at 6-1/2").

Now, holding things in place so the tube doesn't slide around, grab the tubing near its center and bend it in half, forming a "V" shape. The angle should be fairly small, around 45 degrees or less. At this point, the guitar string should not be in danger of slipping out of position.

Next, the top third of each arm of the "V" should be bent. If the V is laid flat, the bend should go

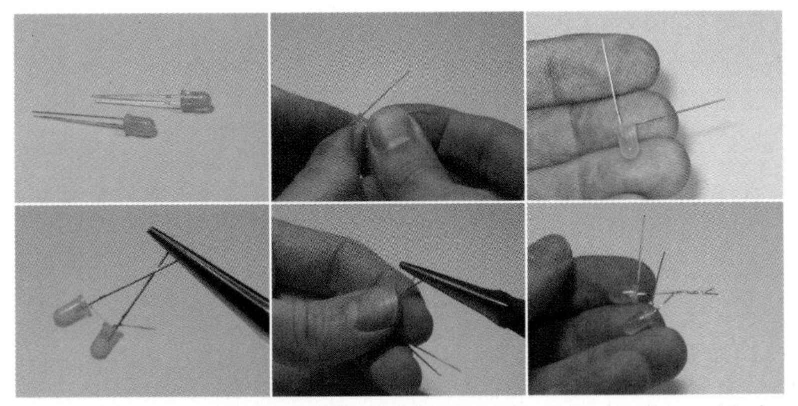

Figure J: Prepare the eyes, part one. The longer LED lead is positive (+, the anode).

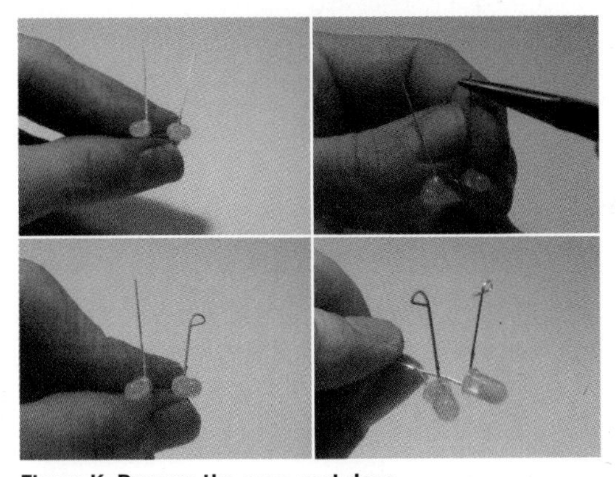

Figure K: Prepare the eyes, part deux

"upward" about 45 degrees. Figure I shows how to do all this. Two pairs of pliers come in particularly handy here.

5. Prepare the "eyes," part one

Grab a couple LEDs of the same color (or not). Take a look and see that one of the wire leads is longer than the other. This is the anode, or the positive (+) side of the LED, and the shorter is the cathode, or negative (-). This is important, as LEDs only work in one direction! If you're not sure that the wires have never been cut (maybe they were salvaged from another project), then look closely at the LED from below. There will be a flat portion of the round plastic package. The lead nearest this flat bit is the cathode.

Bend the anode at the base of the LED so that it is 90 degrees from the cathode. Repeat for the other LED.

Next, cross the ends of each cathode, grabbing the ends with pliers. Twist it like a twist-tie, giving it three or four turns. The LEDs should remain pointing in more or less the same direction, and the anodes should remain roughly parallel. The two LEDs should now be firmly twisted together. If it seems to have a bit of wiggle, give the twisted part a squeeze with the pliers. Mash it good. Figure J shows all the steps.

6. Prepare the eyes, part deux

Next, create a small loop on the end of each anode. Grab the tip of one anode with the end of the needle-nosed pliers, and give it a turn, rotating it outward until a small loop is formed. Leave a small gap in the loop for now, as the guitar string will need to sneak through.

Repeat the same for the other anode, twisting it in the opposite direction. See Figure K for the details.

7. Assemble antenna, eyes, and body

For this step you will need the eye assembly, the battery, and the antenna assembly. Note that the battery has a "+" side (it's marked).

Grab a piece of scotch tape, about 2" long. Hold the lower part of the "V" of the antenna assembly against the "+" side of the battery. Lay the tape across the lower part of the "V," so that equal lengths of tape are extending in either direction. Press the tape firmly on the "+" side so it holds the "V" in place.

Next, hold the twisted leads of the LED assembly along the "bottom" of the battery (opposite from where the antennae are attached). The LEDs should point in the same direction as the arms of the "V," and the looped anodes should extend "upward" (that is, in the direction of the "+" side of the battery). Tightly wrap the tape around the entire battery, so that everything is held firmly in place.

(Figuring out how to fasten this together without soldering was key in simplifying the process... I definitely took some inspiration from the "LED Throwies" Instructable at (www.instructables.com/id/LED-Throwies) and on page 108 in this book.

At this point, the LEDs may light up if the guitar string makes contact with the LED anodes. For now, gently bend the anodes out of the way. Figure L shows the steps.

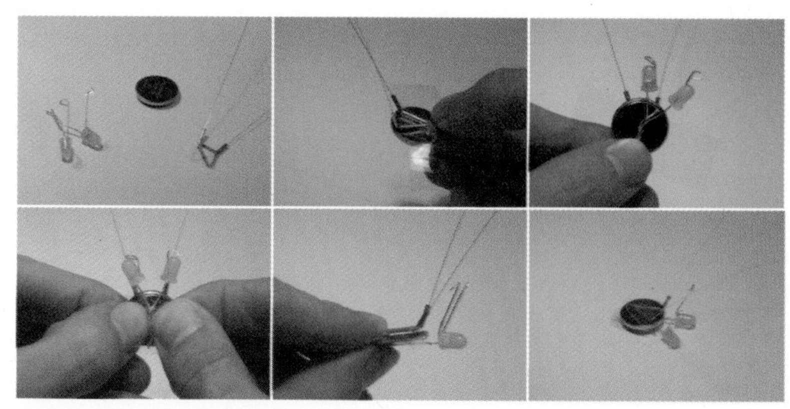

Figure L: Assemble the antenna, eyes, and body

8. Give it legs

The pipe cleaner "legs" are going to be attached to the bottom of the battery (where the LEDs are attached) with hot glue. Plug in your glue gun and let it warm up.

Grab some pipe cleaners and cut them into roughly 3" pieces, and arrange them as shown in the second panel of Figure M.

Figure M: Give it some legs

Have the assembled bug body on hand, and drop a blob of glue onto the center point of the legs. As Figure M shows, I'm pretty sloppy, as hot glue is fairly forgiving. But you might want to be careful to not let too much get onto your work surface.

Quickly stick the bug, bottom-first, onto the glue-blob as shown. Give it a minute or so to cool and gently pick it up from the work surface. (Hopefully it won't be stuck!)

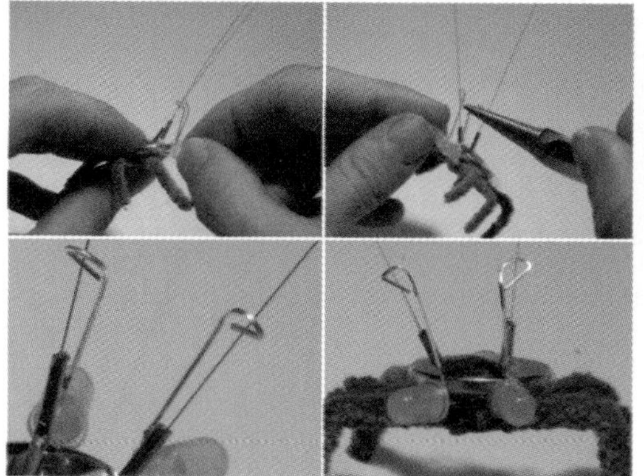

Figure N: Lining everything up; when at rest, the antenna must pass through the loops without touching

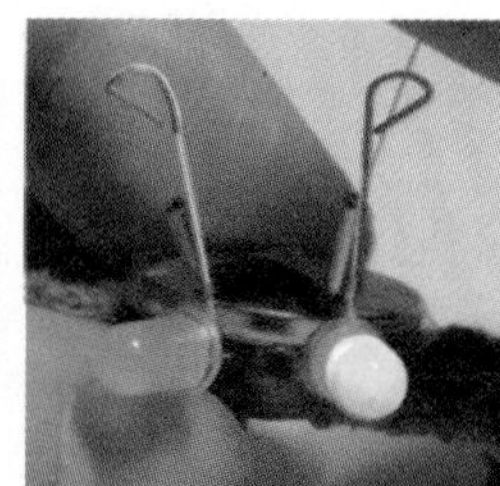

Figure O: When the antenna contacts the loop, the LED turns on

At this point, it's a good idea to apply a little bit of extra glue to the underside of the legs to make sure they're solidly attached.

Give the glue a few minutes to cool down and set. Now bend the legs into a buglike pose!

9. Line things up

Almost done: the next step is to get the antenna (guitar string) to pass through the anode loops. The antenna guide (copper tubes) need to be pointing

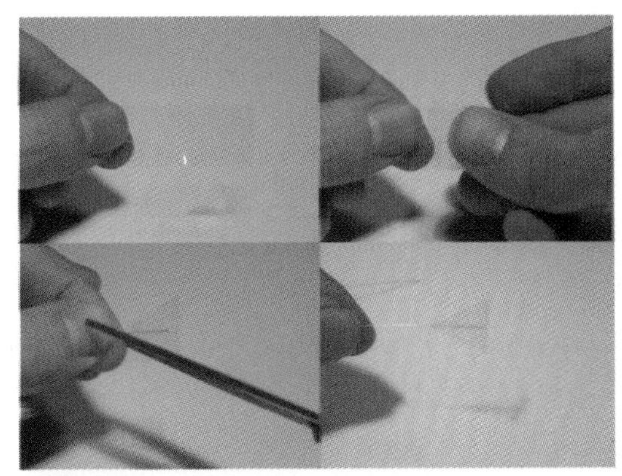

Figure P: Making the antenna flap; even a tiny little flap greatly increases the Blinkybug's sensitivity to air currents

Figure Q: A finished Blinkybug

Figure R: Blinkybug with eyes "blinking"

the antennae in roughly the correct direction.

The objective is to get the guitar string to pass through the anode loops so that when the bug is perfectly still, the antennae don't touch the sides of the loop. However, with any movement or vibration, the antennae will make contact with the loops, close the circuit, and momentarily light up the LEDs.

Using a pair of pliers (or two), try to get everything aligned as described. It helps to grab the entire "loop" and rake it back a bit, so that the guitar wire is perpendicular to the loop. If you left a gap in the loop, you should be able to easily pass the guitar wire in. If not, you may have to pull the loop open a bit with the needle-nosed pliers. Gently squeeze the loop closed once it's in.

You will also want to get the eyes in a position that you like. This all requires a certain amount of futzing until things get lined up. Figure N shows how to futz things into place, and Figure O shows one of the eyes lit up.

As a final step, you might want to attach little scotch-tape flaps on the end of the wire, which makes them more sensitive to wind, and less likely to poke you in the eye. Figure P shows how to do this.

You now have a working Blinkybug as shown in Figures Q-R. The eyes should be "off" when the bug is sitting undisturbed, but should blink rhythmically when you pick it up, move it, or it catches a breeze. If the eyes seem stuck on, you will need to adjust the position of the loops. If the bug does not seem sensitive enough, you might want to try to make the loops a bit smaller.

Enjoy your bug!

Ken Murphy grew up in Massachusetts but has since migrated to San Francisco, where he works on websites for a public broadcasting company. He divides his free time between running off to the mountains, and toiling in his workshop creating strange objects that blink and bleep. He's scored music for movies, done a bit of writing, and most recently has been attempting to turn his living room into a camera obscura.

Build a Water Mortar

Put together your own drill lathe to make this super-soaking water mortar By Carl Morris

This water mortar is made from PVC using a variation on the "drill press lathe" technique from the book *Eccentric Cubicle* (O'Reilly/Make: Books). It launches over a quart of water per shot!

1. Parts and tools

- 30" of 1-1/2" schedule 40 PVC
- 24" of 2" schedule 40 PVC
- End caps for 2" PVC
- Plumber's silicone lubricant
- Size 224 Buna-N O-rings (2)
- 6" of 1" x 3" oak
- Chunks of 2" x 4" for pillow blocks (2)
- 6" of 1/4" x 20tpi threaded rod
- 4" long 1/4" x 20tpi carriage bolt
- 1-1/2" long 1/4" x 20tpi bolt
- 1/4" x 20tpi nuts
- Drill, twist bits, spade bits
- 1-3/4" hole saw (the kind that cuts out a plug)
- Bearings with shoulder bushings from old inline skate (2)
- 8" clamps (2)
- Gouge
- Dial caliper

You can get O-rings from Superior Seals (www.superiorseals.com).

2. Make centers for the inside pipe

Use the hole saw to cut two plugs from the 1" oak.

Figure A: The finished water mortar

From this table for PVC pipe (www.gizmology.net/pipe.htm), we see that 1-1/2" Schedule 40 has an inside diameter of 1.592", so the plug left after cutting a 1.75" hole turns out to be just about the right size to fit the inside of the pipe. If you tighten the collar of the hole saw, you will get a larger plug; if you loosen it, you will get a smaller plug (because the bit wobbles). These pieces should jam tightly into the ends of the 1-1/2" pipe, so it's best if they start out a little too big; you can sand a little to taper the plug if necessary. Use the 1/4" x 20 bolt and nuts to chuck it into your drill and hold a sanding

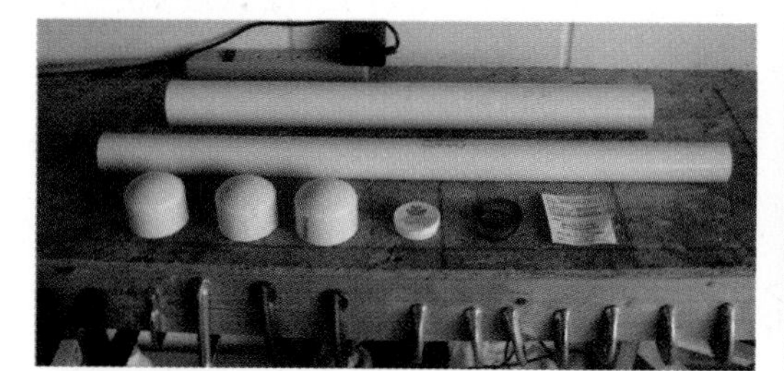

Figure B: Parts

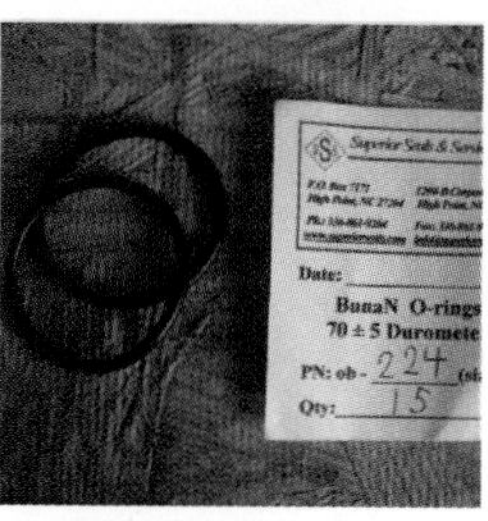

Figure C: O-rings

Figure D: Centers for the inside pipe

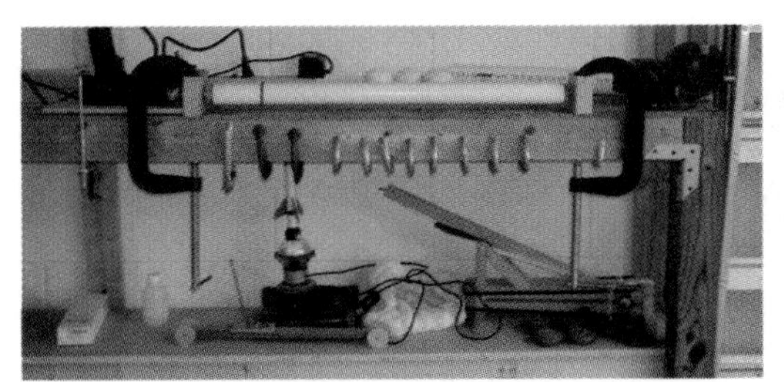

Figure E: Turning set-up—a drill lathe!

Figure F: Mark the grooves for the O-rings

Figure G: Turn and cut the grooves

block up against it to sand evenly. If your pieces are too small, build up their diameter by adding wraps of masking tape. Let the tape hang off the inner edge a bit so it can guide the pipe onto the center without being pushed back.

Once you've got the fit right, assemble the centers (Figure D). The "driven" center is in the foreground, the "non-driven" center is in the background. Using spade bits, drill the 2" x 4" chunks to make pillow blocks for the bearings.

3. Set up for turning

Insert the centers into the ends of the pipe. Roll an O-ring over the pipe. Seat the bearings into the pillow blocks. Check to see that the pipe rotates smoothly. Firmly secure the pillow blocks.

The carriage bolt protruding from the driven center is chucked into the drill.

I haven't tried it, but the whole setup might be more solid with a single piece of 1/4" x 20 threaded rod running from end to end.

4. Mark grooves

Set the drill for a relatively low speed and lock it on. Use a permanent marker to mark positions for the grooves. The first one should be a little over 1" from the end; the second one should be about 4" further down the pipe.

5. Turn grooves

Start with the drill at a slow speed until you get a feel for things, then speed it up for faster cutting action. If you get too much wobble in the pipe, slow down. If you don't have a gouge, experiment with different sharp-edged objects held at about a 90 degree angle to the pipe. I used the bottom end of a round-file for my grooves.

Grooves should be flat-bottomed with smooth, squared-up sides, and enough width to give the O-ring some room to breathe.

6. Check groove depth

Roll the O-ring into the groove and measure the outside diameter of the O-ring with a dial caliper. Rotate the pipe slightly and measure at a couple positions to get an average diameter. The average diameter needs to be about .010" bigger than the inside diameter of the large pipe. The table referenced earlier says the inside diameter of 2" Schedule 40 PVC should be 2.049", but measure yours and add .010" to get your target diameter. If the O-ring fits too tightly, it will bind and not slide smoothly, if it is too loose, water will leak past it (Figure H).

Once the target diameter is reached, remove the pipe from the turning setup. Bevel the inside lip of the outer pipe so the O-rings will slide into it easier. Clean everything before assembly. Put damp, folded pieces of paper towel over the end of the inner pipe and push them through the outer pipe to clean dust and debris out. Put both O-rings into their grooves, liberally lubricate everything, and check the fit.

7. Prepare end plug

Use one of the centers to make an end plug. Insert a 1-1/2" long 1/4" x 20 bolt, put a nut on the other end, and tighten (Figures I-J).

8. Install end plug

Work the end cap into the end of the inner pipe. The head of the bolt should be visible. Once it is started you can turn the pipe over and tap it against the floor to seat the plug tightly. If the plug isn't tight enough it will get sucked out when you draw water into the chamber. You can wrap electrical tape at the base of the plug to make it fit more snugly.

9. Check fit

This is what the end should look like when everything is together (Figure M). It should be nice and

Figure H: Check the groove depth and put on the O-rings

Figure I: Put a bolt through one of the centers to make an end plug

Figure J: Hold the bolt on with a nut

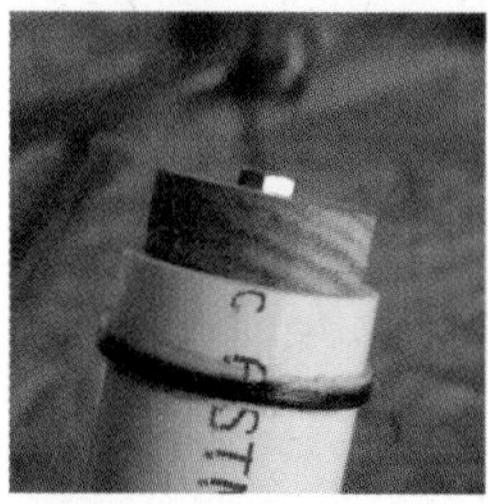
Figure K: Install the end plug

Figure L: Tap the plug in so it fits tightly

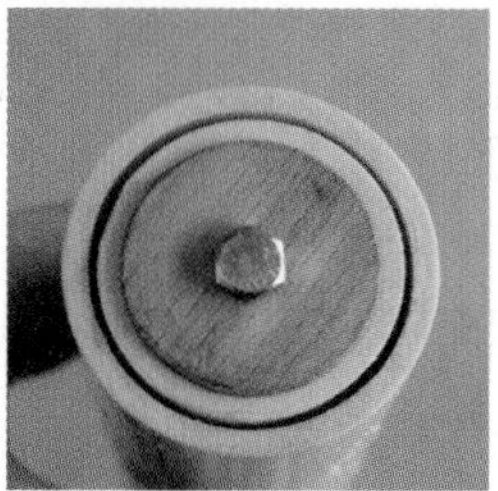
Figure M: Check the fit of the inside pipe and the outside pipe

Figure N: Drill a 1/2" hole in the end cap

Figure O: Assembled water mortar with end cap seated

even, with smooth motion.

10. Prepare end cap

Drill the end cap. A single 1/2" hole seems to work pretty well. Get a few end caps and experiment.

11. Final assembly

Place the end cap onto the outer pipe and seat it by tapping firmly on the floor. Remove the cap by tapping the opposite end of the inner pipe.

12. Try it out

Get a bucket of water, push the inner pipe all the way in, put the end cap under water, and pull the inner pipe back until you can just see an O-ring. Quickly tip the end cap skyward to keep leakage to a minimum, placing the end of the inner pipe on the ground. Tilt in the direction you want to fire, then pull downwards on the outer pipe to fire.

Bonus fun: if you pull the inner pipe back and seat an undrilled end cap, you can launch the end cap with pneumatic pressure, generating an impressive bang in the process. Make sure it's not aimed at anything or anyone. Also make sure you're not inside your garage when you discover how cool this is: a garage door makes a resounding boom when you nail it with an end cap!

Carl Morris is an engineer who likes to build things, solve problems, and learn how other people build things and solve problems. He recently collaborated with MAKE to produce a Magnet Sculpture Kit sold in the Maker Shed.

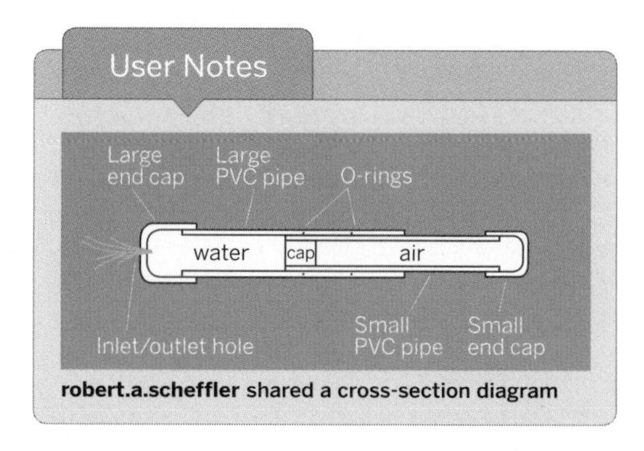

robert.a.scheffler shared a cross-section diagram

An Iron Man–Style Arc Reactor

How to build one part of the complete costume

By Martin Raynsford

Figure A: The finished project

The arc reactor work by Tony Stark before he completed his Iron Man outfit is the inspiration for my own arc reactor. It has 10 well-defined sections and a glowing center as shown in Figure A.

I'm rather pleased with the results and very happy with the segments of light that emanate from it. I'd also like to pay respects to the other arc light reactor on Instructables... imagine my horror as the weekly round up arrives in my inbox only to find out that I had been beaten to the write up for the same project.

1. The materials

I wanted to make something substantial for this project and I remembered I had some suitable plastic in my cupboard. This is the key component for the project. This material is called Polymorph plastic (Figure B) and can be bought from places such as eBay and craft supply shops. It's a thermal plastic that melts around 60°C and it becomes something resembling Plasticine modeling clay. From there it is simple to mold it into the desired shape.

The next thing we will need is a light source. I wanted the whole thing to be quite thin when it was finished, so I opted for some surface-mount white LEDs. Surface-mount LEDs have a very wide viewing angle and being white they produce quite a lot of light, so they are perfect for this application. You can find surface-mount LEDs from electronics suppliers. These LEDs are in a PLCC 2 package which means they are still large enough to be soldered by hand.

You may also want some surface-mount resistors to go with those LEDs. I used the amazing program at LEDCalc (ledcalc.com) to work out exactly what values I need. I am running these LEDs from a 9V battery and wanted 20mA of current to flow throw them. LEDCalc suggested how exactly they should be wired and what values I needed.

Figure B: Polymorph plastic (also sold as Shapelock)

For my LEDs I required 5x 180Ω resistors and 1x 330Ω resistor.

I mounted the LEDs on a piece of plywood, but anything will do as you are gluing the surface-mount components down for ease of soldering. A 9V battery and battery clip are providing the power for the system. These can be bought from any electrical store as required.

Finally you'll need some wire for the detailed decoration. Wire coat hangers could be used but I used tin copper wire of 22 AWG gauge. There is nothing special about the wire, it's just hard finding something chunky enough for the job.

2. Wiring of the LEDs

Some assembly is required for the LEDs. I took my round piece of wood that I was using to mount the LEDs on and I started to glue the LEDs in the desired places (Figures C-D). The circuit from LEDCalc, shown in Figure E, suggested I used 5x 2 LEDs and one single LED in parallel. This ties in well with the arc reactor so I had the single LED in the middle and the pairs arranged around the edges.

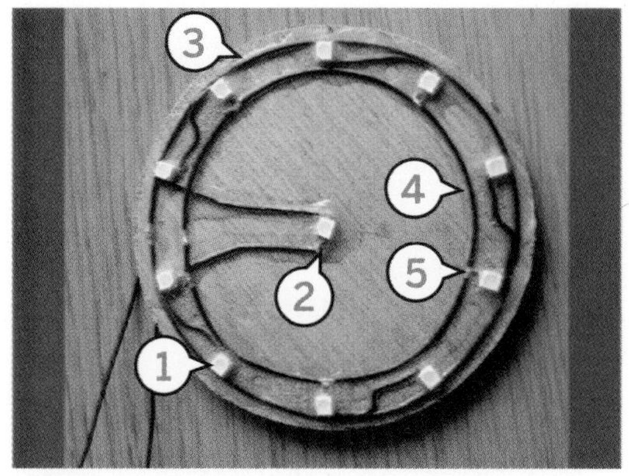

Figure C: 1. SMD LED 2. R2 3. Positive wire 4. Negative wire 5. R1

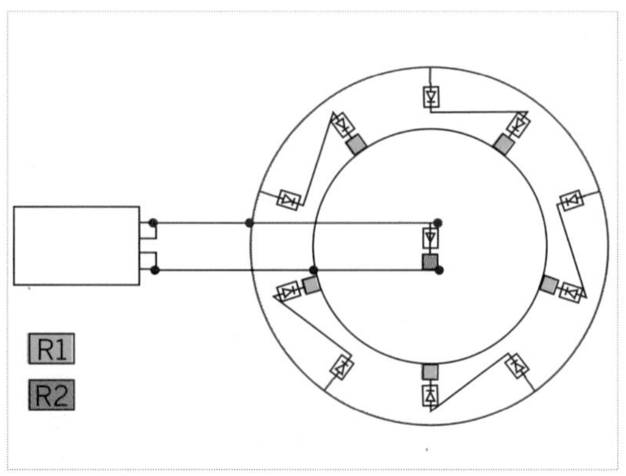

Figure D: Wiring layout

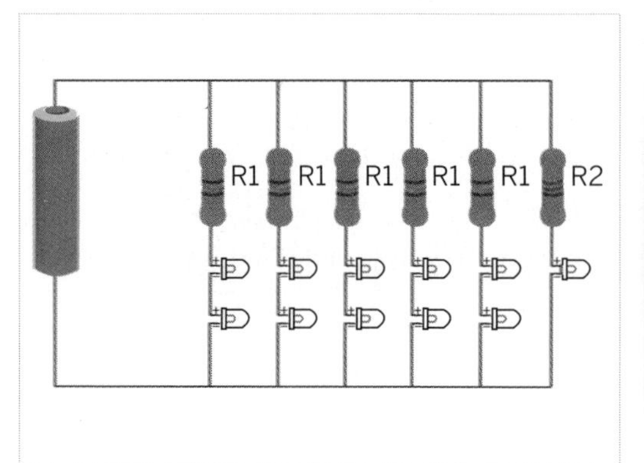

Figure E: Schematic

Figure F: The balsa wood mold

As you can see from Figure C, I made two rings of wire around the edges: the outer wire is 9V and the inner wire is 0V.

The power wires are passed through the back plate through a small hole that will allow me to power the LEDs when they are encased in the plastic.

3. Making the plastic shape

This is the key stage of the build. The Polymorph plastic behaves like Plasticine when it is heated to temperature. This allows it to be pressed into a mold and form the desired shape. As always I wish I had more photos of the stages involved but I don't and it's too late to go back (let this be a lesson for budding Instructable writers).

I formed the mold using balsa wood again on a more solid plywood base (Figure F). I cut the outer circle out of balsa to the required depth of the arc reactor. I used thinner strips of balsa as relief pieces and provide the detail in the plastic. (These are roughly the same depth as the wire I used).

I heated the plastic using water from the kettle. Once ready, it becomes transparent and malleable. I took care to make sure it was pushed right into the mold to reach all the corners of the mold. Once it was fully pushed into the mold, I pushed the LED disk into the back of the plastic. The plastic pushes slightly around the disk that holds it in place, as shown in Figure G. The disk must be aligned with the slots in the mold so that each LED is directly under a raised piece of plastic. (There are no photos of this because it was all done with some haste.)

Figure H shows the plastic once it has been removed from the mold. You can clearly see the raised sections of plastic and the gaps that are soon to be filled with wire. Under each bump there is an

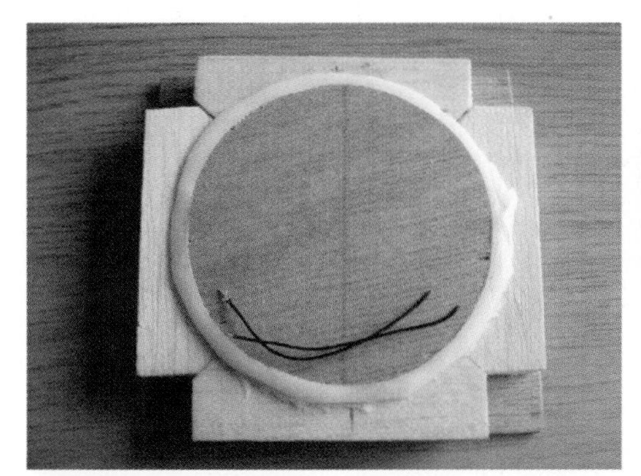

Figure G: Setting the plastic in the mold

Figure H: The plastic, after removing from the mold

Figure I: Adding the wire to the reactor

Figure J: The reactor with all the wires

Figure K: The arc reactor powered up

Figure L: Modeling the reactor

LED; the plastic adds to the diffusion of each LED and really adds to the overall effect.

4. Adding the details

The final step of making the reactor is to add the wire details. I drilled holes in the plastic to hold the wire around the edge of the device. I then bent each piece into a C shape, hooked it into a hole on the edge of the plastic, and again into the holes in the center (Figure I). This was enough to secure them in place. Finally, I shaped four wire rings to go around the center of the reactor, as shown in Figure J. These are held in with PVA wood glue, although any clear drying glue should do the job just as well.

As you can see in Figure K the device lights up very well and looks really good, now onto the final stage to bring it all together.

5. Bringing it together

With the arc reactor now finished, this final stage is about bringing it all together in a costume. I brought a sleeveless T-shirt from the local store for a few pounds. I carefully sewed a pocket on the inside of the shirt to hold the reactor (this proved to be a very good idea due to the number of people that wanted me to take it out and show them during the evening). The wires from the reactor run down the T-shirt and into my back trouser pocket.

After a week's worth of effort I still had nearly zero facial hair so I ended up padding it out with some black shoe polish. I'm particularly proud of the whole chubby Tony Stark thing I had going on but then the photo in Figure L was taken at the wrong end of the evening after quite a lot of good food and drink... normally I'm only half as fat.

I hope you find this useful and encouraging for your own projects!

Martin Raynsford says: "I'm 28, live on the south coast of England, am married to a very understanding wife, and I have a degree in robotics. I don't have kids yet but when I do they are going to win all the fancy dress competitions."

Make an Awesome Magic Wand

Make this realistic-looking magic wand from simple materials (and a wave of your hand)

By Kaptin Scarlet

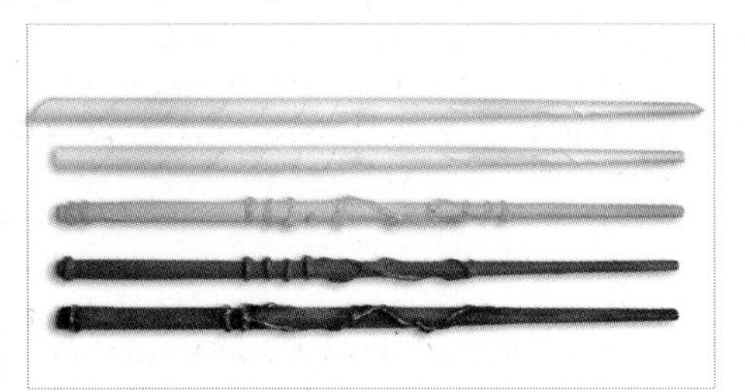

Figure A: The stages of a wand's birth

If you ever wanted to have a go at making some really nice magic wands, this is a simple but effective Instructable for you. With a sheet of paper, some glue and a bit of paint, and about 40 minutes to spare, this Instructable will show you how you can make a magic wand that would not look out of place in a fantasy film. I designed it and I have already made a load for my kids; even my girlfriend and her 18-year-old daughter and her boyfriend wanted one each! For the first few stages I have drawn pictures of what you have to do, and then for the painting part I have taken photos to show you exactly what to do. It's very simple, but the effect is fantastic. For more projects like this visit dadcando (www.dadcando.com), where there are more wizarding projects and a load of other craft printables and templates.

If you like this but want more of a challenge, why not try it with added magic? In the Instructable www.instructables.com/id/EU718C9F3YWV6QY, you get to make a wand with a UV LED at the tip that can reveal secret and otherwise invisible writing. (But I'd try this one first to get the idea.)

1. Prepare your paper

Stick a strip of double-sided tape diagonally across a sheet of A4 or U.S. letter-sized paper.

2. Tightly roll the paper starting in the corner

Roll the paper starting in the corner and roll diagonally, rolling one end slightly less that the other so that the thin paper roll is tapered. Roll until you get to the double-sided tape, then roll over it so that the tapered roll sticks to it.

3. Glue the last third of the roll

Smear the free corner of the paper with a little PVA glue (Elmer's Glue) so that whole corner is covered in glue.

Continue to roll the wand tightly and hold (with fingers) until it's dry.

4. Trim the wand

Wait about 20 to 30 minutes for the wand to dry. As the PVA glue hardens, it should become much

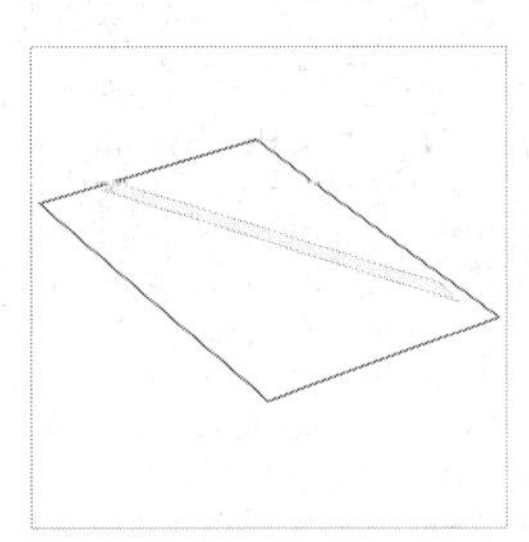

Figure B: Put double-sided tape across a sheet of paper

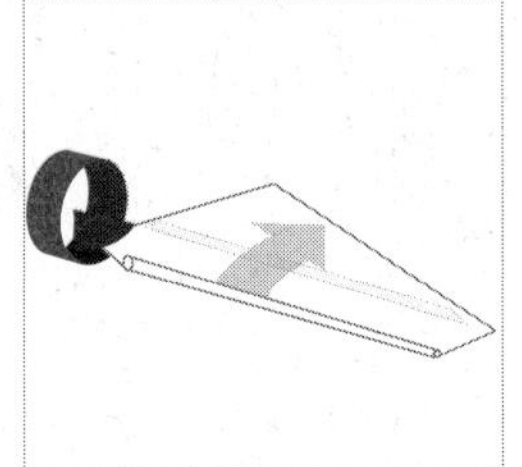

Figure C: Tightly roll the paper diagonally

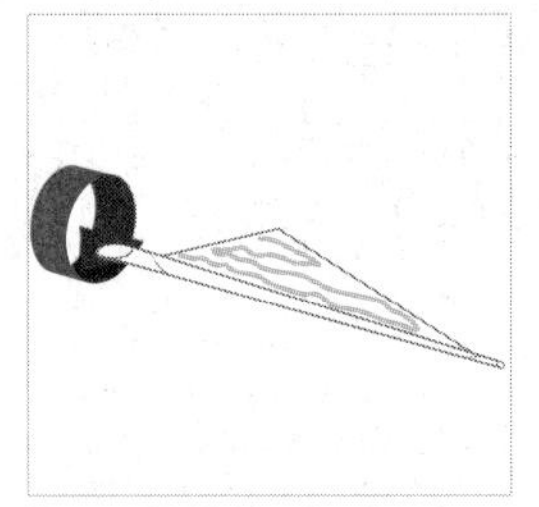

Figure D: Glue the last third of the paper before you finish rolling it

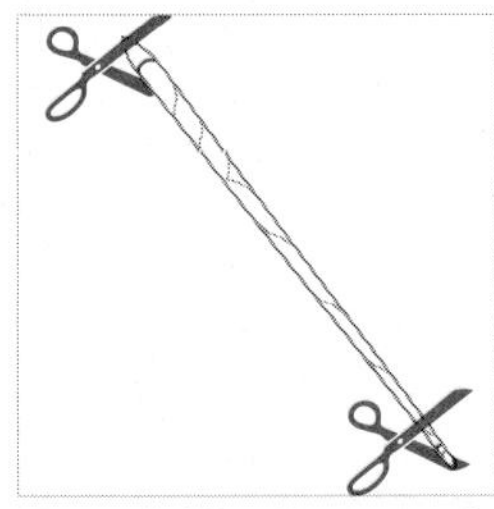

Figure E: Trim the ends of the wand after the glue has dried

For additional information, discussion, and more, please visit the Instructables project page:

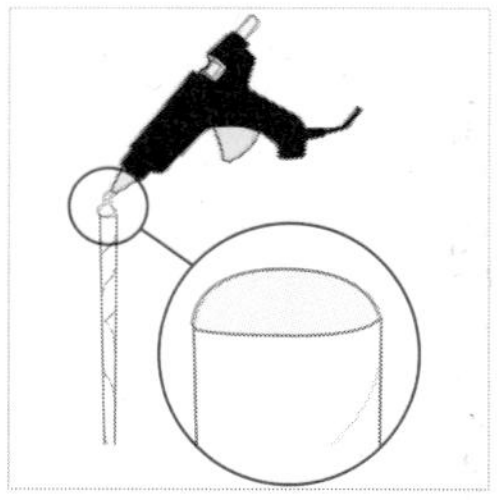

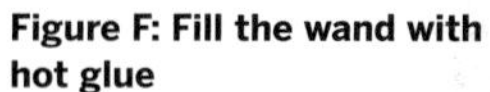

Figure F: Fill the wand with hot glue

Figure G: Decorate the wand with hot glue

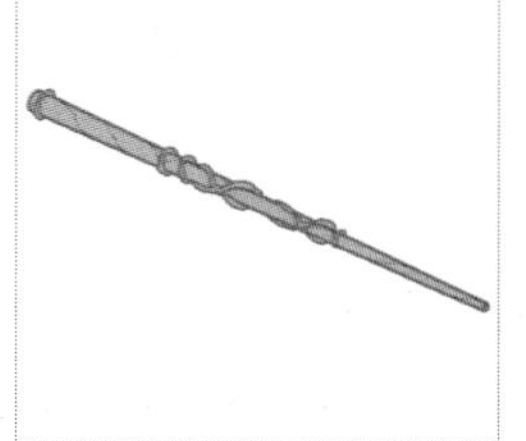

Figure H: Apply a base coat of paint to the entire wand to seal it

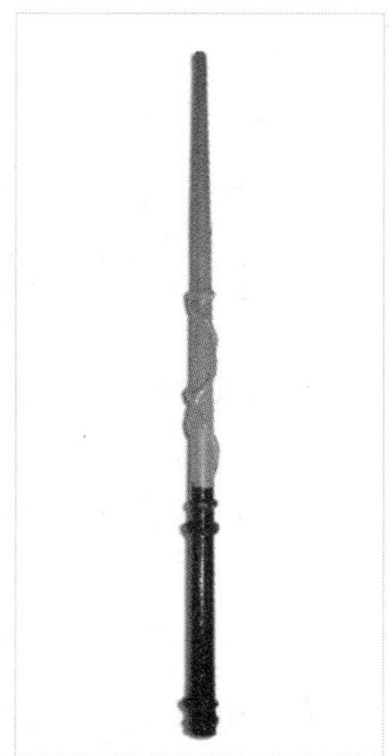

Figure I: Wand half-painted with main color

stiffer. When dry, trim a little bit off both ends of the wand to make the ends straight.

5. Plug the ends of the wand and fill it

Carefully dribble hot glue into both ends of the wand (one at a time, waiting till each end is set). For the bigger of the two ends, you can pack the end with a little crumpled-up tissue pushed down a bit with a pencil so that you don't have to use too much glue. For the bigger end you will probably need to have two goes. If you are careful you can achieve a rounded end; as the glue is setting, make sure that you rotate the wand to stop it slumping to one side or dripping over the edge. The same goes for the little end, although if you have wound the wand tightly enough, you will not need to fill this twice.

Note: if you want your wand to be stiff and very robust, after plugging the little end, but before plugging the big end, you can fill the wand with quick setting, two-part epoxy resin. (This is the sort of glue where you have to mix two equal amounts of glue from glue and hardener tubes.) Use the 5-minute setting version rather than the really fast 90-second version, and carefully dribble the glue down the inside of the wand, making sure not to get it on the outside. Don't worry if you do get a little bit on the outside though, just wipe it off carefully and quickly, you'll be painting over that later.

6. Create the surface detail

Holding the wand in one hand and the glue gun in the other, slowly rotate the wand between finger and thumb as you gently squeeze out glue onto the surface of the wand. Try to keep it even and make a nice pattern. Start with one or two rings at the thicker end, leaving a space for the grip area, then make a criss-cross lattice effect lower down the wand by rotating the wand and moving the glue gun along the wand at the same time.

As the glue sets, rotate the wand in the air to make sure that no uneven drips build up. The glue should be set in about a minute or so, but might be tacky for a couple more minutes so be careful what you rest it on to set properly.

7. Spray with base coat

Apply a base coat to the wand to seal it. You can use spray paint (more or less any color will do); spray paint is good because it dries hard, but you can use latex wall paint if you don't have spray. Make sure that the wand is dry before going to the next step.

8. Paint the main color

I chose brown because I wanted it to look like natural wood, but you could use black, or an off-white for ivory, or any other muted color. Paint the whole surface, but don't worry if it isn't too even; this will make it look more like a natural material.

Important: you must use a type of paint that is waterproof when it dries. Ideally, use acrylic. The next step uses a wash that you have to wipe off while it is wet, and if you have used a non-waterproof paint for this stage, then when you apply the paint from the next stage, it will rub this paint off as well.

To mix brown, use all the primary colors (red, blue, and yellow) in varying proportions depending on what sort of brown you want, or one primary color with any secondary color (orange, purple, or green). Mix in a little black (but not too much) for a darker brown, and allow it to be streaky if you want.

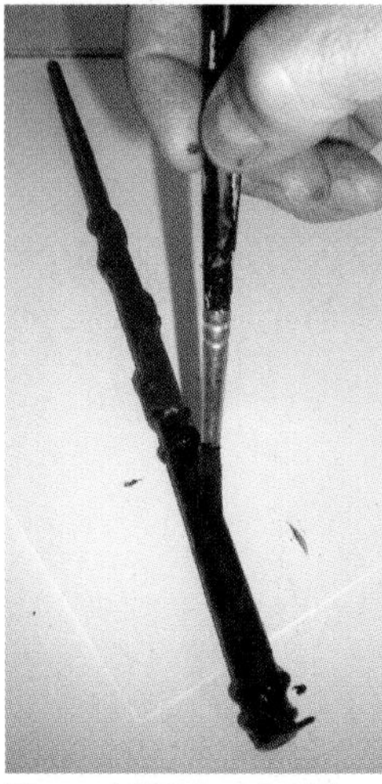

Figure J: Paint part of the wand with a wash of the second color

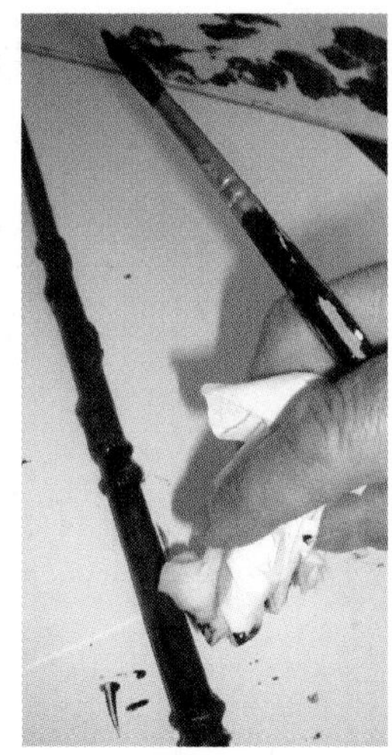

Figure K: Wipe off some of the top coat. Repeat with the rest of the wand.

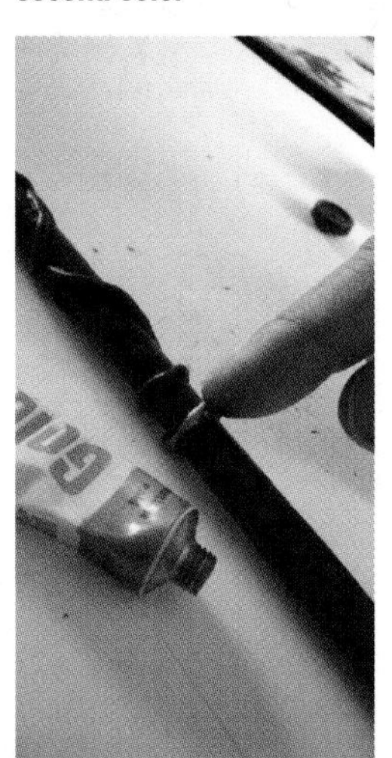

Figure L: Add details in gold or silver

Figure M: A variety of finishes for different types of wizards

9. Start distressing the wand

No, this doesn't mean telling it upsetting news; distressing is the furniture makers' term for making something new look old.

You do this by mixing up a wash of black. Not too washy, but enough so that it remains wet long enough to be able to wipe it off. The best type of paint for this is acrylic. DO NOT paint the whole wand before starting to wipe the paint off, otherwise it will dry and you won't be able to wipe it off.

User Notes

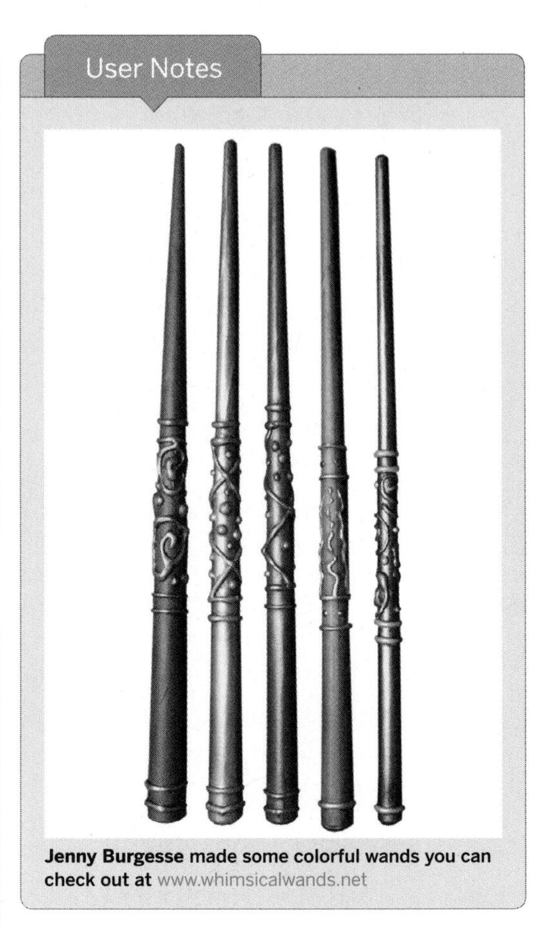

Jenny Burgesse made some colorful wands you can check out at www.whimsicalwands.net

10. Finish distressing the wand

Wipe off the paint as you go along. Use a damp cloth or piece of kitchen towel. Dab and wipe; if you are not happy with the effect, paint over and wipe more. You are trying to achieve the natural look of grime and aging that collects in the cracks and corners.

You won't be able to wipe all the paint off; some will collect in the corners round the glue and this will make it look really old.

Work on the handle area; in real life, handles get worn more so they will be shinier and have less dark areas. Go with the flow, look at the work, and wipe and paint until you are happy with the results.

Don't be afraid to go back a stage and add more light colors and then repeat the distressing until you get the right effect. Always wait till the previous layer is dry. If you use acrylic you can do this as many times as you want, but you really only need to do it once.

User Notes

The Tiny Shocker

By Zot O'Connor

Years ago, my family moved to Louisiana, where my wife and I home school our children. Through home schooling, we soon met a family with whom we became fast friends. Our older sons especially hit it off.

Eventually they relocated to England. While there on business, I stopped in to see them. Their son, Derek, who is always taking things apart and making new things, had gotten into electronics. I wasn't sure if he was doing what he claimed he was—it all seemed pretty advanced for a 14-year-old—but at least his terms matched with my aging electrical engineering background.

When I visited recently, Derek showed me a circuit diagram. It looked similar to one I was trying (and failing) to build, called a Joule Thief [found on Instructables and MAKE, BTW —Ed.]. "Oh, I've built a couple of these," he said dismissively." I mention MAKE and he's heard of it, so I mention Instructables. "Oh, you know about that?" he asks. I say "Why, yes," and am about to mention that I'm going to be in the upcoming book, writing a sidebar story about "The Marshmallow Shooter" (page 232) when Derek cuts me off: "I've posted some projects. Ever hear of the 'World's Smallest Electronic Shocker'?" (No, but now my spidey sense is tingling.) "How about a Marx Bridge?" he continues. (You mean the high-voltage spark-gap thingy I spent 30-45 minutes reading about this past week!?) "You built one?" I ask. "Uh, no—I posted the Instructable! You want to see the real thing?" "I would love to!" I say, trying to bring the conversation back to my point. "You know they're publishing *The Best of Instructables*?" "Yeah, I'm in it, for the tiny shocker..." (See *The Best of Instructables* Community Contest winners, page 304.)

After seeing Derek last year, I thought maybe I should become or find him a mentor. This year, I shot video of his Marx Bridge, shockers, and the other Instructables he's posted. I was thinking, he still should have a mentor. And then, as I was leaving, he offered to show me how to make a Joule Thief—this time, without screwing it up.

11. Apply the gold detail

Using your finger tip, apply some gold rubbing paste to the raised bits of the wand. You can use a gold marker, or a gold gel pen, or gold modeling paint. The best stuff is the art store product that is designed to be rubbed on with the finger tip and then burnished a bit, but any gold paint will do.

For one of my kid's wands, I used silver leaf to make some highly reflective parts metallic-looking. It was a brilliant effect but a bit more complicated. Paint a thin layer of gold size (a model makers' and artists' liquid glue that dries slightly tacky) onto the raised surface and let dry for two hours. Then apply the gold or silver leaf with great care (it is sooo thin), remove the excess with a very fine brush, and burnish with a soft cloth. The effect is truly amazing, like real polished gold or silver (which is what it is, of course). Gold or silver leaf is cheap, but you need to get it from a good art shop or from the Web.

12. My (but could be your) finished wands

There are photos of my wands (each took about 20 minutes to make) on my website dadcando (www.dadcando.com), just one of the 100s of free projects available there. Many have printables and templates of the highest quality, and are designed for kids to be able to do with their dads or mums.

Have fun making the Magic Wand, and send me a picture of any you make so that I can put them up on my website (in the Your Models section). Send a message through Instructables and I'll give you my email address.

A single father of four, Chris Barnardo is the founder of Dadcando.com, a website full of craft projects, tips, and advice for dads and single parents. An inventor and designer at heart, Chris holds 25 patents.

Post Office Drawings

Let the bumps and grinds of the shipping process generate art By Brandon Sullivan

Figure A: The stylus was made with a Super Ball at its core and pencil leads hot-glued to it

Figure B: The materials you'll need

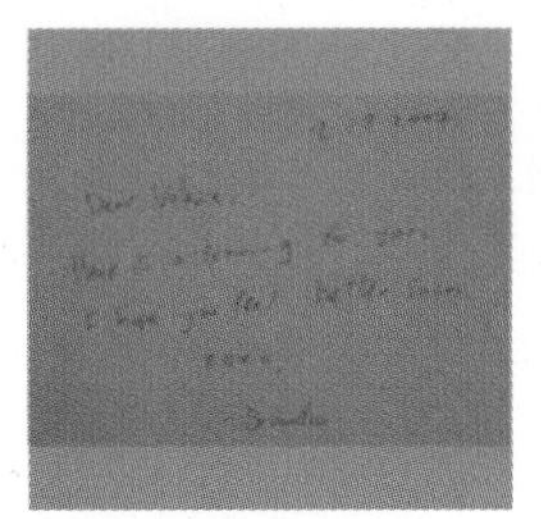

Figure C: Letter to the recipient, an ill friend

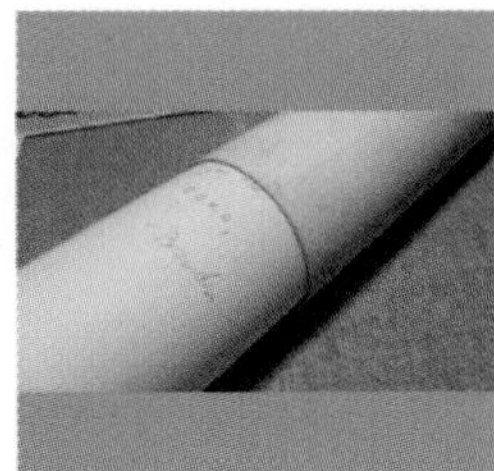

Figure D: Make sure your note is on the outside of the paper roll!

This project was the genesis for an Instructable I posted earlier called Drawing Machine (www.instructables.com/id/Drawing-Machine).

Post Office Drawings takes a more mellow approach and turns our friends at the postal service and the shipping process itself into co-creators of art!

1. Gather your materials

- A shipping tube
- A suitably sized piece of paper
- A "stylus" of your own making
- Packaging tape
- A marker
- Money for postage
- An understanding friend

Choose a shipping container that will accommodate your stylus. In this project I am using an old stylus from my Drawing Machine Instructable. It fits nicely inside this cardboard tube, allowing for free movement and mark-making abilities.

Cut a piece of paper to fit inside the shipping container. It should be as long as the tube is tall (allowing for the lid to fit, which actually squeezes into the end of the tube). The paper should be as wide as the inside circumference of the tube in order to not have any overlap, which would yield a blank spot where the paper wrapped over itself.

2. Assembly

I will be sending this to a friend who is ill, so I wrote a little note on the back of the drawing wishing her

Figure E: Shipping tube with rolled-up paper and stylus inside

Figure F: Ocho the Cat likes projects, too

Figure G: Address the package

Figure H: Include words of encouragement. Who doesn't like to shake it?

For additional information, discussion, and more, please visit the Instructables project page:

a speedy recovery. Here's how to prepare your Post Office Drawing machine for mailing:

- Roll up the paper with your note on the outside and insert it into the tube
- Drop the drawing stylus into the tube
- Seal package with packing tape
- Write opening instructions near the lid
- Address the package
- And just for giggles, write some words of encouragement on the package

3. Post

Take your package to the post office and purchase the required postage. It's probably best to purchase the slowest, longest route to your destination to give the package plenty of time to get jostled and tossed, creating a more "storied" piece of art. Encourage the nice woman at the counter at the post office to shake up your package—tell her she's helping make a unique gift for your recipient.

4. Other ideas

A tube is not the only way to go here. You can use a long piece of paper rolled up inside of a square box. You can use multiple styli. How about sending two sheets of paper, so your friend can remove the inner piece to keep, then repackage and send the tube back to you, so you can each have a drawing too? Or they can send it on to someone else. Oh, the mind reels with the possibilities...

Brandon Sullivan is a sculptor from Jamaica Plain, Massachusetts. He has been working on a body of works he calls "auto-drawings." Post Office Drawings is one in the series.

User Notes

Brandon has done a number of similar drawing machines that use human and mechanical movement to create randomized works of art.

Revenge of Post Office Drawings
www.instructables.com/id/Revenge-of-Post-Office-Drawings
In Revenge of Post Office Drawings, he uses a rectilinear stylus matrix, a box, and flat pieces of paper.

Vibrobot Paintings
www.instructables.com/id/Vibrobot-Paintings
Why should humans have all the art-generating fun? Here, Brandon outfits little jumpy, motorized robo-critters with bristles dipped in paint.

Pocket Drawings
www.instructables.com/id/Pocket-Drawings
For Pocket Drawings, the drawing machine fits into an Altoids Tin and your pocket becomes the place where the magic happens.

Drawing Machine
www.instructables.com/id/Drawing-Machine
An old drill shakes up a host of drawing styli to create Brandon's original Drawing Machine.

Revenge of Post Office Drawings

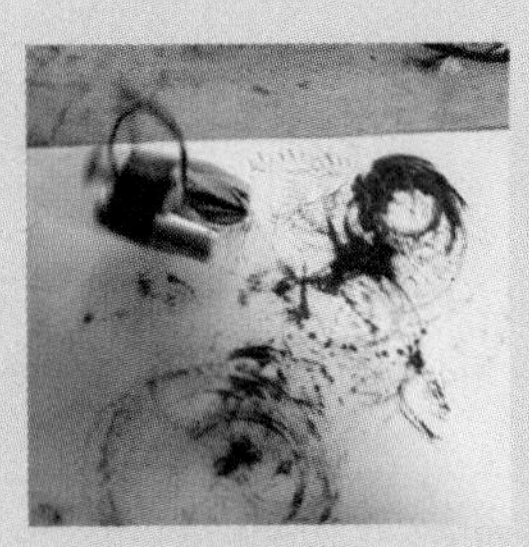

Vibrobot Paintings

Pocket Drawings

Drawing Machine

Jar of Fireflies

Simulate the behavior of fireflies with a microcontroller and some surface-mount LEDs By Xander Hudson

Figure A: The jar of fireflies

This project uses green surface-mount LEDs along with an AVR ATtiny45 microcontroller to simulate the behavior of fireflies in a jar.

1. About this project

The inspiration for this project comes from having never lived in an area where fireflies were common and being deeply fascinated whenever I encountered them in my travels.

The flash patterns have been digitized from firefly behavioral research data found online and were modeled in Mathematica so that variations of speed and intensity could be generated. The final output was transformed by a lightness (www.poynton.com/notes/colour_and_gamma/GammaFAQ.html#lightness) function and written into header files as 8-bit PWM (pulse-width modulated) data.

The software is written in avr-gcc C (GNU C for the Atmel AVR chip) and source code is provided along with a pre-compiled *.hex* file for convenience. You can download both here: www.makezine.com/go/fireflysrc.

The code has been significantly optimized for efficiency and to minimize power consumption. Crude runtime estimates predict a 600mAh 3V CR2450 battery should last between 4 to 10 months, depending on the song pattern used. The source comes with two patterns, song1 and song2, with song2 as default. Song2's estimated runtime is 2 months, song1's is 5 months.

This project involves a fair amount of surface-mount level soldering. However, the circuit design is trivial and the fact that we're able to use an off-the-shelf SMD prototyping board rather than having a custom PCB made greatly saves on cost.

It would be very simple to create a non-surface-mount version using the PDIP version of the ATtiny45 and through-hole LEDs.

The cost of the electronic components comes in at around $10-$15 (after shipping) and assembly time is on the order of two hours.

2. Parts

In this section I list the parts I used in the construction of this project. In many cases, the exact part is not required and a substitute will suffice. For instance, it isn't required that you use a CR2450 battery to power the circuit, any 3V power supply will suffice and CR2450s just happened to be the cheapest battery that I found that fit the size and capacity requirements I was looking for. Figure B shows many of these parts.

- One AVR ATtiny45V microcontroller, 8-pin SOIC package, DigiKey part number ATTINY45V-10SU-ND (see note A)
- One Surfboard 9081 SMD prototyping board (www.capitaladvanced.com/9081.htm), DigiKey 9081CA-ND
- Green LEDs (6), DigiKey 160-1446-1-ND (see note B)
- One 22.0K 1206 resistor (see note C)
- 100Ω 1206 resistors (2) (see note B)
- One CR2450 battery holder, DigiKey Part# BH2430T-C-ND
- One CR2450 battery (any 3V power supply will do)
- One spool of #38 Magnet wire from http://ngineering.com, part number N5038
- Six inches or so of bare thin wire. I used stripped wirewrapping wire but about anything will do.

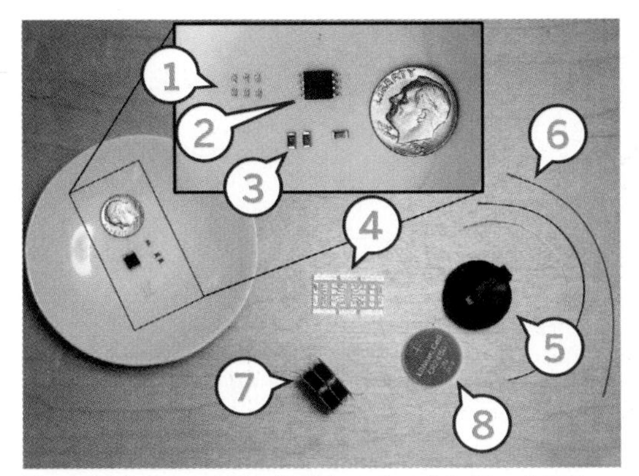

Figure B: 1. Green LEDs (0603 package) 2. Atmel AVR ATtiny45V microcontroller 3. Resistors in 1206 package (2x100Ω, 1x22.0kΩ) 4. Surfboard 9081 SMD prototyping board 5. CR2450 battery holder 6. Bare wire for jumpering pads on SMD board, red and green wires for attaching battery holder 7. #38 Magnet wire 8. CR2450 3V battery

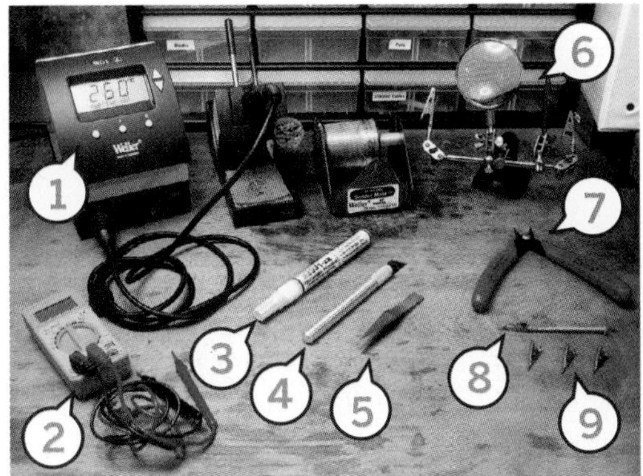

Figure C: 1. Weller soldering station 2. Multimeter for testing the board 3. Water-soluble flux pen 4. X-Acto knife 5. Tweezers 6. Two Helping Hands with magnifier 7. Flush-cut shears 8. "Clip-on-a-stick" 9. Micro clips

Notes:

- **A.** The difference between an ATtiny45V and an ATtiny45 is that the ATtiny45V is spec'd to run on voltages between 1.8V–5.5V while the ATtiny45 wants 2.7V–5.5V. For this project, the only implication is that the ATtiny45V can possibly run for just a little bit longer as the battery dies. In reality this probably isn't the case and the ATtiny45 can be considered interchangeable with the ATtiny45V (guess which one I happened to have on-hand when I started?). Use whatever one you can get your hands on. Also, the ATtiny85 will work just fine too, just for a little bit more money.
- **B.** Substituting a different model of LED with different current-draw characteristics will have implications on what resistor you use. See the Circuit Schematic appendix on the Instructables site (www.instructables.com/id/Jar-of-Fireflies) for more information and be sure to check the spec sheet for your LEDs.
- **C.** This is only a pull-up resistor, so the specific value is not important. It just needs to be big enough without being too big. See the Circuit Schematic appendix online for more information.

3. Tools

These are the tools I used—some are in Figure C

- RadioShack #270-373 1-1/8" Micro Smooth Clips (4)
- One "clip-on-a-stick"—one of the Micro Smooth Clips mounted on a nail or other sort of stick for manipulating and twisting thin wires
- One temperature-regulated soldering iron with a fine tip (I use the Weller WD1001 digital soldering station with 65 watt iron and 0.010" x 0.291"L micro tip). On a budget, however, a 15-watt Radio Shack style soldering iron should be fine. 260°C is a good temperature to solder LEDs without damaging them.
- One set of helping hands; the third hand was cannibalized from another set
- One multimeter (for circuit testing)
- One set of Xcelite 170M flush cut wire shears, DigiKey part number 170M-ND
- Some flux. I like the Kester water-soluble flux pen, DigiKey part KE1808-ND. It works like a magic marker and rinses off with warm water.
- Some solder (the thinner the better)
- Light tweezers with a fine sharp point (Gingher model G-7600, 3-1/2")
- X-Acto knife or razor blade
- An AVR programmer such as the USBtinyISP programmer from Adafruit Industries (see the "Dance Messenger" or "Programmable LED" Instructables in this book for information on AVR programmers)
- Pomona 5250 8-SOIC test clip (available from HMC Electronics for around $10)

Figure D: Applying the flux

Figure E: Applying solder to the pad

Figure F: Solder the bare wire to a pad, then cut it to length after the solder has cooled

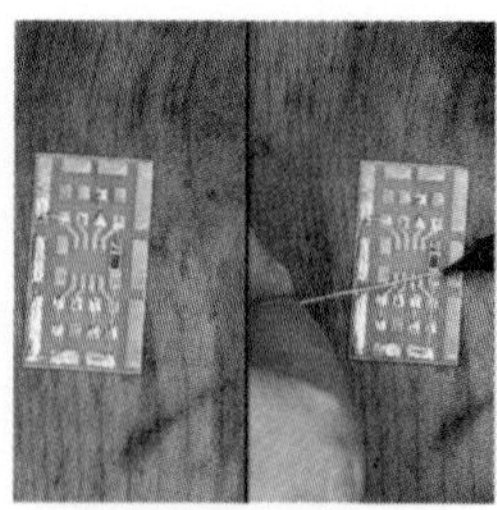

Figure G: Solder the resistor to one pad first, leaving the other pad dry

4. Circuit board assembly—part 1 of 3

Now it's time to prepare the circuit board and attach the resistors.

Flux the pads (Figure D)
I tend to flux everything, even when using solder that already contains flux. This is especially true when I'm using the water-soluble flux pen since cleanup is so easy and the pen makes it easy to not get flux everywhere. You can scribble over all the pads on the board and once you're done soldering, simply rinse it off under hot water.

Apply solder to the pads (Figure E)
There are a number of good guides on the Internet with examples of how to solder surface-mount components. In general, you want to start by putting a little bit of solder on one pad as shown.

Solder jumper wire across pads (Figure F)
The consequence of not having a custom PCB made for this project is that we have to add our own bus wires. Note the bus wires on PIN_C, PIN_D, and PIN_E (see Figure H for the pin assignments). These are not strictly necessary but it looks cleaner this way and gives us some elbow room when attaching a clip to the microprocessor for programming.

Solder resistors to the board (Figure G)
Holding the component in a pair of tweezers, heat the solder that you applied to the pad and hold one side of the component in the solder until it flows onto the pin. You want to keep the component flush with the board while you're doing this. Then, solder the other side. Only add solder to the other pad after the first pad has cooled and is holding the component flush to the board.

5. Circuit board assembly—part 2 of 3

Now it's time to solder the microcontroller to the board.

Bend the pins on the microcontroller
Another consequence of not having a custom PCB made is that we have to deal with the unusual width of the ATtiny45 chip, which happens to be slightly wider than will comfortably fit on the Surfboard. The simple solution is to bend the pins inwards so that the chip stands on the pads instead of sitting on them (Figure I).

Solder microcontroller to board
Again, there are many SMD soldering guides online but the executive summary is this:

- Apply flux to the pins of the chip (I find this makes it *much* easier to get a good solder joint, especially with the weird surface topology of these bent pins).
- Hold the chip to the pad and draw solder down from the square pad and onto the first pin of the chip (add more solder if there isn't enough on the square pad but you'll typically have enough already).
- Make sure that the solder actually flows up and *onto* the pin. The soldering motion is sort of like "pushing" the solder onto the pin (Figure J).
- Once the first pin is soldered, go to the pin on the opposite corner of the chip and solder that down as well. Once those two corners are tacked down, the chip should remain firmly in place and the remaining pins become simple to complete.

Also, be very careful that you solder the chip to the board in the correct orientation! If you look closely on the chip you'll see a little round in-

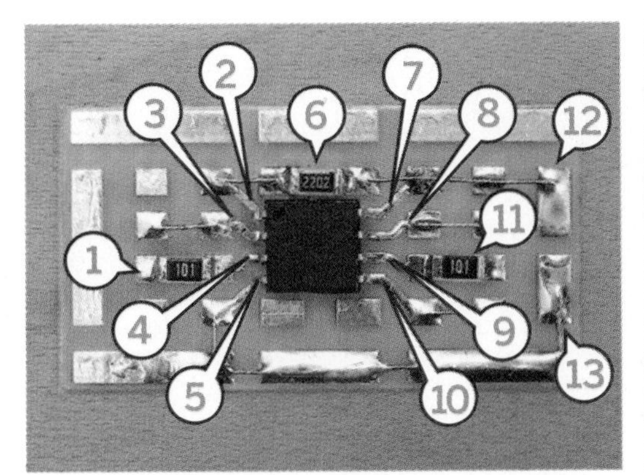

Figure H: 1. 100Ω resistor 2. Reset 3. PIN_C 4. PIN_A 5. GND 6. 22.0kΩ resistor used as pull-up 7. VCC 8. PIN_D 9. PIN_B 10. PIN_E 11. 100Ω resistor 12. VCC 13. GND (jumper across from here to the chip's GND)

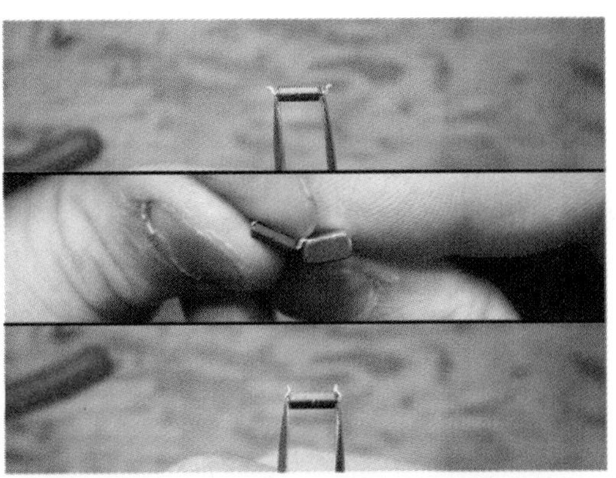

Figure I: Bending the pins to fit the board using a piece of metal (the wirestripper from a wirewrap tool)

Figure J: Draw the solder from the square pad and push it onto the pin.

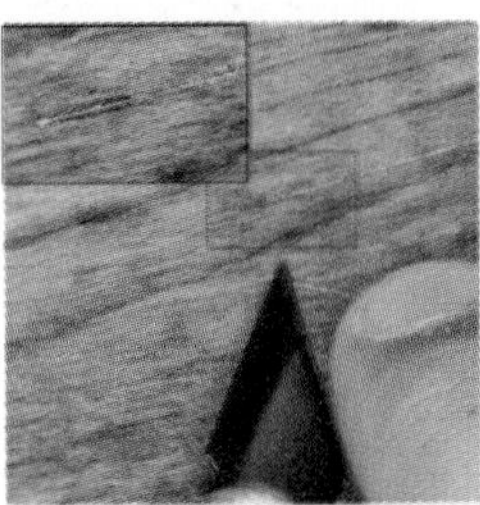

Figure K: Gently scrape away about a millimeter of insulation (don't damage the wire)

dentation on the top in one of the corners. That indentation marks pin #1 that I marked as Reset in Figure H. If you solder it down in the wrong orientation, I promise you that it won't work. :-)

6. Circuit board assembly—part 3 of 3

Now you need to test all the connections.

Since everything is fairly small here, it's quite easy to make a bad solder joint that looks fine to the eye. That's why it's important to test everything. Use a multimeter and test all of the pathways on the board for connectivity. Make sure to test everything; for example, don't touch the probe to the pad that the pin of the chip looks soldered onto, touch the pin itself. Also test the resistance values of your resistors and make sure they match with their expected values.

A small problem now is easy to correct, but it will become a big headache if you don't discover it until after all the LED strings have been attached.

7. Making a firefly LED string—part 1 of 4

Now it's time to prepare the wires.

Ngineering.com has a good writeup (www.ngineering.com/Wire_wiring_tips.htm) of how to work with the magnet wire and covers tinning as well as twisting it which are two key steps of making a firefly LED string. However I've never been satisfied with the results of burning away the insulation as they describe in the guide and have instead settled on gently scraping the insulation away with a razor. It's quite possible that I simply wasn't doing the tinning steps right (despite many attempts) and your own mileage may vary.

Cut red and green wires to the desired string length. I prefer to use different lengths of wire for each firefly string so that once assembled they don't all hang at the same "altitude." Generally I calculated the lengths that I was going to use by figuring out the shortest string (based on measuring the jar I was going to use), the longest string, and dividing the interval between them equally into 6 measurements. The values I ended up with for a standard widemouth jelly jar are: 2-5/8", 3", 3-3/8", 3-3/4", 4-1/8", 4-5/8".

Strip one end of each wire exposing a millimeter or less. Using the razor method, gently scrape the insulation away by softly dragging the blade over the wire (Figure K). Turn the wire and repeat until the insulation has been removed. Using this method I find it hard to only strip a millimeter of wire so I simply trim the length of the exposed wire to one millimeter after I strip it.

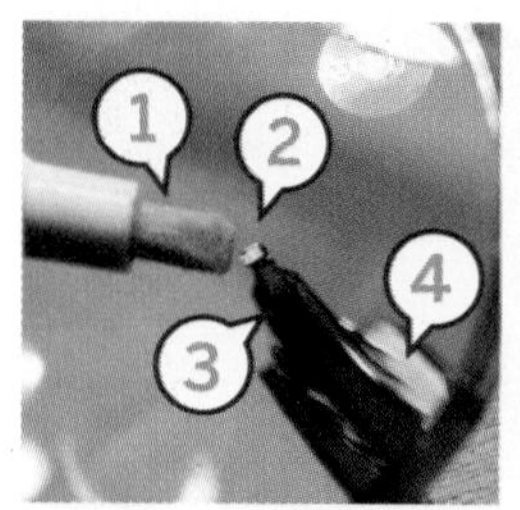

Figure L: 1. Flux pen 2. LED 3. Microclip 4. Helping hands

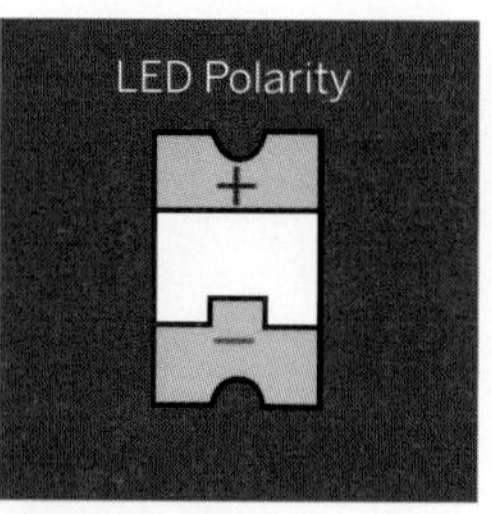

Figure M: Surface-mount LED polarity

Figure N: 1. The green wire is being held in place by another microclip (out of view) held by one of the helping hands. Don't try to hold the wire in place with your fingers 2. The LED gently resting on the pad

Figure O: The wires, soldered to the LED

8. Making a firefly LED string—part 2 of 4

Next, you need to prepare the LED.

Using a microclip, pick up an LED so that the bottom side is facing out, exposing the pads. Mount the microclip and LED in the helping hands and apply flux to the pads on the LED as shown in Figure L.

9. Making a firefly LED string—part 3 of 4

Here's where you'll solder the LED.

Using another microclip, pick up the green wire first and mount it in the helping hands.

Now comes the hardest part of the project, soldering the LED. Manipulate the helping hands so that the exposed part of the green wire is resting gently on the cathode pad of the LED (Figure M shows the orientation; the—side is the cathode). This is the time-consuming part that requires patience and cannot be rushed. Plan your moves in advance and act slowly and with deliberation. This is basically ship-in-a-bottle type delicate work and shouldn't be underestimated. However you don't have to be the favorite son of a watchmaker in order to pull this off either, it *is* within the realm of mortals. I find it considerably easier to manipulate the arms of the helping hands rather than the wire itself or the microclip.

Rest the exposed part of the wire on the cathode pad and arrange your magnifying gear and lighting to make sure you can perfectly see what you're doing in preparation for soldering.

Using a soldering iron set to around 260°C, pick up a very small blob of molten solder onto the tip of the iron and, very gently, touch the tip of the iron to the cathode pad on the LED as shown in Figure N. A small amount of solder should instantly run off of the tip and onto the pad (thanks to the flux), securing the wire to the pad in the process. Be careful not to burn the LED by holding the iron to the pad too long (3 seconds max, when done right you need less than 0.10 seconds of tip contact, it's very fast).

Unfortunately what tends to happen here is that you knock the wire off the pad with the tip of the iron, forcing you to go through setting it all up again. For that reason you have to be *very* slow and gentle with the iron. I tend to place my elbows on the workbench on either side of the helping hands and hold the iron with both hands in a seppuku (en.wikipedia.org/wiki/Seppuku) type grip, gently bringing the iron down towards the pad. This grip is sometimes the only way I can get enough control. Another tip: don't drink a pot of coffee before attempting this.

This does get easier with practice.

Very gently tug on the green wire to test that it's firmly secured. Release the wire (not the LED) from the microclip and, without changing the orientation of the LED, repeat the process with the red wire, only this time soldering it to the anode (+) pad of the LED. Since the red wire will be flying over the cathode (green) pad, it's important to not have too much exposed red wire, lest it come down in contact with the cathode pad and create a short. Figure O shows an LED with green and red wires attached.

10. Making a firefly LED string—part 4 of 4

Next, twist the wires and test the LED.

Once both wires have been attached to the LED, it's time to twist the wires. Twisting the wires results in a cleaner look, greatly adds durability to the LED

Figure P: Twisting a pair of wires

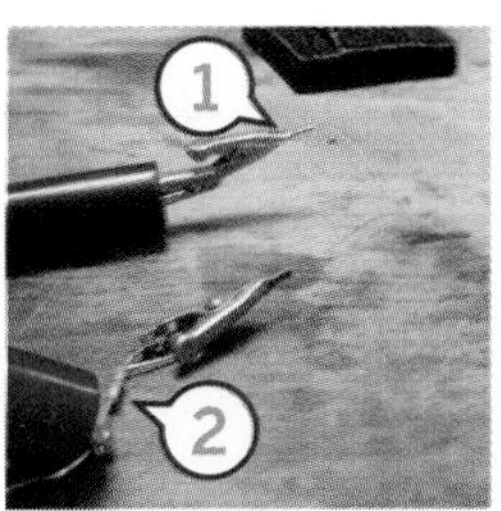

Figure Q: 1. Using microclips to get a more reliable connection to the little wires 2. 100Ω resistor

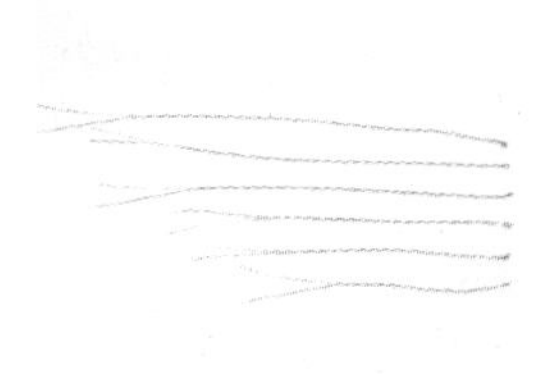

Figure R: An hour or two of your life that you'll never get back

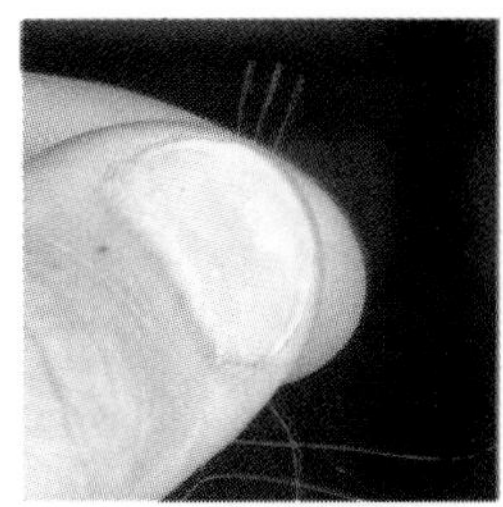

Figure S: Grab three red wires between thumb and forefinger. Clip them together using a microclip and mount the microclip on your helping hands

string, and also reduces the number of delicate free-flying wires you have to deal with when working with the board later.

To twist the wires, begin by mounting a microclip in your helping hands and clip it to the two wires right beneath the LED. Now, using another microclip (I have it mounted on a nail to make this process easier), grab the other end of the string about 1.5" from the end (you'll leave that length untwisted). Gently twist the microclip while applying just enough tension to keep the wires straight until the wires are sufficiently twisted together. I tend to prefer a somewhat tight twist as this results in a string that's easier to keep straight (Figure P).

Once the string has been twisted, strip about 2-3mm from the free end of the wires and test by putting 3 volts through a 100Ω resistor and into the ends of the wires. I've found it very difficult to make a good connection by pressing probes into the bare ends of the magnet wire so I clip microclips onto the ends and touch those with the probes instead (Figure Q). You don't have to get a good solid "on" from the LED for the string to pass the test, since even with the clips it's hard to get a good connection. Even a few flickers are enough to pass. When soldered, the connection will be much better.

Set the LED string aside in a safe place. Repeat this process for each of the six strings. Figure R shows six strings in a variety of lengths.

Note: Figure Q shows a 100Ω resistor being used to protect the delicate LED. LEDs are a hair-trigger when it comes to voltage sensitivity and it's possible to burn them out by driving them just a few tenths of a volt higher than their rated specification unless you use a resistor to protect them. This is doubly true for SMD LEDs since they're so small.

11. Attaching LED strings to the board—part 1 of 2

Once you've completed all six of the LED strings and the circuit board, it's time to attach the strings to the board.

Sort the LED strings into two groups of three. For each group, we will twist and solder the three red wires together into one and then solder that to the board.

Grab three of the red wires between your thumb and forefinger as shown in Figure S. After taking special care to ensure that the stripped ends of the three wires all line up, microclip the three wires close together and mount the microclip in the helping hands.

Twist the exposed parts of the wires together as shown in Figure T. This is to prevent them from coming apart while you solder them to the board.

Tin the twisted ends of the wires with solder. Use flux first to ensure a good contact between the wire tips (the last thing you want to do is have to untwist these three wires to get at one that's not making good contact) (Figure U).

Carefully solder the red-wire bundle to the far side pad of PIN_A (Figure V), so that the resistor separates the bundle and the microcontroller. Repeat the process with red wires from the other three LED strings, soldering the bundle to the far side of the resistor on PIN_B. You should now have both 3-string red wire bundles soldered to the board with the green wires flying free.

12. Attaching LED strings to the board—part 2 of 2

Using a similar process to how you made the red 3-wire bundles, join the green wires together into

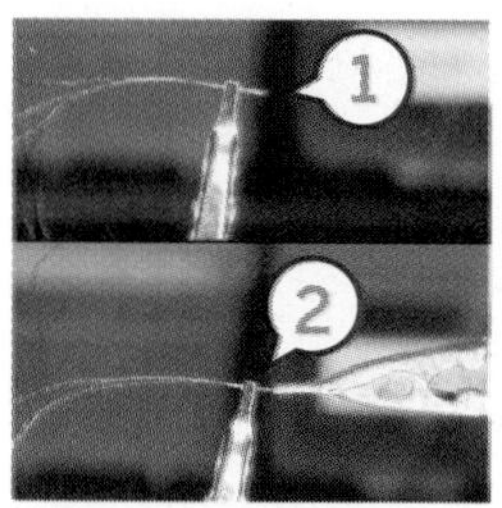

Figure T: 1. Clip right behind the start of the exposed wire 2. Twist to ensure wires don't come apart

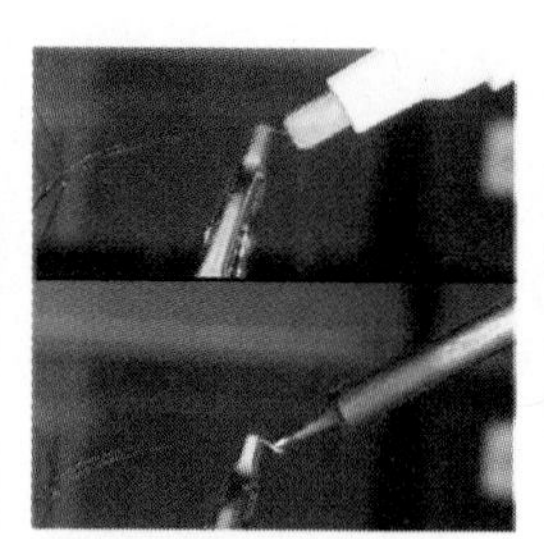
Figure U: Fluxing and soldering

Figure V: Solder the red-wire bundle to the pad on the far side of the resistor

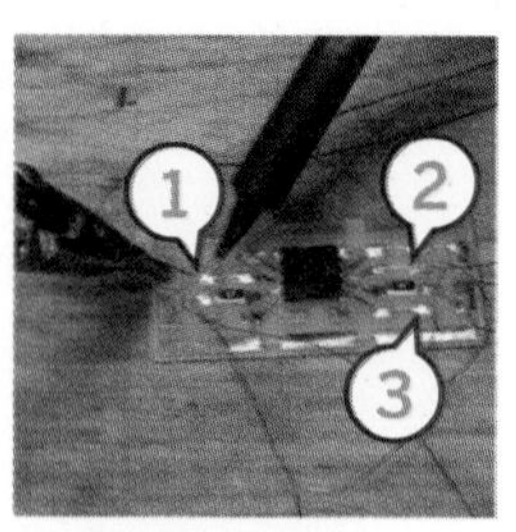

Figure W: 1, 2, and 3: Solder green 2-wire bundles to these pads

2-wire bundles and solder them to PIN_C, PIN_D, and PIN_E (Figure W). By not soldering the bundles to the pad closest to the microcontroller, we give ourselves more elbow room should we need to do any touchup soldering work on the microcontroller or attach a programming clip to the board.

Once all the LED strings have been soldered to the board, it's a good idea to test them as shown in Figure X. With a 3V power source, test the strings by placing a positive voltage on either PIN_A or PIN_B, being careful to place it *behind* the resistor since 3V will damage these LEDs without it, and moving the negative voltage between PIN_C, PIN_D, and PIN_E. Each combination of pins should result in an LED lighting up when probed.

Note: If your chip happens to be already programmed at this point then simply applying power to the board (VCC and GND) should be enough to test all six LEDs in one go. The provided program cycles through all the LED's on boot.

13. Preparing and attaching the battery holder

Take the wires that you're going to use to attach the battery holder with and cut them to length. I tend to use the following lengths:

- Red Wire: 2"
- Green Wire: 2-3/8"

Strip a little bit off both ends of the wires and solder one end of the wire to the battery holder and the other end to the circuit board, being careful to get the polarities correct as shown in Figures Y and Z.

Also, once you have soldered the wires to the battery holder, you may want to snip the pins on it short so it's not quite as awkward to attach to the lid of the jar.

14. Final assembly

By this point you've completely assembled the circuit board and attached the LED strings and battery holder.

All that's left is to program the chip and affix the board assembly to the lid of your jar (Figure ZA).

As to how to program the chip, I'm afraid that's a bit beyond the scope of this document and is heavily dependent on what platform (FreeBSD, Mac OS X, Linux, Windows, etc.) of computer you're using and what development environment you're working with. I've provided the source code (written for GCC) as well as compiled binaries but figuring out what to do with them is up to you.

Note: The "Dance Messenger" or "Programmable LED" Instructables elsewhere in this book discuss how to program this family of chips, but you will need to use a special clip, described next, to physically connect the chip to the programmer.

Thankfully, there are loads of good resources out there for getting started with AVR programming—here are a couple:

- AVRFreaks (www.avrfreaks.net)—This is the ultimate site for AVR. The active forums are indispensable.
- AVR Wiki (www.avrwiki.com)—I found this site quite helpful when I got started.

The file firefly.tgz contains the source code and compiled .hex file for this project. You can download this file from www.makezine.com/go/fireflysrc.

This project was built using avr-gcc 4.1.1 (from the FreeBSD ports tree) along with avr-binutils 2.17 and avr-libc-1.4.5. For instructions on setting up avr-gcc on Windows, Mac OS X, and Linux, see the "Dance Messenger" or "Programmable LED" Instructables elsewhere in this book.

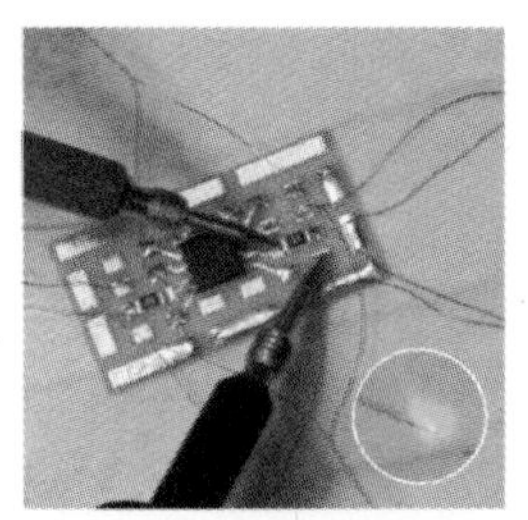

Figure X: Be sure to test from behind the resistor so you don't damage your LED

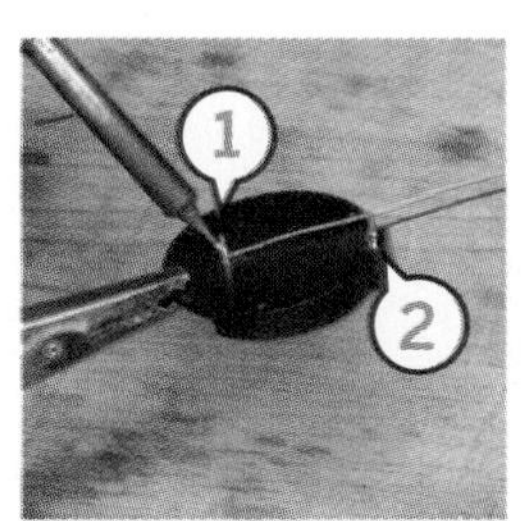

Figure Y: 1. Negative terminal 2. Positive terminal

Figure Z: Connecting the positive and negative terminals to the board

Figure ZA: The assembled jar

Figure ZB: 1. The Pomona 5250 8-SOIC test clip 2. Make sure the two battery contacts aren't touching when you program the chip, or it will short out

Figure ZC: Inside the lid of the Jar of Fireflies

User Notes

In the Comments to this project, **Instructables member ccarlson** presented a problem he ran into while trying to reprogram a chip that was already in a working circuit. The problem was encountered while using Ladyada's USBtiny-ISP to program the ATtiny45V for this project. Ladyada herself offered the solution: don't let the programmer (in this case the USBtinyISP) power the process via USB power, remove the jumper and use a battery to power the circuit instead.

The Pomona 5250 8-SOIC test clip, shown in Figure ZB, is the only test clip I've used which works well with the 8-SOIC AVR processors like the one in this Instructable. Other clips tend to have a problem due to the fact that the AVR chips are a little wider than normal and the other clips weren't designed for that.

As for attaching the board and battery to the lid, there are probably a million ways to do this but I'm not confident that I've found the best one yet. The methods that I've tried have been to use either epoxy or hot glue. I've already had a few instances of epoxied boards pop off so I wouldn't recommend using that. Hot glue seems to work ok but I have little faith that after a few hot/cold cycles it'll fare much better than the epoxy. Figure ZC shows the board attached to the lid.

So, I leave figuring out how to attach the board and battery holder to the lid up to you as well. However I will offer a couple tips:

- Be careful when you attach the battery holder that the two pins don't short out due to the metallic lid. Some lids are insulated, and others aren't.
- This to That (www.thistothat.com) is a website that offers glue recommendations based on what you're trying to glue. For glass to metal (the closest approximation I can think of for a silicon circuit board) they recommend "Locktite Impruv" or "J-B Weld." I haven't ever used either.

Note: Be sure to check out the web page for this Instructable at www.instructables.com/id/Jar-of-Fireflies for four appendixes, including descriptions of the circuit schematic and possible areas for improvement.

Xander Hudson runs Synoptic Labs in Sacramento, California, and maintains the Synoptic Labs weblog (www.synopticlabs.com).

The Best of Instructables Tools

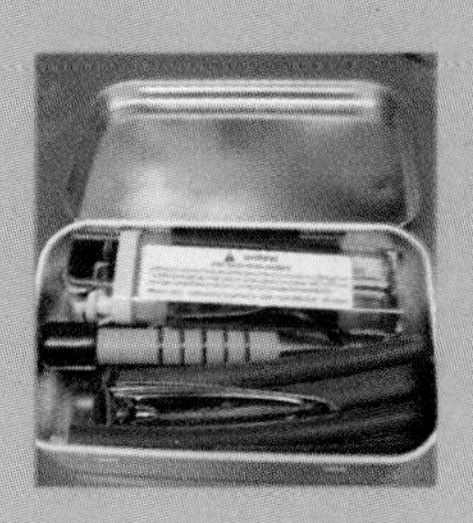

First you learn to work with your hands, draw with crayons, work with clay.

Then you learn to work with tools, starting with the basics like scissors and rulers.

Eventually we graduate to ever more capable and specialized tools: lathes, mills, computer-controlled machines like laser cutters, even things like electron microscopes.

This gradual mastery of tools is the monk-like apprenticeship of the committed builder and hacker.

Every tool's strength is also its weakness.

Eventually we need to do something beyond the specialized capability of the tool, in fact, beyond the capability of any tool.

This is when you graduate from apprentice to master, the creator of new tools, capable of building new things never before possible.

The highest art of the committed craftsman, hacker, and builder, is to become a meta-craftsman: a designer and builder of tools.

To these great masters, I tip my hat and vow my thanks.

Innovation rides on the back of the capacity of our tools.

—Saul Griffith (saul)

Third Hand++

How to build a multi-use helping hand for electronics and other delicate work By Ryan Straughan

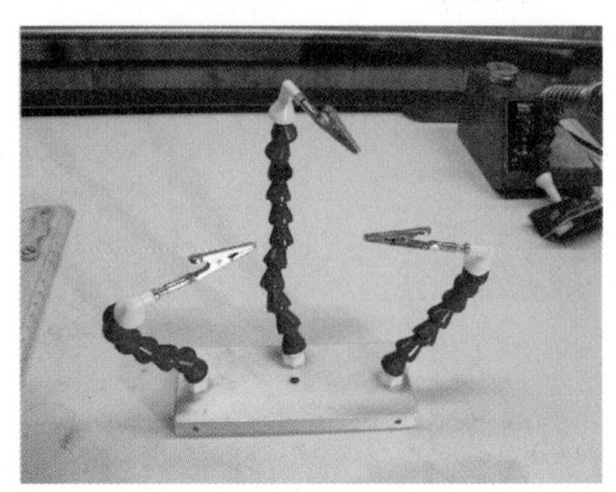

Figure A: The basic Third Hand++, made from machining coolant hose

In the past, I've used the third hands/helping hands available at chain electronics shops and been frustrated with their usability. I could never get the clips exactly where I wanted them, or it took more time than it really should to get the setup right. I also wanted the ability to hold small circuit boards and alligator clips just don't do a very good job of that.

I was familiar with the adjustable coolant hose systems used to spray coolant at cutting tools in the machining industry and thought that would be a great place to start. I ordered various nozzles and hose segments from my favorite online machine tool supply company and started experimenting. This is what I came up with. While it still has plenty of room for improvement, it has served me well over the last 3-4 years.

These arms can be placed into pretty much any position and they will stay there. Another nice feature is that you can make all sorts of attachments for holding whatever you need to work on. So far, I've made a circuit board holder, a clamp, a mount for an LCD, and an extraction fan for keeping fumes out of my face.

All you really need are some simple hand tools, a couple of taps, a drill bit, and a drill to make the basic version. If you have all the tools you need, it can be made for $20 or less.

1. Getting started

The first step is to gather everything you'll need.

Tools

- Drill (a hand drill will work, but a drill press would be better)
- 3/8" drill bit
- 1/8-27 NPT tap
- 6-32 tap

Figure B: The Third Hand++ in action

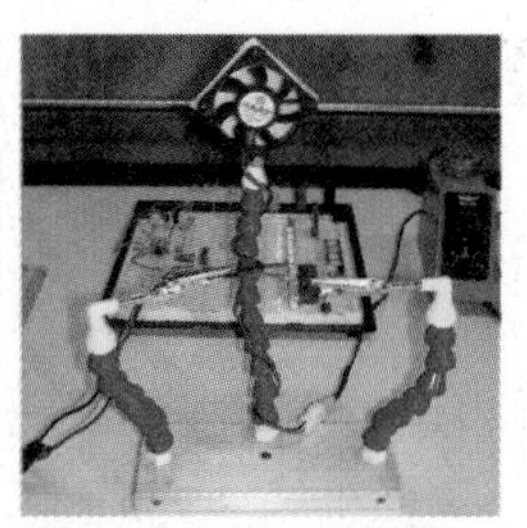

Figure C: Middle arm with a fan for extracting the fumes produced while soldering

Figure D: Circuit board holders

Figure E: Circuit board holder close up. Note how the board slots in place.

For additional information, discussion, and more, please visit the Instructables project page:

- A tap handle
- Ruler
- Center punch
- Don't forget safety glasses!

Parts

The base: I used a block of 1/2" thick aluminum (5.75" x 2.5" x 0.5").

Aluminum is heavy enough to be stable and is easily tapped. You can use whatever you want as long as it is at least a 1/2" thick and can be tapped (plastic, wood, MDF, steel, etc.). The lighter the material, the larger the base needs to be in order to remain stable. If the material is too soft, the threads will wear out and the arms won't stay in. If you don't have a local source for the aluminum, you can order it from an online metal sales company cut to length for about $6 plus shipping. I have used onlinemetals.com for other projects in the past.

The arms: The arms are made from coolant hoses and nozzles used in the machining industry to keep cutting tools cool and lubricated. I used the Snap Flow brand coolant hose system that I bought from use-enco.com. They sell a "Male NPT Hose Kit" that has 13" of hose and an assortment of nozzles and connectors. That gets you most of what you need to make a two-handed Third Hand.

I'd recommend purchasing two kits and a few extra nozzles and connectors. For around $12, you will have more than enough parts to make four arms.

For each arm, you will need:

- A 1/8 NPT connector
- 4-5" of hose
- A 1/8" 90 degree nozzle

You may want to consider buying the hose assembly pliers for $23. They are a little difficult to snap together by hand. I didn't buy the pliers but I kind of wish I had.

The hands: Each hand is made out of a banana plug threaded into the 90-degree nozzle and an alligator clip. I chose the "Flexible Banana plugs (2-Pack)" from RadioShack because it has 6-32 threads that will thread into the nozzle. The alligator clips are the standard 2" size.

Figure F: Center arm sports a holder for an LCD display

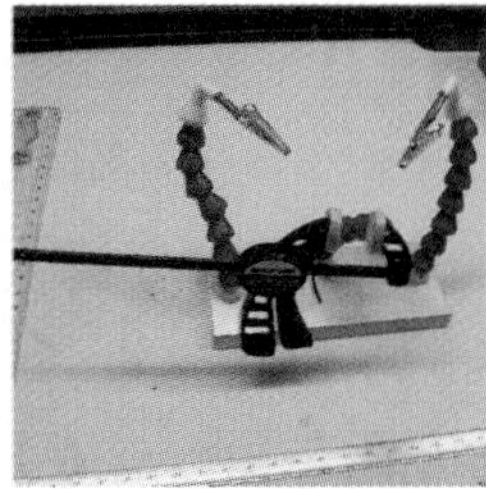

Figure G: My first version of the Third Hand++, with a Quick Grip attachment

Figure H: A 2.5" diameter x 9.5" long solid aluminum round bar that weights 5lbs 1 oz, to show the load capacity of the arms

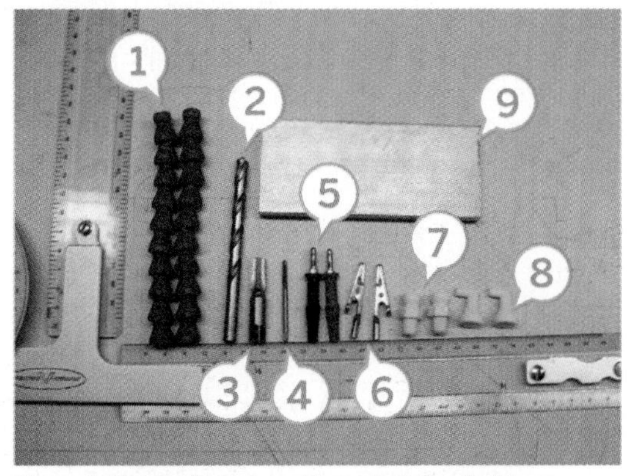

Figure I: Some of what you'll need—1. Coolant hose 2. 3/8" drill bit 3. An 8-27 NPT Tap 4. A 6-32 tap 5. Banana plugs 6. Alligator clips 7. 1/8-27 NPT coolant hose connectors 8. 90-degree 1/8" coolant nozzles 9. Aluminum block for the base

2. Building the base—lay out

Once you have chosen your base material, you will need to cut it to size, if it has not been done already.

Figure J: Laying out where you want your "arms" to go

Figure K: I used a spring loaded center punch to mark the holes, but I suppose even a nail and hammer would work

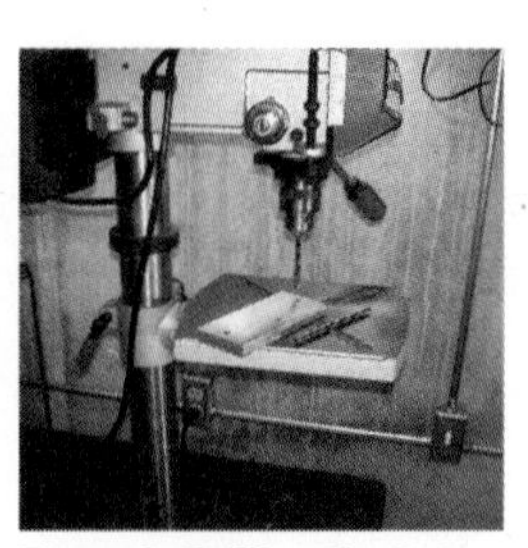

Figure L: Drilling the holes for the base

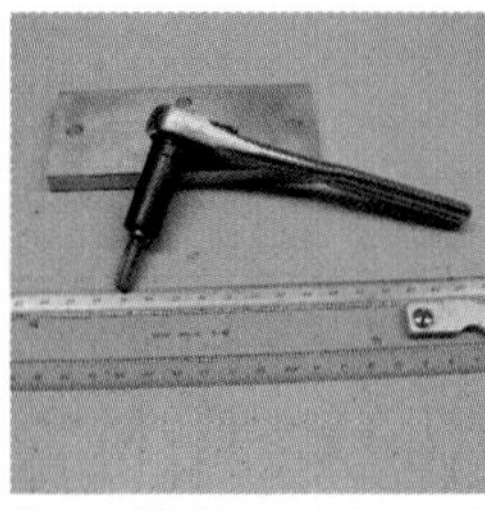

Figure M: A tap socket and a 1/8-27 NPT Tap

Figure N: Tapping the holes in the base

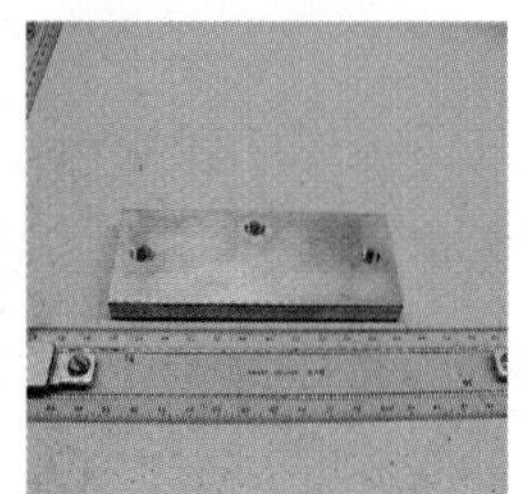

Figure O: The finished, tapped base plate

Figure P: Finish the base surface by sanding. I used 80 and 220 grit and finished with a coarse Scotch-Brite pad (shown).

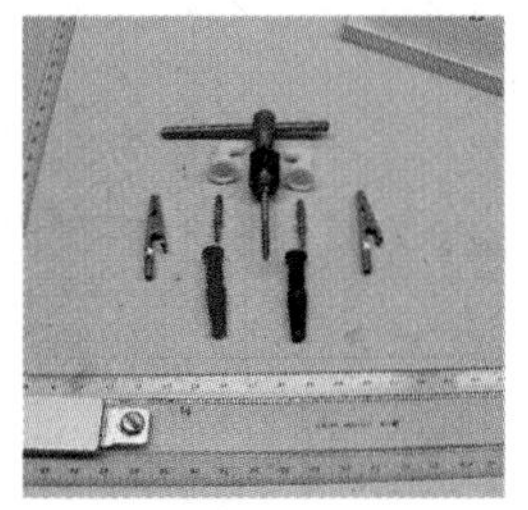

Figure Q: The parts to build our "hands"

I used a block of 1/2" thick aluminum (5.75" x 2.5" x 0.5").

Next you need to lay out the location of the hole for each arm. In this case, I'm using three arms. The arm locations are not critical, they just need to be close enough that the hands will be able to reach each other and be symmetrical so they look nice. It will also depend on the shape and size of your base material. A triangular base might also be a good way to go if you plan on using three arms.

Use a center punch to mark the center of each hole for drilling.

3. Building the base—drilling the holes

I usually start with a smaller drill bit to get it started and then finish with the 3/8" bit. Make sure you drill all the way through the material so it can be tapped. You want the hole to be perpendicular to the base so the hose connector will be flat on the surface when threaded in. This can be done with a hand drill but a drill press would be easier.

4. Building the base—tapping the holes

Tap the holes for the arms using the 1/8-27 NPT tap. Remember that pipe thread is tapered so you will need to tap it deep enough that the hose connector screws all the way in. But tapping it too deep will cause it to be loose and potentially strip the threads on the hose connector. Also remember to keep the tap perpendicular to the base.

I don't have a tap handle large enough for the 1/8-27 NPT tap so I used one of my sockets designed to hold taps.

If your base is metal, I recommend using thread cutting oil or any lubricant like WD-40 that you may have on hand. I use Tap-Magic thread cutting oil.

5. Building the base—surface finish

Now that the holes are drilled and tapped, you can smooth the surfaces and round the corners using sand paper. I started with 80 grit, then used 220 grit and finished it off with coarse Scotch-Brite. The Scotch-Brite gives it a nice satin finish.

6. Building the hands

Remove the red and black covers from the banana plugs and discard. We only need the metal parts. Using the 6-32 tap, start threading the 90-degree

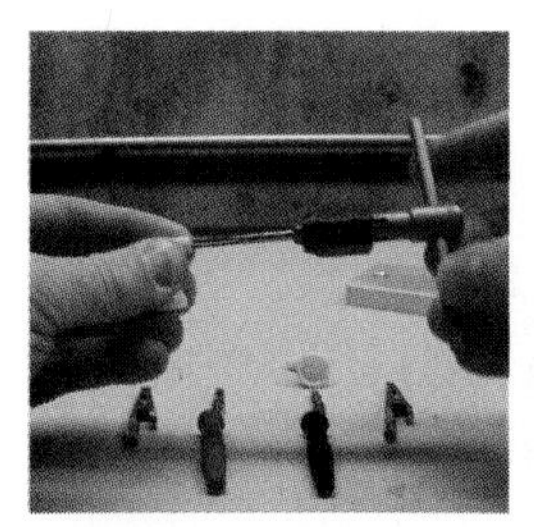

Figure R: Threading the 90 degree nozzle using the 6-32 tap

Figure S: Attaching the nozzle-banana clip assembly to the alligator clip

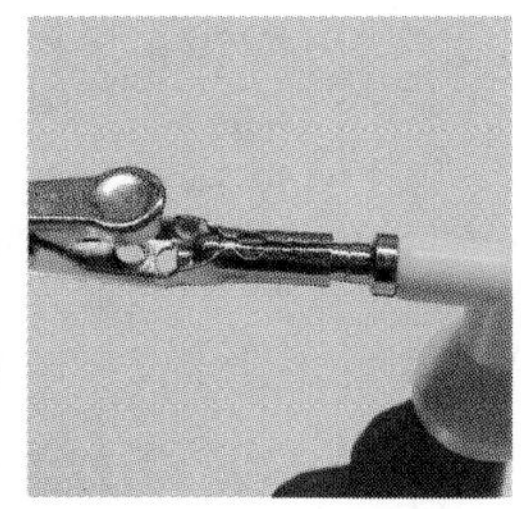

Figure T: The gap in the alligator clip can cause the clip to rotate

Figure U: The top two clips are modified, the top with a metal sleeve, the middle with a wire wrap. The bottom clip in unmodified.

Figure V: All the parts to an arm

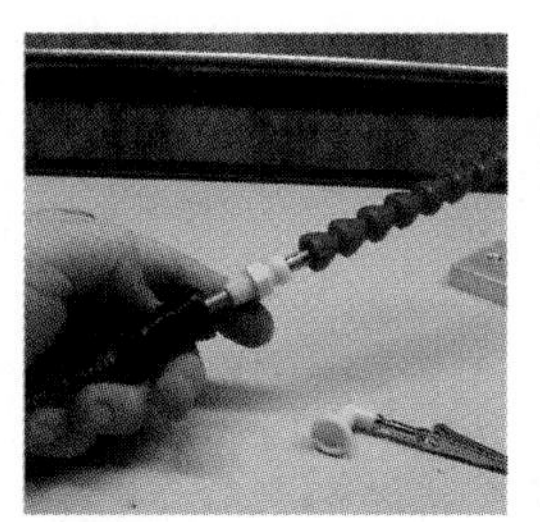

Figure W: Using a #2 Phillips head to thread the parts for easier assembly

Figure X: Tapping all of the components to snap into place

Figure Y: The finished, basic Third Hand++

nozzle. The banana plug threads aren't actually 6-32 but are close enough and provide a nice tight fit. Once the banana plugs are installed you can just slide the alligator clips onto the banana plugs.

The alligator clips work pretty well as is, but they do have a tendency to rotate when holding longer or heavier items. In the next step I'll show you how to improve them.

7. Improving the hands—optional

As I said in the previous step, the hands have a tendency to rotate on the banana plugs. While this is a feature I wanted, the ease at which they rotated was a problem in some situations. This is in part due to the alligator clips expanding when installed. You can see the gap in Figure T. To fix this, I came up with a couple of solutions. You can certainly skip this step and be happy with the results, but doing this will make it a lot easier to use.

Metal sleeve

I had some stainless steel tubing sitting around that was the perfect size. OD: 1/4" ID: ~3/16" (0.192"). I cut a 3/8" long section of tubing and using a hammer, lightly tapped the alligator clip into the sleeve. This is the best fix in my opinion (Figure U).

Wrapping with wire

I found some thin solid core wire, wrapped it around the clip, and soldered it in place. This is the easiest and cheapest solution to the problem (Figure U).

8. Assembling the arms

Unless you bought the assembly pliers ($23) when you ordered the coolant hose parts, assembling the arms can be a little tricky. I didn't buy it, but here is how I figured out how to put them together with ease. Slide the parts you want to join onto a #2 Phillips screwdriver. This will keep everything aligned and all it takes is a sharp tap to get the parts to snap together. Just hold onto the hose and tap it on the work surface in the direction of the part you want to attach.

Although theses pictures show 10 (Figure V), I found that about 7 hose segments per arm to be about the right length. Of course that is my preference and you can use as many as you want.

Figure Z: The circuit board holder in action

Figure ZA: The circuit board holder made out of 1/2" x 1/4" aluminum

Figure ZB: The newer holder made from black Delrin plastic—an easy to work with, electrically benign material

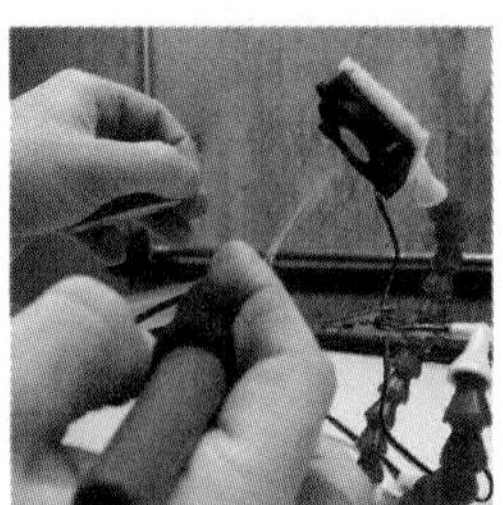
Figure ZC: A fume extraction fan made from a PC fan and a Scotch-Brite pad

9. Finishing it up

Now all you need to do is thread the arm assemblies into the base and you're done!

Next I'll show you some of the attachments I've made.

10. Attachments—circuit board holder

The circuit board holder has been one of the best attachments I've made for the third hand. I've used it to hold boards as small as an inch wide all the way up to ~8" wide (Figures Z-ZA).

Making these might be beyond the capabilities of most people because of the tools needed. First, my first set of holders. I used two pieces of 1/2" x 1/4" aluminum, each about 2.5" long. In the end of each, I drilled a 5/32" hole around 3/4" deep. You could do that with a hand drill and a vise, but there isn't a lot of room for error. A drill press or a mill would be best.

To make the slot, I used a slitting saw in my mill. The slot is a little over 1/16" wide and 1/8" deep running the entire length of the rail. I suppose you could do this with a hack saw or Dremel, but it would be difficult, and I suspect the results would be pretty rough.

Figure ZB shows a set of newer circuit board holders I made out of black Delrin plastic. These were a bit easier to make since I was able to use a 1/16" end mill instead of a slitting saw. Plastic is also a safer material, since many circuit boards have components and traces right up to the edge of the board and could be shorted by the aluminum.

11. Attachments—fume extraction fan

I made an extraction fan using an old CPU cooling fan, a bit of filter material, another 1/8" 90-degree nozzle, and a couple screws—all stuff I had lying around the house (Figure ZC).

For the filter, I cut a piece of white Scotch-Brite to the shape of the fan and attached one corner with a screw. The opposite corner I used a long screw to go through the fan and filter into the nozzle. Connect it to a 12-volt source and solder without having fumes in your face.

In my next version, I plan to add white LEDs to provide extra light in addition to the extraction functionality.

12. Attachments—LCD mount

I made this mount to hold a graphics LCD when I was playing with a BASIC STAMP II (Figure ZD).

I used a mill to build this, but I'm sure it can be done with hand tools. I milled a black bracket out of Delrin plastic, drilled and tapped the appropriate holes, and screwed it to a large straight nozzle. I'm not going to get into the details as it is fairly self explanatory and not everyone is going to have the same LCD. I mostly wanted to show you the variety of attachments that can be made for the helping hands.

13. Attachments—clamp

This clamp can be used to hold larger items than what will fit in the alligator clips. All I did was drill a couple holes in the bar near the stationary end of a Quick Grip clamp and attached it to a 90-degree spray bar nozzle with two screws. I used a right angle adapter to make positioning the clamp easier and #4-40 screws thread right into the spray bar nozzle without having to tap them. However, drilling through the hardened metal bar of the clamp took some effort (Figure ZE) .

Figure ZD: An LCD display mount

Figure ZE: The Quick Grip mount using the 90-degree spray bar nozzle part

Figure ZF: The Third Hand++ holding 2.2lbs of round bar

Figure ZG: The Third Hand++ with an O-scope probe attachment

14. Other attachments and ideas

ESD (electrostatic discharge): When I made the first version, that I've used for years, I was concerned with ESD and so I soldered a wire to the banana plug that ran inside the arm and was grounded to the base. I also drilled holes in the front and back so I could plug in a static wrist strap in the front and in the back and connect it to the static ground at the workbench. I probably should have soldered in a 1MΩ resistor to the wire going to each hand for more protection.

Powered hands: I've also thought about adding binding posts to the base and wires going up to the banana plugs so that voltage can be applied to the hands for powering circuits or loads being held by the alligator clips. The down side to this is that the left and right hands are wired to the base so changing attachments, that requires a nozzle change, means you have to disconnect the wire. However, most of the attachments I made that require a nozzle change are primarily used on the center arm. The alligator clip hands and the circuit board holder just slide onto the banana plugs, so no nozzle change is required.

DMM/O-scope probe: I am also working on an attachment to hold voltmeter leads or an O-scope probe (Figure ZG). I always seem to run out of hands when measuring signals on a circuit board.

Magnifying glass: Although I never used the magnifying glass on the old helping hands I had, I'm sure many people would use one. It would be easy to adapt one to use on the center arm.

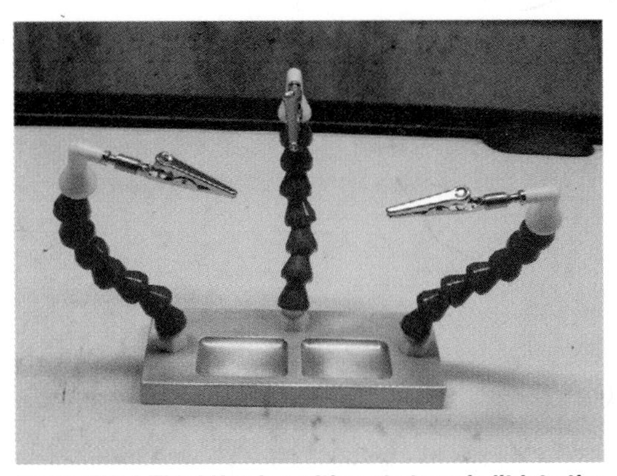
Figure ZH: A Third Hand++ with parts trays built into the base

LED light: A little extra light would be helpful as well. I plan to combine the extraction fan and an LED light.

Small parts tray: Figure ZH shows a base that I made for a friend that has a pair of parts trays milled into the aluminum.

Other thoughts: There are a ton of different nozzles and connectors available for the coolant hose systems. I'm sure there is no end of attachments and accessories that can be made for this third hand.

Ryan Straughan works as an airline flight simulator technician in Denver, CO. His hobbies include RC aircraft (both planes and helicopters), electronics, metal working/machining, an amateur radio (N0RYN). He has been fixing, making, and inventing since he was very little. He lives in Centennial, CO, with his wife and two cats.

Guide to Field Soldering

Do many soldering tasks in the field with this pocket-size kit

By Brian Cochran

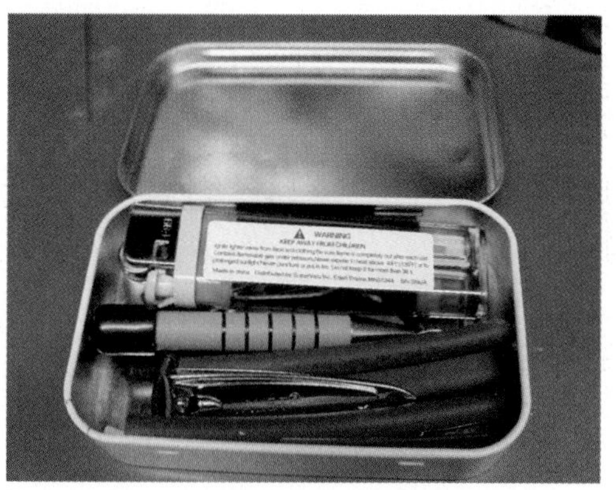
Figure A: The completed field soldering kit

This is a kit that allows you to do many soldering tasks in the field. It costs only about $8 to put together and it all fits in an Altoids tin! I've used this same kit for years and was inspired to share it based on a recent Instructable on soldering (an outstanding one). This Instructable goes one step further in building a portable helping hand, solder dispenser, and assembling all the other tools that you will need to get things done in the field.

Let me be clear at this point that I know this is not the best way to solder and this method has its limitations. It is, however, the best way to fix surveillance equipment at 3 a.m., in the back of a van, in the dark. I can testify to that. Good times, good times. I was a tech for an undercover narcotics unit for five years and this method came through for me many times.

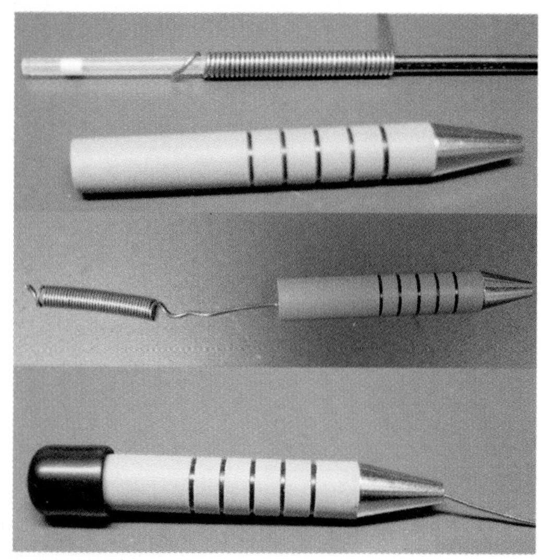
Figure B: Our portable solder dispenser made from an ink pen

1. Building the dispenser

I hate having to carry a whole spool of solder around with me. When I first started having a need to do these tasks in the field, I would just cut off a couple of feet and throw it in the bag. It soon became a tangled mess. So I came up with this approach.

Materials

You will need:

- An ink pen (metal-tipped is the best)
- 2' of thin solder wire

First, take the pen apart and cut it down so that it fits in the mint tin. Next, wrap the solder around the ink cartridge that you just removed from the pen. Pull the solder off the cartridge and insert it into the portion of the pen that you've retained. Cap off the end with the pen cap or whatever you have. I got a ton of these little plastic caps someplace and they are my favorite. I like the metal tips the best because they don't melt if you use this with a regular soldering iron. The result is a compact solder dispenser! I'm always amazed just how much solder you can fit in one of these suckers.

2. Helping hand

If you've soldered very much, you know that a

For additional information, discussion, and more, please visit the Instructables project page:

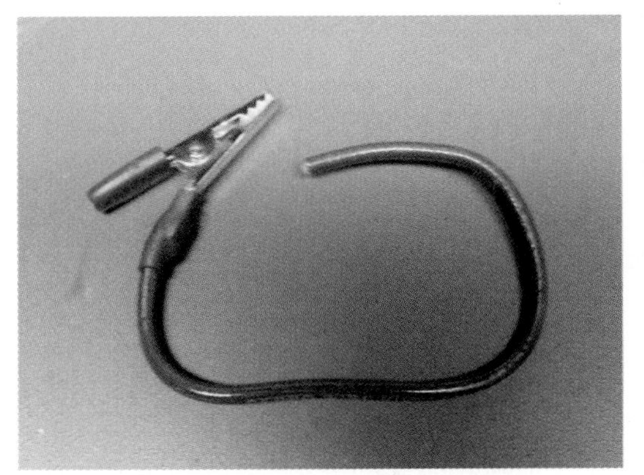

Figure C: A portable helping hand that folds flat

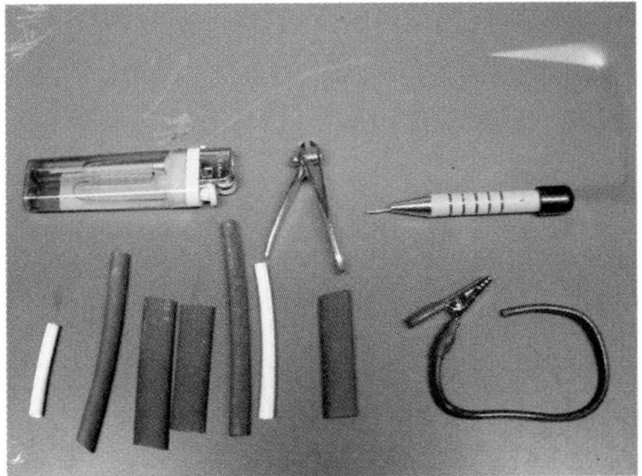

Figure D: Adding some heat shrink tubing (bottom left), a lighter, and clipper

helping hand tool is a necessity. There isn't much instruction here. There have been many Instructables on making helping hands, this one folds flat. Very simple. Heavy-gauge copper wire with an alligator clip on the end. It's very easy to use, just bend the clip up over the base (Figure C).

3. Rounding out the kit

You will need some additional items to complete the kit (Figure D).

Materials

- A pair of nail clippers (great foldable small gauge wire cutters!)
- A disposable lighter
- Various lengths and diameters of heat shrink tubing

The nail clippers are the cheapest and most compact flush wire cutters that you can buy. After 9/11, I was in a surveillance engineering class and they took all my tools for the course at the gate at the airport. A nice airport cop held them for me, but I was still without tools for the course. A quick trip to Wally World yielded a cheap multitool and some nail clippers for $6. I've been a fan ever since.

4. Using the kit

First you have to twist the wires to be connected. A good electrical connection is imperative before soldering. Next, place the connection in your helping hand. I like to wrap the solder around the connection prior to heating. This makes the heating process very easy in the field and yields a great distribution of solder over the entire joint. Fold the joint over and heat shrink it. I couldn't have done it better in the shop (Figure E).

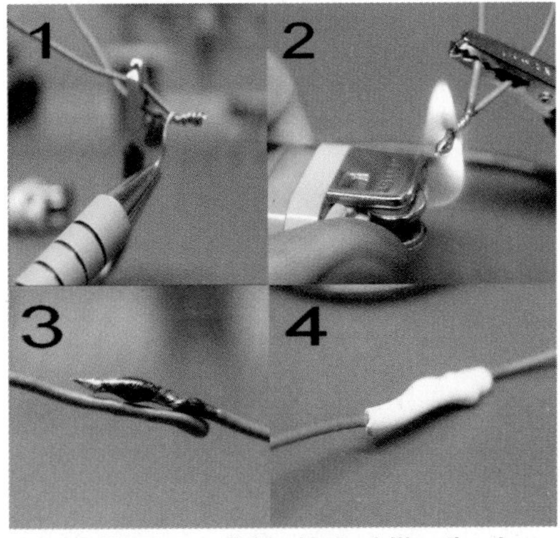

Figure E: The steps to field soldering 1: Wrap the wires and the solder around the wire wrap 2: Heat the solder 3: Bend the wires over 4: Heat shrink the joint.

Brian Cochran is a 35-year-old detective in a law enforcement agency in Kentucky with responsibilities for surveillance equipment and crime scene processing.

Flashlight Business Card

Give away a business card that recipients will never throw away

By Tom Ward

Figure A: The finished flashlight business card

If you've read my other business card projects on Instructables, you know what this is all about: make a business card that's actually useful, or too cool for people to want to throw away, and you have a successful piece of advertising. This is a variation on my previous flashlight card, here made to look a bit more flashy, and to be much easier to construct. No printed circuit board is required, just some self-adhesive copper tape. The finished design costs less than a dollar to make, and they're simple enough to make up a small batch in an hour or two.

1. Materials

What you will need:

- A CR2032 battery (I got them for about 16 cents on eBay when I bought 100)
- A 3mm high-intensity white LED (eBay again! I got 100 for about $16, so 16 cents each)
- Blank PVC ID cards (2) (Again, I got these for about 16 cents—find a specialized ID card printing store on the Web near you—I used www.digitalid.co.uk)
- Some double-sided foam adhesive tape (I got mine from an office supply store for a couple of dollars—you just need tape that is thicker than the battery you are using—mine was 4.5mm thick)
- Some self-adhesive copper tape (eBay! Mine cost a few dollars for a long roll of 1" thick tape, and I cut it into thinner strips.

Tools: You will also need a soldering iron (plus solder), a cutting knife, some spray adhesive, and a way of printing the front of your card—you can use a color laser or inkjet printer. I printed on paper and laminated it, but have had success before with printing in reverse on OHP transparency film which can look good as well.

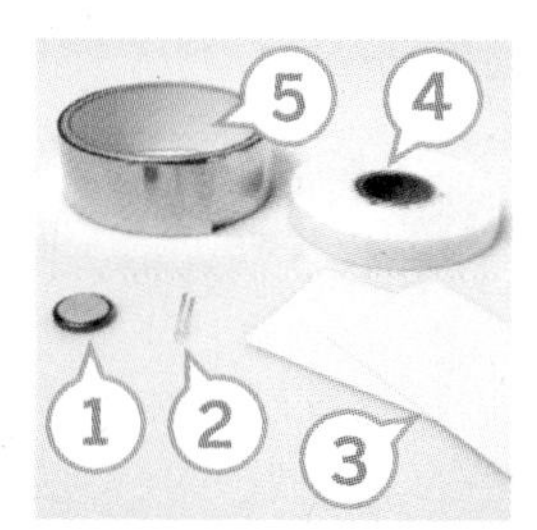

Figure B: Parts—1. CR2032 2. High-intensity LED 3. PVC ID cards 4. Double-sided tape 5. Self-adhesive copper tape

Figure C: One LED lead soldered to the first copper strip

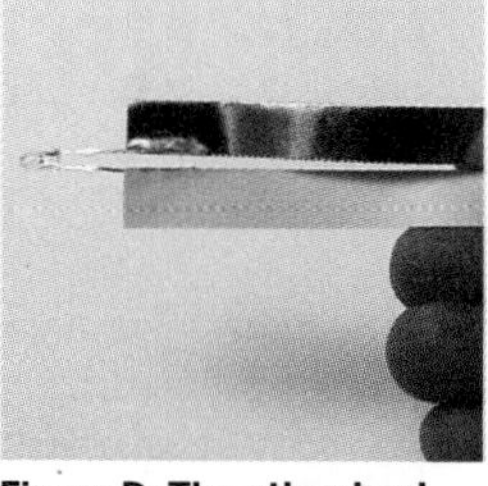

Figure D: The other lead of the LED soldered to second copper strip (the copper sides should face away from each other)

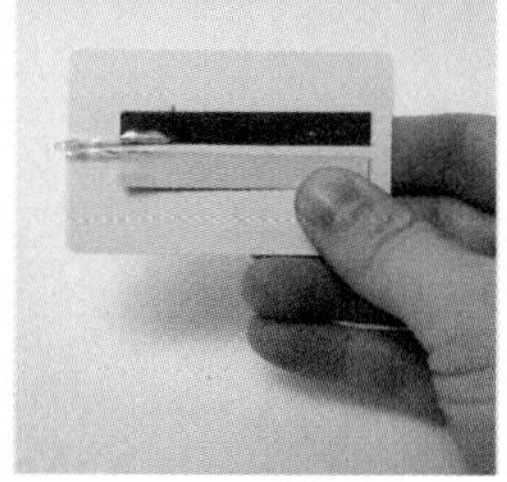

Figure E: Affixing the LED/copper tape to the card (with the LED bulb protruding slightly)

For additional information, discussion, and more, please visit the Instructables project page:

Figure F: Applying foam tape around the components

Figure G: The card art printed and laminated

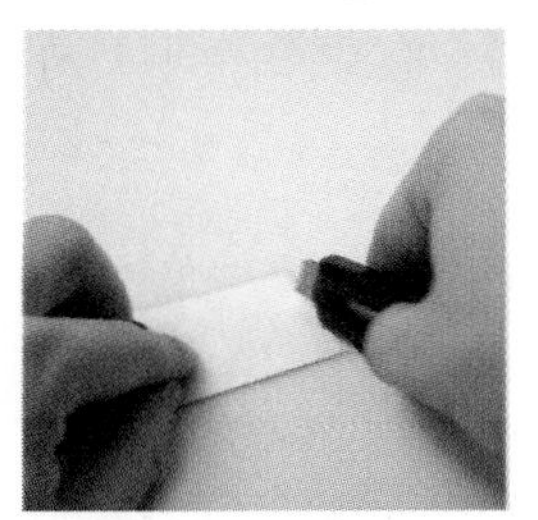

Figure H: Cutting the excess around the edges after mounting with spray adhesive

Figure I: The business card flashlight in action

2. Soldering the LED to copper

Cut a couple of strips of copper tape about 1/4" wide and 2-1/2" long. Solder one end of the LED to one of the copper strips, starting about half way down the lead of the LED. Make the connection so that the soldering joint is close to the edge of the copper strip, and don't use too much heat or the self-adhesive backing paper will peel off. Turn the LED over and do the other side (Figure C).

3. Attaching to the card

One of the leads of the LED will be longer than the other. This will connect to the positive side of the battery. Peel off the backing on the copper strip attached to this lead, and apply it to the middle of one of the blank ID cards. Try and get it so that the LED is poking out just a bit from the edge of the card (Figures D-E).

4. Attaching the foam tape

Apply some self-adhesive foam tape around the perimeter of the card and around the battery to stop it from moving. Make sure that the unattached copper strip is free to go above the battery. It does not need to be attached to anything (Figure F).

5. Attaching the overlay

Design the layout of your card on the computer and print it out. I just printed on plain paper and then laminated it. You will want the image to be a little bit bigger than the card (called a "bleed"), so that you can cut it exactly to size later (and not have a white border around it). Spray some adhesive on the back of the printout, and attach it to the unused ID card. Then turn the card over, and cut neatly around the outside of the card. Finally, attach this ID card to the front of the electronic assembly (Figures G-H).

6. Changes

If you are making a few of these, you can ditch the double-sided tape and use foam sheeting attached with adhesive spray. By using one of the blank cards as a template and cutting out around the outside, you can get a neater edge on the finished design. You can then just cut holes for the battery and LED, making sure there is a vertical cut halfway down the card for the top copper strip to poke through. You might also want to fix the LED a bit more securely so that it doesn't get knocked around in use. A blob of 5-minute epoxy does the job well, and dries transparent.

If I produced these en masse, I would probably change a couple of things. First I would change the CR2032 cell to a CR2016, as this is thinner. I'd also design a custom self-adhesive foam cut-out for the LED and battery to replace the double-sided tape. Getting the cards printed professionally would then allow them to be assembled in seconds each, rather than minutes. If anyone is interested in getting commercial versions of this card or another design produced for corporate use in quantity, please let me know.

Tom Ward (t.ward@lightboxdesigns.com) is a an electrical engineer who has a short attention span and has never really grown up. He spends some of his time designing weird and wonderful gadgets, and the rest teaching science and running technology camps for children.

Dot Matrix Business Card

If your business card is looking too 20th century, how about one with a graphical display on it? By Tom Ward

If my flashlight business card wasn't advanced enough for you (see "Flashlight Business Card," page 290), how about one with a full graphical display that can be customized with any number of scrolling messages? This card could be made in quantity for about a $5 parts cost, and is only a little bit more expensive if you're only making a few. I won't kid you, this is not an easy design to make—don't try it unless you have good soldering skills and some experience in electronics. Some of the components are smaller than grains of rice, so it would be useful to have good eyesight as well! Like the flashlight card, it is more of a proof of concept than something you can churn out in quantity, but it might at least give you an idea of what can be achieved, and where business cards might be in just a few years time.

1. About the design

This is the sort of card that would suit any high-tech business, or those involved in high-value contracts, where an innovative image is all-important. I would never suggest that it replace a conventional business card, but to impress that all-important prospective client, there would be more than a few companies who'd be happy to spend just a few extra dollars. Like the flashlight card, the aim here is to design a card that people just can't throw away!

The design is really quite simple for what it does. It uses a matrix of 5 x 15 LEDs, connected to a single-chip PIC microcontroller. A handful of resistors and switches complete the design (a PDF schematic is available on the project page found via the web address at the bottom of this article). By keeping the microcontroller in sleep mode unless the buttons are pressed, the battery can last several years, and still have enough juice for a couple of thousand displays of your messages.

Figure A: The dot matrix display business card in operation

2. Materials

You will need:

- A CR2032 battery (I got them for about 16 cents on eBay when I bought 100)
- A CR2032 battery holder (I used part 18-3780 from www.rapidonline.com. This costs around 14 cents in quantities of 100—these are a common type of holder that you should be able to find at places like www.mouser.com if you are on the other side of the Atlantic than I!)
- A PIC16F57 (Order code 1556188 from www.farnell.com—these cost 66 cents each in 100+ quantities. Again, you can find them at www.mouser.com).
- Surface mount switches (4) (Part 78-1130 from www.rapidonline.com at 20 cents each)
- Some miscellaneous resistors and capacitors in "0805" surface-mount packages—you will need 5 x 100Ω resistors, 2 x 10kΩ resistors, 1 x 47kΩ resistor, 1 x 47pF capacitor, and 1 x 100nF capacitor. Any of the suppliers mentioned above do these, and they cost almost nothing!
- "0603" LEDs (75)—as bright as possible, and as

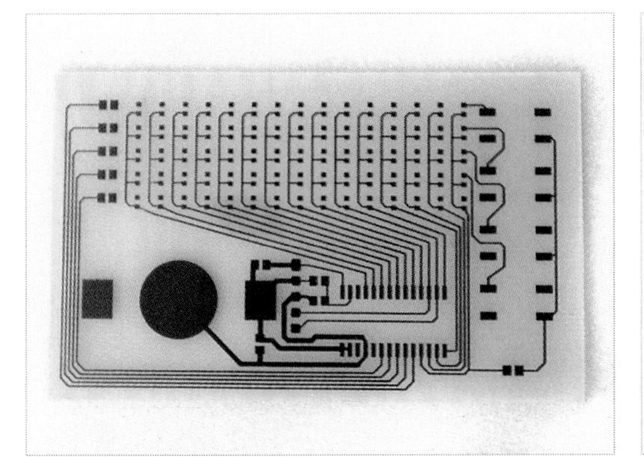

Figure B: The printed circuit before components are added

Figure C: First round of soldering complete. Whew.

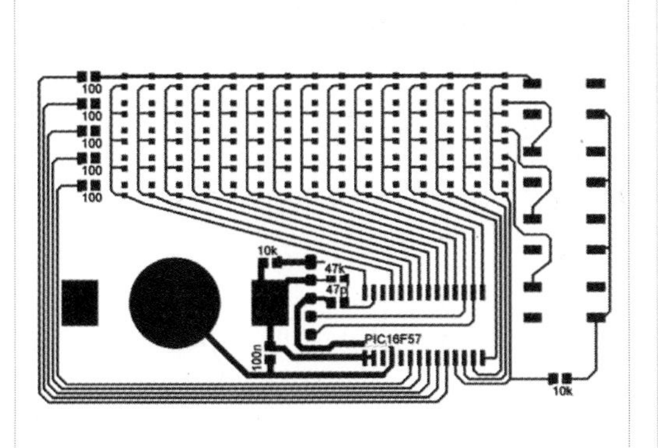

Figure D: The layout with component labels

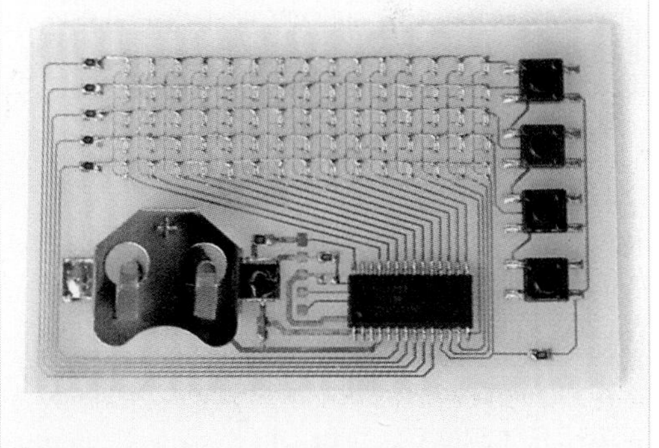

Figure E: Our LEDs connected to the 100Ω surface-mount resistors with fine copper wire

cheap as possible! I used item 72-8742 at 6 cents each from Rapid, but again, you should be able to get them at other suppliers. In quantity, you can get these down to about 3 cents each.

- Some double-sided foam adhesive tape that is slightly thicker than the battery you are using—mine was 4.5mm thick
- A printed circuit board (PCB) for the project—instructions for producing your own are beyond the scope of this article, but you may have some success with the iron-on or photographic technique (my preferred technique). You can find instructions for making your own printed circuit boards elsewhere on Instructables and other sites. The PCB layout for the card can be found in a PDF file on this project's Instructable page.

Tools: You will also need a soldering iron (plus solder), a cutting knife, some spray adhesive, and a way of printing the front of your card—you can use a color laser or inkjet printer. I printed on OHP transparency film. You will also need a way of programming the PIC microcontroller. I use the PICKit2 which is part number 579-PG164120 from www.mouser.com, and available at around $35. A strip of 5 x 0.1" PCB pins (such as 22-0510 from Rapid) can be pushed into the programmer to act as an interface with the board.

3. Soldering starts!

Solder the components onto the board, starting with the smallest first (see Figure C). A pair of tweezers is useful here. By putting a blob of solder on a pad, and then re-melting it while positioning the resistors or capacitors with the tweezers, you can neatly add these smaller components. It doesn't matter which way around these components go, but it does for the PIC (which should read with the writ-

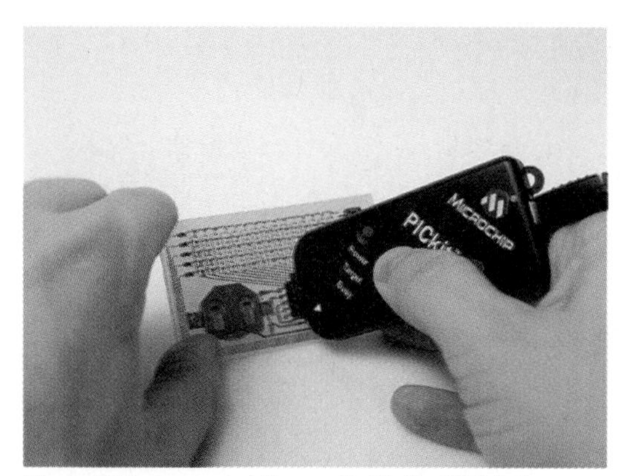

Figure F: Programming the card with the PICkit 2 programmer

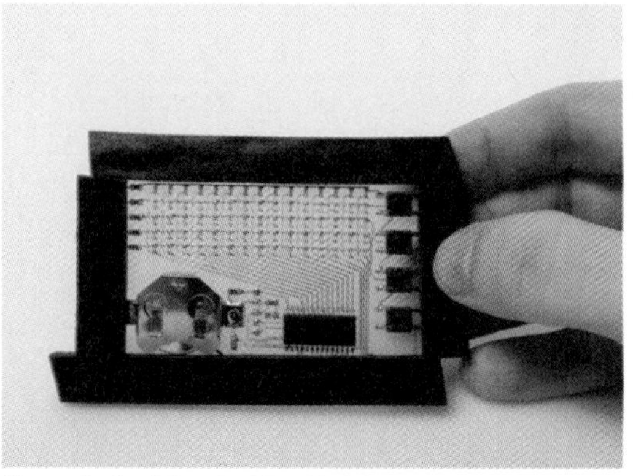

Figure G: Applying double-sided foam tape

ing the correct way up as shown in these photos), and also the LEDs must be put in the correct way. It is harder to tell with the LEDs which way around they should go—the top connection should be the positive (or "anode"). You can tell by consulting the datasheet for the LED—one of the two leads will usually be marked in some way. An easier way is sometimes to test one of them by attaching a couple of wires to a 1.5V battery, and then touching the leads on the ends of the LEDs. If it is the right way around, you should see a glow. If you're using a single 1.5V battery, it will be extremely faint, so you'll have to observe carefully. Again, a tutorial on soldering is not within the scope of this article. Note that the LEDs are only initially soldered on their bottom lead (see Figure C). We will use some wires later to connect their top leads.

4. An ad-hoc double-sided board

Lay some fine strips of "invisible tape" down along the vertical PCB traces next to each column of LEDs. This will stop the wires we are about to solder from touching them. Next, solder some fine tinned copper wire along the top of each row of LEDs, to reach all the way to the resistor as in Figure E. Note that you will need only four wires. The top one will not be needed if you use the PCB layout found on the project page, as it uses a PCB trace to connect the components.

5. Programming

The next step is to put the display program into the chip. If you have bought the PICkit 2 programmer, it has everything you need. Download the MatrixCode.zip file from this article's Instructables page, unzip it, and put it in a directory somewhere on your computer. Then from within the MPLAB IDE, go to the Project menu, select Open, and navigate to the "main.asm" file. Change the stored messages (around line 115 in the code) to your contact details rather than mine(!). The messages are spelled out with a series of "1"s and "0"s—a "1" means the LED is on. If you look closely, you will see my name spelled out with "1"s. (You might need to turn your head 90 degrees to see this!) You have complete freedom to make your own characters or symbols, so you could have, for example, a simple animation of a car moving to the left if you wanted. Note that there are four messages—one for each button—you will need to specify the length of each message by stating the number of columns it takes up in the "MSG1LEN,MSG2LEN..." definitions.

Go to the Project menu again, and select Quickbuild—check that there are no errors, and you are now ready to program. I use a simple technique of inserting a broken-off strip of five pins from a strip of 0.1" header pins into the programmer, and then just touching the five pins while programming. This is a little fiddly, but as the erase or program cycle only takes a second or so, it is quite manageable. The arrow on the end pin of the programmer should align with the top pin of the PCB (NOT as shown in Figure F—Whoops!) If you are experimenting, it is well worth soldering the strip of five pins onto the board until you have finished your changes. When you are ready to program, you will have to use the separate PICkit 2 utility supplied with the programmer, as for some reason, the MPLAB IDE doesn't support programming of the PIC16F57 directly.

Figure H: Cutting the excess from the tape

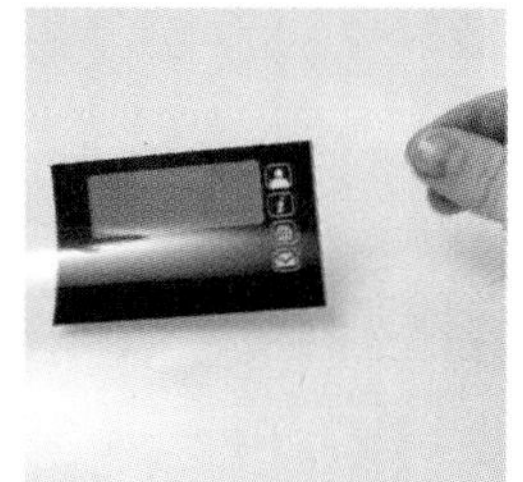
Figure I: The transparency sheet (with white icons)

Figure J: White icons created by using a printing label on the back of the transparency

Figure K: Attaching a sheet of polypropylene to the overlay

To do this, you need to specify the family of PICs ("baseline"), and the particular part (16F57), before loading the Hex file created in the previous step, and then finally, programming the chip. If all is successful, you should be able to insert the battery (positive side down), and press one of the buttons to see your message scroll along!

6. Finishing off

To encapsulate the prototype card, I applied some double-sided foam tape to the board, turned it upside down, and then cut the excess off. I then reverse-printed the graphic overlay on an OHP transparency sheet. By turning the sheet over, and attaching a white printer label, you can get the clear icons on the transparency to show up white. I also attached a sheet of thick polypropylene (made as a cover for binding documents) to the overlay using some adhesive spray before attaching it to the front of the card and trimming off the excess. If you wanted to use the same graphic as mine, it is also available on this article's page as a PDF.

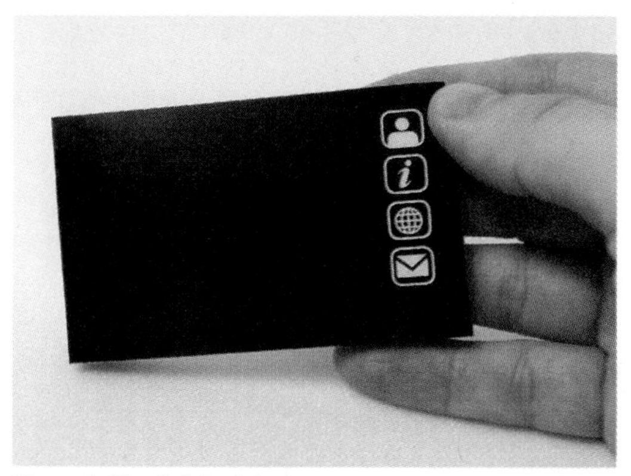
Figure L: The finished design

7. The finished product

You can now relax, content that you have the world's most advanced business card (at least until I make my next one which will have a color OLED screen!).

8. The future

If I produced these commercially, I would probably change a couple of things. First, I would change the CR2032 cell to a CR2016 as it's thinner, and then embed it within a space cut in the PCB. By using lower-profile components, the thickness of the card could probably be reduced to about 1/8th of an inch (rather than the current 1/4"). By using some of the new thin-film batteries, it might even be possible to make a flexible card, albeit at a higher price. A professionally printed overlay and a custom die-cut replacement for the foam tape would see the cards assembled much more quickly, and look slicker too. Of course, the PCBs would be manufactured professionally as well, and populated by a "pick and place" robot to allow assembly to be sped up even further.

Next, I would like to work on a high-resolution version using a color OLED display—I'm thinking photographs and animations. The sky is the limit. Almost any electronics could be put into business cards: wireless links, audio soundtracks—if anyone is interested in using these ideas or other related ones commercially, then let me know!

Tom Ward (t.ward@lightboxdesigns.com) is an electrical engineer who has a short attention span and has never really grown up. He spends some of his time designing weird and wonderful gadgets, and the rest teaching science and running technology camps for children.

Gallery

Notebook Mods

Nine ways of making your Moleskine (or other notebook) your own

Three Clever Mods

Pauric O'Callaghan

/moleskine-notebook-mods

Some simple mods to make your notebook more functional

Create a pencil holder by wrapping a pencil in tape to form a long tube, sticky side out. Tape over the tube to attach it to the spine of the notebook.

Make a little pad from the sticky parts of Post-it Notes and store them in the back of your notebook—fold them over the edge of a page to make a tag

Keep your pencils sharp! Glue two pieces of sandpaper into the back cover; coarse for the wood, fine to sharpen the lead.

Pen Holder sl4sh3r

/Another-Moleskine-Pen-Holder

Sew a pocket for your pen. Cut a 1" wide piece of elastic to a length that will wrap around your notebook, with an overlap of about 5". Pin into a loop.

Sew the elastic on three sides, creating a pocket for your pen

Place the loop around your notebook and put your pen in the pocket!

Pen Clip Mod TeNeal Metcalf

/Moleskine-Pen-Clip-Mod

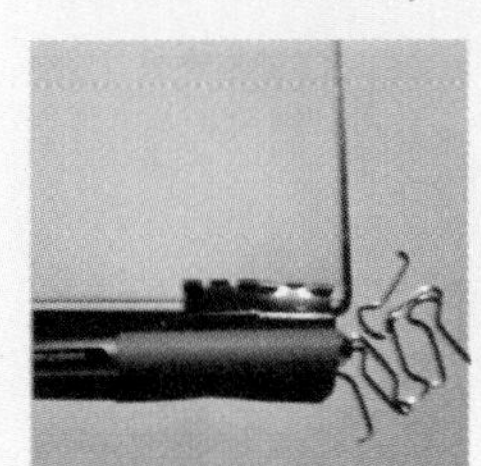

Never lose your pen again. Clip two binder clips onto the edge of the back cover; flip over the handles so they're out of the way. You can leave the handles on or remove them. Slip the pen clip into the binder clips and you're good to go!

Paper Cover Mod

Jillian Soh

/Moleskine-Paper-Cover-Mod

Here's a simple mod for decorating your notebook cover. Cut origami paper into desired width, then glue together to make a long band to wrap around your notebook and under the edges.

Fold the strip around the cover and glue into place. Easy!

Pocket Diary Hipster PDA

Michael Mauzy

/How-to-make-a-hipster-PDA-from-a-Moleskine-pocket-

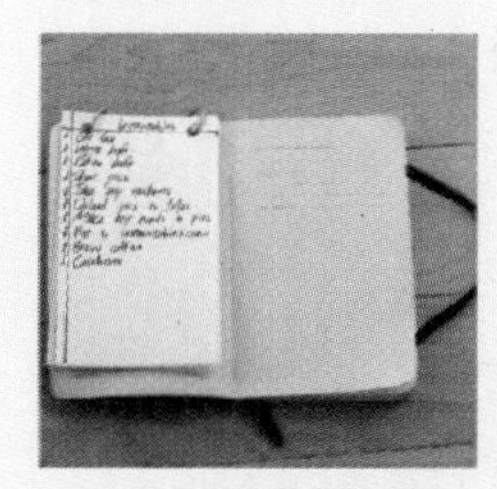

Here's a way to "hipify" your Moleskine. Position an index card inside the front cover. Punch 2 holes through the card and cover and holes in the rest of the cards. Hold it all together with 1/2" binder rings.

Poke the binder rings through the holes in the cover and the cards, then close the rings. Voilà! You have a planner.

Mini Moleskine

Jordan Running

/Mini-Moleskine

Make a tiny Moleskine by cutting one in half; mark where you want the cut and take it to Kinko's for the surgery. Prevent fraying by running a thin bead of glue along the cut edge and the ends of the thread binding.

How to Carry a Pen

Marco Poponi, Italy

/how-to-carry-a-pen-with-your-MOLESKINE

Here's another clever way to keep your pen with your Moleskine. Thread the cloth bookmark through the pen clip, with the clip facing up. Slide the pen up the bookmark to the top of the book.

Close the notebook and pull down on the bookmark so the pen is held snugly against the top of the book. Loop the elastic band around the notebook and the pen to hold the pen securely.

To view these Instructables, and read comments and suggestions from other notebook users, type in www.instructables.com/id followed by the rest of the address given here. Example: www.instructables.com/id/Mini-Moleskine

Gallery

Top Tool Tips

Because sometimes your most valuable tool *is* a tip

Sharpen a Chisel

Eric Seidlitz

/Tool-Tip-How-to-Sharpen-a-Chisel

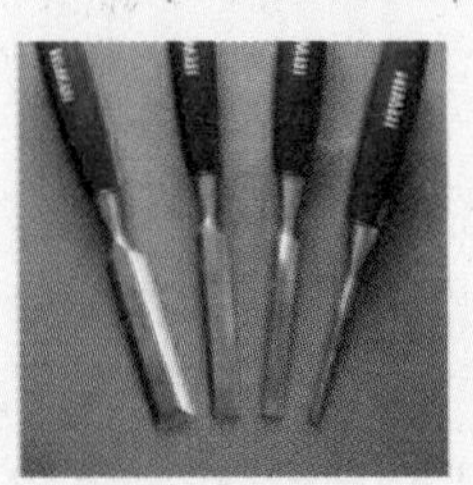

Getting a good point on a chisel is essential to it doing its job. Getting that precious point is best done with a "honing jig" and the right kind of sharpening stones. This Instructable demonstrates the proper use of both.

Recycle Old PCB Components

Patented

/Recycle-old-PCB-components

Every high-tech maker needs a "techno-junk" box of reusable parts, from resistors and capacitors, to switches, voltage regulators, LEDs and integrated circuits. This tutorial shows you how to safely remove all of these circuit board treasures for later use.

How to Solder

Noah Weinstein

/How-to-solder

People are far more intimidated by soldering than they need to be. It really only takes a few tools, following a few basic do's and don'ts, and a couple hours of practice to get the hang of it. This Instructable patiently walks you through the process.

Belt Sander Tool Rest

Robert H. Dutton

/Tool-Rest-for-a-Belt-Sander-for-Sharpening

Building a jig like this—this one built from the parts of an old homebuilt router table—allows you to invert a belt sander to use for tool-sharpening. Holes in the base allow the sander to be held in place with zip ties.

To view these Instructables, and read comments and suggestions from other tool users, type in www.instructables.com/id followed by the rest of the address given here. Example: www.instructables.com/id/How-to-solder

Introduction to Files and Rasps TechShop

/Introduction-to-Files-and-Rasps

If you're not a seasoned wood or metal worker, all of the available files and rasps, with different types of teeth, working surfaces, and shapes, can be confusing. This Instructable, by the folks at TechShop (www.techshop.ws), will brief you on the differences and offer tips for using and maintaining these tools.

Wire Twisting Jig

Chris Palmer

/Free-Wire-Twisting-Jig

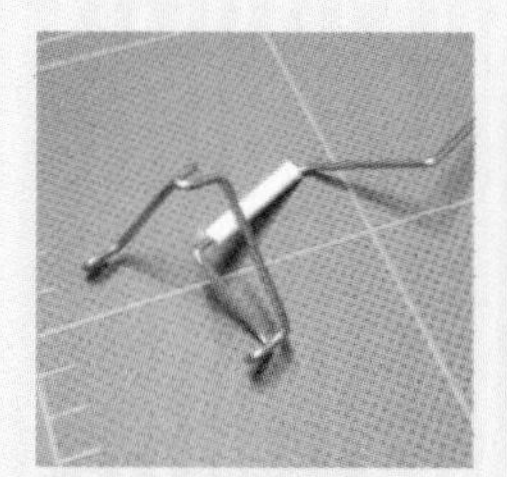

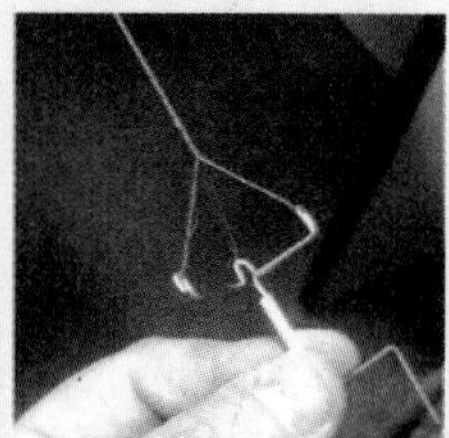

This ingenious little device will allow you to spin tight runs of neatly twisted wire. To build it, all you need is a paper clip, a straw/coffee stirrer, pliers, and some patience. The resulting jig will twist three strands of wire.

How to Maintain Your Multi-tool LftnDbt

/How-to-maintain-your-Multi-tool.

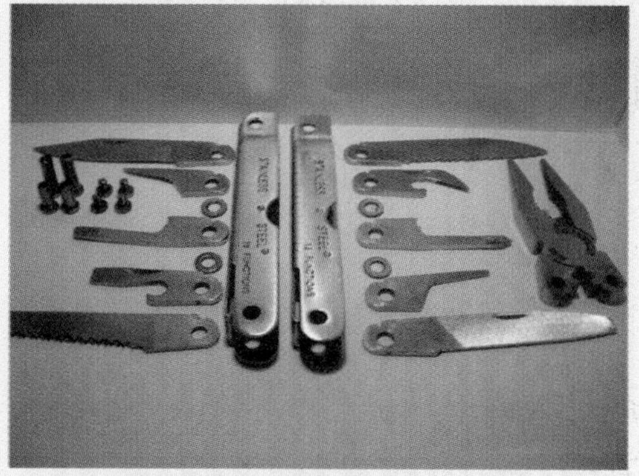

Look on the belt of any cop, emergency/rescue worker, firefighter, or anyone else who relies on tools to stay alive, and you'll find a multi-tool. As the name implies, there are many tools (knives, scissors, pliers, drivers, and specialty tools) all packed into these belt-borne wonders. If you want to know how to take one apart, clean it, lubricate it, and put it back together, hit this link.

Get the Most out of your Dremel

Connor Kaufmann

/Get-the-Most-out-of-your-Dremel

A Dremel, or "rotary tool," is basically a universal spinning shaft with a slew of attachments that can turn it into everything from a cutting and drilling device to a carving tool, a finishing tool, an engraving tool—all sorts of gadgets. This Instructable gives you a rundown of the most useful features of this versatile tool.

contest WINNER!

Easy to Build Desktop 3-Axis CNC Milling Machine

This DIY computer-controlled milling machine puts the power of computer controlled machining at your disposal

By Tom McWire

Here's how to make a computer-controlled milling machine. This puts the power of computer-controlled machining into the hands of the average human. Small enough to set on the desk but scalable to any size. As inexpensive as possible without sacrificing accuracy (too much). Almost all the parts can be purchased in local retail stores. And above all, it's CHEAP, so you can be up and running for well under $200. With it, you can do 2-dimensional engraving and PC board etching, and 3D milling and modeling in foam, wood, plastic, and other soft materials.

Note: You should also check out the YouTube movie that shows how to build this Instructable: www.youtube.com/watch?v=6drMZqmyXQc.

Note: Be sure to check out my "Easy-to-Build CNC Mill Stepper Motor and Driver Circuits" Instructable for details on how to make this machine go: www.instructables.com/id/EJ3KFVBF5R8QRL3.

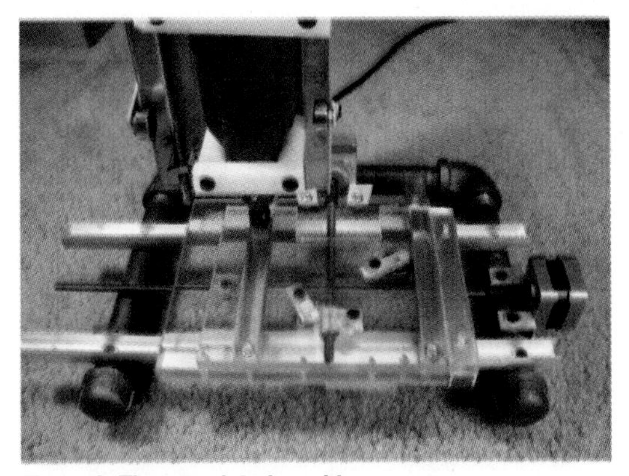

Figure A: The completed machine

1. The frame

The frame needs to be two things:

1) A flat base that you can mount everything on horizontally
2) A goose neck of some kind to hold the Z axis (the part with the motor tool that goes up and down) firmly in place.

For my frame, I used 1" pipe just for fun, but as it turned out it was pretty handy too. When I needed to make adjustments I could just tap it with a hammer. As you can see in Figure B, the post that holds the Z axis doesn't have to be in the center. It just needs to be firm and the water pipe does a good job of that. Later, after you are sure all the pipe joints are in the right place, you can add a drop of thread sealer to the joints and it will be a good solid structure.

Figure B: The frame **Figure C: X-stage rails**

2. The x-stage rails and motor

It's time to add the rails for the x-axis stage. These rails are 3/4" U-channel aluminum that you can get from the hardware store. Put a washer under each end to space the rail off the pipe just a bit. Don't worry about the rails being perfectly parallel. You'll see why later.

Next, mount the stepper motor with a bracket like you see in Figures C and D. Connect a length of 1/4" by 20" threaded rod to the motor shaft with a short piece of rubber hose (1/4" inch fuel line). Now you're ready to set the movable part of the x-axis (the stage).

For additional information, discussion, and more, please visit the Instructables project page:

Figure D: Closeup of the rails

Figure E: Plexiglas mounted to a U-channel

Figure F: Bearings mounted to a piece of aluminum

Figure G: Another piece of aluminum wrapped around a coupling nut

Figure H: Plexiglas, bearings, and coupling nut in place

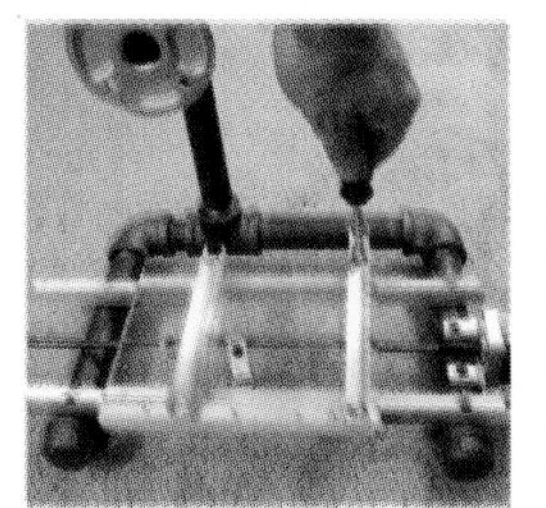

Figure I: Making the y-stage

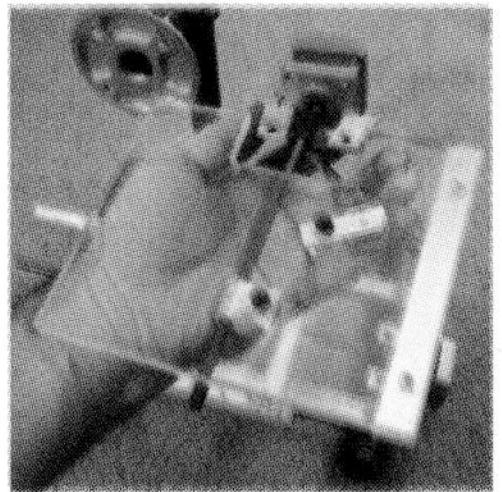

Figure J: Assemble the Plexiglas, bearing, and coupling nut

Figure K: The y-stage

3. X-it stage right

Take a piece of plastic (I used Plexiglas, shown in Figure E) or metal, something strong and flat and mount a piece of the U-channel to it.

Now comes the tricky part. The round object shown in Figure F is a bearing. You can get them out of motors or buy them at a hardware store. Mount it to a short piece of aluminum as shown.

Next, take a 1/4" coupling nut (a long nut) and wrap it with aluminum as shown in Figure G. The bearing will hold the x-stage to the x-rail, and the coupling nut will allow the motor to run the stage back and forth (Figure H). It wouldn't hurt to grease the skids and the nut a little too.

Note: You can see a video on making the bearing fixture at www.youtube.com/watch?v=Y6vDllrRSDw.

4. The y-stage

The y-stage is just like the x-stage but turned 90 degrees. Mount two rails and a motor on the x-stage as shown in Figure I and then take another piece of flat material and a U-channel and make the moving y-stage. Make the little bearing assembly and a coupling nut for it too (Figure J). When you're done it should look like Figure K.

5. Zee z-axis

Again, we are going to reproduce the x- and y-stage to create the z-axis stage. But first, take a flat piece of material; I used a piece of white Plexiglas. Mount some rails and a motor to it (Figure L). Then make a moving stage piece with a U-channel and a roller bearing. We'll do something a little different with the nut (Figures M-N). The four posts you see on the stage will hold the motor tool. Since this stage is going to move up and down the weight of the motor tool will make it want to come off of the rails so lets add a few more roller bearings to each side to keep it together (Figures O-P).

6. Get it together

Now we slap the motor tool into the z-stage (Figure Q). Then it's time to mount the stage to the frame as shown in Figures R and S. And there you have it. This is the mechanical structure. From here we will need to hook up the stepper motors to a controller and get some software running on the computer (see my "Easy to Build CNC Mill Stepper Motor and Driver Circuits" Instructable for details on how to make this machine go: www.instructables.com/id/EJ3KFVBF5R8QRL3).

Figure L: Making the z-stage

Figure M: Adding the posts

Figure N: Another view of the posts

Figure O: Attaching roller bearings

Figure P: The z-stage

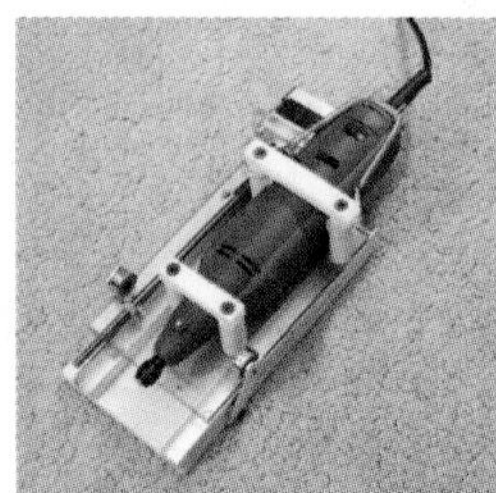

Figure Q: Attaching the motor tool to the z-stage

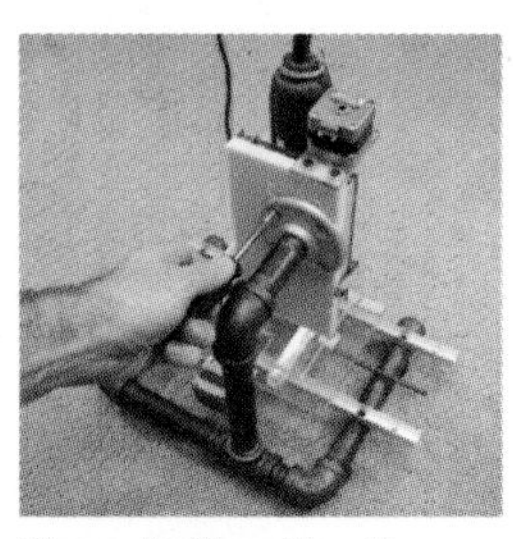

Figure R: Mounting the tool to the frame

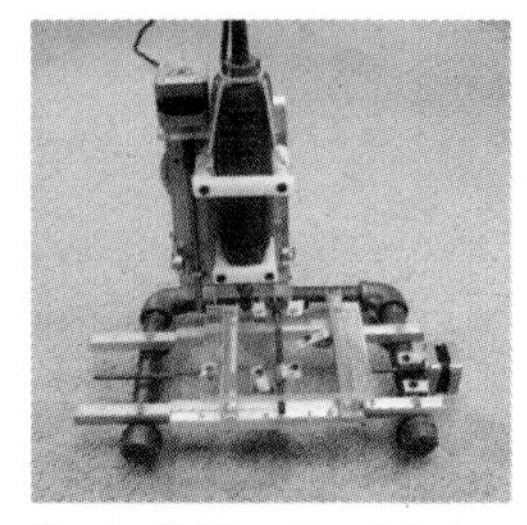

Figure S: The completed mechanical structure

Figure T: Milling a PC board

Figure U: Carving a 3D object out of foam

7. What does it do?

If you were interested in this project it's likely you have already seen what can be done with an 3-axis (XYZ) computer-controlled milling machine. What is surprising is what kind of accuracy you can get out of this thing after you tinker with it a little bit. Make sure all the rails are held firm and straight. Tighten the roller bearings so the stage doesn't shift.

I use it to make PC boards (see Figure T). It's also good for engraving name tags and signs. And it's pretty exciting to see it carve a 3D object out of a block of foam (see Figure U) or plastic.

Note: There's a lot to learn about the software. Some venders offer package deals of motors, drivers, and software. That makes it easier but you pay for it. I have been using KCAM from www.kellyware.com. You can run the demo for 30 days or buy it for $100. I think it works pretty good for the price. Look there, and you will get some idea of what it takes to run the motors from a computer.

8. Engraving

I put the machine back together after making the Instructable and I did some engraving and made a PC board.

Cutting plastic is no problem but when I tried a PC board, the bit went a little too deep on the left side of the board and took out all the finer traces. This is when you start tweaking it. Just take some aluminum foil and put it under the rail of the y-axis. As the stage travels left to right, the height of the bit should stay the same.

Notice in Figure T I'm just holding the material down with masking tape. What I like about this Instructable is that it's easy to fix these kinds of problems because it's all made from simple elements.

Tom McWire is an engineer, inventor, and artist who loves to work with machines and inspire people to have fun with technology. See more of him at www.tommcguireart.com.

Tim's Tool Tricks

Nifty nuggets from an obsessive tips collector By Tim Anderson

Figure A: Fix your broken blender bottom

Figure B: C-clamp lock for casters

Figure C: Hanging bins for parts

Figure D: Landfill consciousness photo

Figure E: Water bottle mic stand

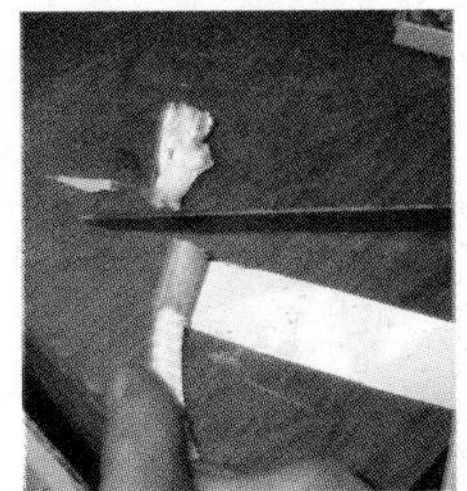

Figure F: Clean ends to synthetic rope

Instructables über-author Tim Anderson loves to collect tips, tricks, simple hacks, and ingenious workarounds. He shares these in a series of Handy Tricks Instructables. Here are just a few to whet your whistle. (Gee, I wonder if he has any tips on whistle-whetting?)

Jar-lid blender fix: A canning jar lid happens to be the same size as the bottom of many blender jars. If your blender bottom breaks, you can use one of these lids as a replacement. And if your blender glass breaks, you can use a canning jar in its place! (Figure A)

Lock casters with clamps: Vincent put a boat hull on a big dolly to move it around. The only problem is it tends to move around when you're sanding or drilling on it. Solution: attach some C-clamps to the wheels! (Figure B)

Hanging parts bins: Here's a handy way to organize small parts without using up shelf or bench space. Just screw a bunch of lids to your ceiling and put the parts in the jars (Figure C).

Landfill consciousness wastebasket: This photo is pasted on the lid of the wastebasket at Instructables HQ, SF, CA, USA (Figure D).

It reminds people that each piece of trash is a nail in the coffin of our beloved mother earth.

Water bottle mic stand: I couldn't find my mic stand and needed to type while talking, so I tied the mic to a water bottle with a jumper cable. It works so well I'm still using it this way (Figure E).

A perfect ending: To create a neat, unravel-able ending to any synthetic cord, tape the end, slice off the fray, heat the end to melt strands together, and remove the tape (Figure F).

For dozens more of Tim's awesome tool tricks, search for "handy tricks" at instructables.com.

The Best of Instructables Community Contest

To give the Instructables community a voice in the content for *The Best of Instructables*, we ran a contest on the site. Users were invited to submit their projects and then the Community voted for projects they thought should be included in the book. From almost two thousand Instructables entries, here are the top 75, with the winners ranked by users' votes. You can find quick links to all of the winners here: www.instructables.com/contest/bookcontest.

—Eric J. Wilhelm

	Instructable	Member Name	URL
1	LED Throwies p.108	Q-Branch	/LED-Throwies
2	Invisible Book Shelf p.16	dorxincandeland	/Invisible-Book-Shelf
3	Laser Flashlight Hack!!	Kipkay	/Laser-Flashlight-Hack!!
4	Minty Boost! Small battery-powered USB charger p.180	ladyada	/MintyBoost!---Small-battery-powered-USB-charger
5	DIY Compact Survival Kit	ledzeppie	/Compact-Survival-Kit
6	How to Make a Cardboard Costume Helmet	Honus	/How-to-make-a-cardboard-costume-helmet
7	Turn Signal Biking Jacket p.158	leahbuechley	/turn-signal-biking-jacket
8	Audio Visual Art....FOTC Style	scooter76	/Audio-Visual-Art....FOTC-Style
9	Wall-E Robot	4mem8	/Wall-E-Robot
10	Paper Wallet	theRIAA	/Paper-Wallet
11	Screen Printing: Cheap, Dirty, and At Home p.192	tracy_the_astonishing	/Screen-Printing:-Cheap,-Dirty,-and-At-Home
12	Munny Speakers p.222	fungus amungus	/Munny-Speakers
13	Knex Heavy Cannon	I_am_Canadian	/Knex-Heavy-Cannon
14	How to Make Playdough (Play-Doh)	canida	/How-to-Make-Playdough-Play-doh
15	LED Chess Set	Tetranitrate	/LED-Chess-Set
16	Sew your own Instructables Robot Plushie!	jessyratfink	Make_your_own_Instructables_Robot_Plushie
17	DIY Vinyl Wall Art	britsteiner	/DIY-Vinyl-Wall-Art
18	Bluetooth Handgun Handset for your iPhone: iGiveUp	ManaEnergyPotion	/Bluetooth-Handgun-Handset-for-your-iPhone-iGiveUp
19	Easy to Build Desktop 3-Axis CNC Milling Machine p.300	Tom McWire	/Easy-to-Build-Desk-Top-3-Axis-CNC-Milling-Machine
20	Build Your Own Butler Robot!!!	Erobots	/Build-Your-Own-Butler-Robot
21	How to Build a 96-Volt Electric Motorcycle	Kentucky-bum	/How-to-build-a-96-Volt-Electric-Motorcycle

To view these Instructables, type in www.instructables.com/id followed by

	Instructable	Member Name	URL
22	How To: Make Bath Bombs	SoapyHollow	/How-To-Make-Bath-Bombs
23	Steampunk Dystopian Sniper Rifle (Mercury Bow)	gmjhowe	/Steampunk_Dystopian_Sniper_Rifle_Mercury_Bow
24	Save $200 in 2 Minutes and Have the World's Best Writing Pen	kingant	/Save-$200-in-2-minutes-and-have-the-worlds-best-wr
25	Concrete Light Bulb Wall Hook p.39	whamodyne	/Concrete-Lightbulb-Wall-Hook
26	Light Bulb Lamp	bumpus	/Light-Bulb-Lamp
27	Light Bulb Greenhouse p.82	LinuxH4x0r	/Lightbulb-greenhouse
28	Instructables Robot—Paper Model	SMART	/Instructables-Robot-Paper-Model
29	How to Build a Robot—The BeetleBot v2 (Revisited) p.210	robomaniac	/How-to-Build-a-Robot-The-BeetleBot-v2-Revisite
30	Wallet Made from a Computer Keyboard	zieak	/Wallet-made-from-a-computer-keyboard
31	How to Add EL Wire to a Coat or Other Garment	enlighted	/how-to-add-EL-wire-to-a-coat-or-other-garment
32	Gandhi: 17' Tall Cardboard Avatar	delappe	/Build-a-17-Tall-Cardboard-Papercraft-Gandhi
33	Ture Trigger, 10 Round, Auto-Loading, Knex Concept Rifle	bannana inventor	/10-Round-Auto-Loading-Knex-Concept-Rifle-by-banna
34	How to Build Your Own Jet Engine	russwmoore	/How-to-build-your-own-Jet-Engine
35	Shake It Like a Tic-Tac!	MrMunki	/Shake-it-like-a-Tic-Tac!
36	Cyber/Steampunk Futuresque Sci-Fi Hand Gun p.232	gmjhowe	/Steampunk-Futuresque-Sci-Fi- Hand-Gun
37	How to Make a Marshmallow Shooter	ewilhelm	/Marshmallow-gun
38	How to Build a 72Volt electric motorcycle	Stryker	/How-to-build-a-72Volt-electric-motorcycle
39	Friends Are Easy To Make	cuteaznprincesss	/Doll-Friends-are-easy-to-make
40	Building Small Robots: Making One Cubic Inch Micro-Sumo Robots and Smaller	mikey77	/Building-Small-Robots-Making-One-Cubic-Inch-Micro
41	Portable 12V Air Conditioner --Cheap and easy!	CameronSS	/Portable-12V-Air-Conditioner---Cheap-and-easy!
42	How to Make a TRON Style Lamp: The MADYLIGHT	Greg Madison	/How-to-Make-a-TRON-Style-Lamp-The-MADYLIGHT
43	Giant Match	Tetranitrate	/Giant-Match
44	Magnetic Rubik's Dice Cube	burzvingion	Magnetic-Rubik_s-Dice-Cube
45	How to Build a Wood Fired Hot Tub	veloboy	/How-to-build-a-wood-fired-hot-tub
46	Iron Man Helmet	pmaggot	/Iron-Man-Helmet
47	Home-made Sun Jar	cre8tor	/Home-made-Sun-Jar
48	Grow Your Own Bioluminescent Algae p.84	ScaryBunnyMan	/Grow-Your-Own-Bioluminescent-Algae

the rest of the address given. Example: www.instructables.com/id/LED-Throwies

	Instructable	Member Name	URL
49	Construction of Two Portuguese Style Dinghies (Small Boats)	rook999	/Construction-of-Two-Portuguese-Style-Dinghies-Sma
50	Uni-Directional WIFI Range Extender	tm36usa	/Uni-Directional-WIFI-Range-Extender
51	"1UP Mushroom" Mushroom Burger! p.54	momo!	/1UP-Mushroom-Mushroom-Burger
52	Open Any Padlock	Tetranitrate	/Open-Any-Padlock
53	Coilgun Handgun	rwilsford07	/Coilgun-Handgun
54	Electroforming an Iris Seed Pod	MaggieJs	/Electroforming-an-Iris-Seed-Pod
55	100 Ways to Reduce Your Impact	Brennn10	/50-Ways-to-Reduce-Your-Impact
56	How to Make a Three Axis CNC Machine	Stuart.Mcfarlan	/How-to-Make-a-Three-Axis-CNC-Machine-Cheaply-and-
57	Airgun with eXplosive air-Release Valve	chluaid	/Airgun-with-eXplosive-air-Release-Valve
58	Lego USB Stick p.97	ianhampton	/Lego-USB-Stick
59	ChapStick LED Flashlight	BCat	/ChapStick-LED-Flashlight
60	Awesome LED Cube	AlexTheGreat	/Awesome-led-cube
61	Solar Powered Trike p.172	dpearce1	/Solar-Powered-Trike
62	Barbie Doll Electric Chair Science Fair Project!	jessyratfink	/Barbie-Doll-Electric-Chair-Science-Fair-Project!
63	DIY 3D Controller	kylemcdonald	/DIY-3D-Controller
64	The Stirling Engine, Absorb Energy from Candles, Coffee, and More!	thecheatscalc	/The-Sterling-Engine-absorb-energy-from-candles-c
65	Build the World's Smallest Electronic Shocker!	Plasmana	/How-to-build-the-Worlds-Smallest-Electronic-Shock
66	Simulated Woodgrain for Metal Boxes	amz-fx	/Simulated-woodgrain-for-metal-boxes
67	How to Grow Pineapples	woofboy111	/How-to-Grow-Pineapples
68	Creating a 3D Effect with Image Editing Software	Andrew546	/Creating-a-3D-effect-with-image-editing--software-
69	S.P.R.E.E. (Solar Photovoltaic Renewable Electron Encapsulator)	charlitron	/SPREE_Solar_Photovoltaic_Renewable_Electron_
70	How to Get a T-shirt for GoodHart	RocketScientist2015	/Tshirt-for-GoodHart
71	Grow a Square Watermelon	watermelon	/Grow-a-square-watermelon
72	Build a 4-Color T-Shirt Printing Press	Progfellow	/Build_a_4_Color_T_Shirt_ Printing_Press
73	Electromagnetic Floater	J_Hodgie	/Electromagnetic-Floater
74	The Accidental Pocket Jet Engine	killerjackalope	/The-accidental-pocket-jet-engine...
75	Digital Picture Frame	micahdear	/Digital-Picture-Frame

Thank You!

A big Thank You to the following Instructables members who were cut from the book due to space limitations.

James Allen, Earbud Cord Wrapper (photo)
Dale Amann, How to Make Your Very Own Rainstick!
Matthew Beckler, ATX → Lab Bench Power Supply Conversion
Benjamin Bustard, How-to-Build a 72Volt-Electric Motorcycle
Laura Cesari, Fire Skirt
Glenn Chandra, Home-Made Sun Jar
Ben Crist, Magnetic Rubik's Dice Cube
Jérôme Demers, BeetleBot Revisited
Gary Fixler, Magnetic Acrylic Rubik's Cube
Paul Gussack, Concrete Countertops for the Kitchen
Hunter Frerich, Paper Wallet
James Harrigan, IKEA Aquascape on the Cheap $12
David Hayward, Looty
Jonathan Hodgins, Electromagnetic Floater
Aaron Hoffman & Paul Gussack, Concrete Countertops for the Kitchen
Ineluctable, Knobby All Terrain Rubber Wallet
Paul Jehlen, Hydroponics at Home and for Beginners
Avram "AviK" Kaufman, Vespa ET4 iPod Speaker System
Kip "Kipkay" Kedersha, Laser Flashlight Hack!!
Jerome Kelty, How to Make a Cardboard Costume Helmet
LaFabricaDeCosasBonitas, Hungarian Shelves
Ken Lloyd, Hidden Door Bookshelf
Gilad Lotan, DIY Touch Sensor
Greg Madison, How to Make a TRON Style Lamp: The MADYLIGHT
Nicole Magne, Six-armed-Hindu-Goddess-Kali-Costume
Jonathan Mayer, 60-Minute Bookcase
Stuart Mcfarlan, How to Make a Three Axis CNC Machine
Brano Meres, How I Built a Carbon Bike Frame at Home...
Travis Odegard, Build a Tetris DVD or Bookshelf
Billy Robb, Solar Heater
David Rogers, Iron Man Helmet
Michael Roybal, Conductive Fabric: Make Flexible Circuits Using an Inkjet Printer.
Brian Sawyer, Handbound Book
Jim Shealy, The Stirling Engine
Star Simpson, Build a Microwave Transformer Homemade Welder
Neel Sutton, Theater-Effects—Lacerations
Alyson Vega, NoSew USB Electronics Organizer
Sam Wirshup, aka imanalchemist, Floppy Disk Bag

Contributors

Eric J. Wilhelm (ewilhelm)
Christy Canida (canida)
Ed Lewis (fungus amungus)
Rachel McConnell (rachel)
Ryan McKinley (ryan)
Eric Nguyen (nagutron)
Cloude Porteus (lebowski)
Randy Sarafan (randofo)
Noah Weinstein (noahw)
Saul Griffith (saul)
Tim Anderson (TimAnderson)
Dan Goldwater (dan)
Kenny Jensen (argon)
Robert H. Dutton (Tool Using Animal)
Mark Langford (Kiteman)

To learn more about the Instructables.com team and the Squid Labs Community, go to www.instructables.com/about.

Every week we put together a newsletter of the best new Instructables. Get inspired and see what everyone else has built by signing up. Enter your email address in the Subscribe box at the bottom of any page at www.instructables.com.

How to Contact Us

We have verified the information in this book to the best of our ability, but you may find things that have changed (or even that we made mistakes!). As a reader of this book, you can help us to improve future editions by sending us your feedback.

If you have a comment, question, or suggestion for a specific Instructable, please visit the project page for that Instructable and add your voice to the community by posting in the comments there. To find the project page, look at the bottom of the first or second page of each Instructable.

To comment on the book, send email to: bookquestions@oreilly.com, and let us know about any errors, inaccuracies, misleading or confusing statements, and typos that you find anywhere in this book or what we can do to make this book more useful to you. We take your comments seriously and will try to incorporate reasonable suggestions into future editions. You can write to us at:

Maker Media
1005 Gravenstein Hwy N.
Sebastopol, CA 95472
(800) 998-9938 (in the U.S. or Canada)
(707) 829-0515 (international/local)
(707) 829-0104 (fax)

Maker Media is a division of O'Reilly Media devoted entirely to the growing community of resourceful people who believe that if you can imagine it, you can make it. Consisting of MAKE magazine, CRAFT magazine, Maker Faire, as well as the Hacks, Make: Projects, and DIY Science book series, Maker Media encourages the Do-It-Yourself mentality by providing creative inspiration and instruction.

For more information about Maker Media, visit us online:

MAKE: www.makezine.com
CRAFT: www.craftzine.com
Maker Faire: www.makerfaire.com
Hacks: www.hackszine.com

The Make: Books website for *The Best of Instructables* lists examples, errata, and plans for future editions. You can find this page at: http://makezine.com/instructables.

For more information about this book and others, visit the O'Reilly website: http://www.oreilly.com.

Index

F

G

H

Q

R

S

User Notes

Maker SHED

DIY KITS + TOOLS + BOOKS + FUN

SAVE 15% WITH COUPON CODE INSTRUCT
OFFER EXPIRES AT MIDNIGHT ON 12/31/2009

makershed.com

ARDUINO STARTER KIT

Arduino lets you connect the physical world to the digital world. Contains everything you need for your first Arduino project, including our *Making Things Talk* book.

FASHIONING TECHNOLOGY

Get the top book on smart crafting, because LEDs are the new sequins.

BLINKYBUG KIT

These simple, little electro-mechanical insects respond to movement, wind, and vibrations by blinking their LED eyes. Great for all ages!

MOUSEBOT KIT

Herbie the Mousebot is a very quick, easy-to-build, light-chasing robot kit perfect for beginners.

BEST OF MAKE

If you're just catching on to the MAKE phenomenon and wonder what you've missed, this book contains the best DIY projects from our first ten volumes — a surefire collection of fun and challenging activities going back to MAKE's launch in early 2005.

MONSTER KITS

Make a friendly neighborhood mini-monster, as seen in CRAFT, Volume 06. No two mini-monster kits are alike. Good for all skill levels.

DISCOVER DIY KITS, TOOLS, BOOKS, AND MORE AT makershed.com

User Notes

Make:

technology on your time®

"The kind of magazine that would impress MacGyver."

—*San Francisco Chronicle*

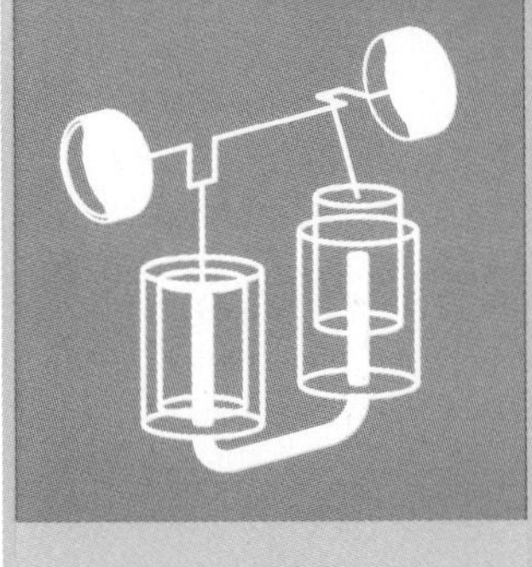

In the pages of MAKE magazine, we teach you how to:

» Snap aerial pictures with a disposable camera, a kite, and a timer made of Silly Putty

» Launch your own water rocket

» Build a backyard zip line

Subscribe today and get a free issue:
» makezine.com/subscribe

Enter promo code TINSTR

If you can imagine it, you can MAKE it. » makezine.com

User Notes

User Notes

User Notes

User Notes

User Notes

User Notes

User Notes

User Notes

Need more space for your notes? Check out the Maker's Notebook: www.makezine.com/notebook